Hecht, Jeff
The laser guidebook

DATE DUE

OCT 3 1988			
MAR 1 1994			
FEB 13 1995			
MAR 29 2003			
DEC 10 2008			
DEC 1 2008			
DEC 11 2008			

THE LASER GUIDEBOOK

THE LASER GUIDEBOOK

JEFF HECHT

Contributing Editor, *Lasers & Applications*

McGRAW-HILL BOOK COMPANY

New York St. Louis San Francisco Auckland Bogotá Hamburg Johannesburg London Madrid Mexico Montreal New Delhi Panama Paris São Paulo Singapore Sydney Tokyo Toronto

Library of Congress Cataloging-in-Publication Data

Hecht, Jeff.
The laser guidebook.

Includes index.
1. Lasers—Handbooks, manuals, etc. I. Title.
TA1683.H43 1986 621.36'6 85-23203
ISBN 0-07-027733-8

1234567890 DOC/DOC 8932109876

ISBN 0-07-027733-8

The editors for this book were Richard Krajewski and Dennis Gleason, the designer was Naomi Auerbach, and the production supervisor was Sally Fliess. It was set in Primer by Williams Press.

Printed and bound by R. R. Donnelley & Sons Company.

This book is dedicated to the scientists and engineers whose pioneering work made the whole laser industry possible: Theodore H. Maiman, Charles H. Townes, Arthur L. Schawlow, Gordon Gould, Nikolai Basov, Aleksandr M. Prokhorov, Peter Sorokin, Ali Javan, William Bennett, Donald Harriott, Kumar Patel, Robert N. Hall, Nicolaas Bloembergen, William Bridges, William Silfvast, and many more.

contents

preface

I wrote this book because it is something I have long wanted to have on my bookshelf. Several excellent textbooks describe the physics of lasers. Handbooks tabulate laser lines reported in the scientific literature. But I have never found any other book devoted to the functional characteristics of commercial lasers—information vital to those of us who work with lasers.

This book is intended both as tutorial and as reference. Readers seeking to learn the basics of lasers and optics should start by reading the first few overview chapters, then dig into the descriptions of specific lasers to see how they really work. Those who are looking up information on lasers may start with the index, specific chapters, or the tabulation of laser characteristics in Appendix B. The chapters on specific lasers share a common structure to make reference use easier.

Earlier, shorter versions of many chapters originally appeared in *Lasers & Applications* magazine. I am grateful to the magazine's publisher, Carole Benoit, for sharing my belief that this material would be useful to the laser community. I also received valuable help from members of the magazine's editorial staff, particularly Richard Cunningham, Breck Hitz, Jim Cavuoto, and Tom Farre, and invaluable encouragement from some of the magazine's readers.

Many people have given graciously of their time to check for errors and misunderstandings in earlier drafts. Alex Jacobsen read the whole thing with his careful eye. Many other members of the laser community checked parts of the manuscript and gave me the benefits of their deeper understanding. Thanks particularly to Bob Anderson, Stephanie Banks, Tony Bernhardt, Dan Botez, Gerry Bricks, Joan Bromberg, Carl Burns, Evan Chicklis, Paul Crosby, Brian Davis, Mark Dowley, Gary Forrest, Horace Furumoto, Bob Goldstein, John Grace, Tim Grey, Hans-Peter Geieneisen, Don Heller, Chuck Higgins,

Jim Higgins, Bill Hug, Bill Jeffers, Tony Johnson, Andrew Kearsley, Paul Kenrick, Gary Klauminzer, Phil LadenLa, Kurt Linden, Arlan Mantz, Peter Moulton, Ed Neister, Bob Pitlak, Dick Roemer, Bob Rudko, Roger Sandwell, Dick Steppel, Jim Stimson, Tom Stockton, Tim VanSlambrouck, Bill Vaughan, Colin Webb, Sicco Westra, Dave Whitehouse, and Bill Young. If any mistakes have slipped through it's my fault, not theirs.

I also owe thanks to my editors at McGraw-Hill, Harry Helms for getting the project going, Roy Mogilanski for keeping it alive, and Rich Krajewski and Dennis Gleason for putting it together at last. Finally, thanks to my wife Lois, for running innumerable errands and applying encouragement when needed.

Jeff Hecht

THE LASER GUIDEBOOK

CHAPTER 1

introduction

The word *laser* is an acronym for "light amplification by the stimulated emission of radiation," a phrase which covers most, though not all, of the key physical processes inside a laser. Unfortunately, that concise definition may not be very enlightening to the nonspecialist who wants to *use* a laser but has less concern about the internal physics than the external characteristics. The laser user is in a position analogous to the electronic circuit designer. A general knowledge of laser physics is as helpful to the laser user as a general understanding of semiconductor physics is to the circuit designer. However, their jobs require them to understand the operating characteristics of complete devices, not to assemble lasers or fabricate integrated circuits.

This book is written and organized with the needs of the laser user in mind. The first few chapters describe the basic ideas behind lasers, how they work, and the characteristics that are most important from a user's standpoint. The following chapters focus on individual laser types, first outlining operating principles, then describing specific characteristics such as output power, wavelength, and input power requirements. Major laser industry buzzwords are defined in the glossary in Appendix A, and Appendix B tabulates the most important characteristics of major lasers. The chapters on individual types of lasers follow the same basic outline and are structured both to be read as chapters and to be used for looking up specific data. Inevitably, such a structure brings with it some redundancy; wavelength and power range, for example, may be mentioned a few times in the same chapter. However, what may seem repetitive when reading the chapter also makes the information easier to find when searching out specific data in the book.

Because this book is a book, it cannot hope to keep up with the continual changes in the product lines of laser manufacturers. Specific models are mentioned in places, but only as examples of the types of characteristics to expect in commercial products. Companies also come and go in the field. The best way to locate specific laser products is first to find manufacturers through an industry directory, then to get product literature from the companies. All three laser optics trade magazines publish industry directories. My favorite—only partly because I helped organize it—is the *Lasers & Applications Buying Guide.* The others are the *Optical Industry & Systems Purchasing Directory* and the *Laser Focus Buyers' Guide.* All three are updated annually.

Definition and Description of a Laser

From a practical standpoint, a laser can be considered as a source of a narrow beam of monochromatic, coherent light in the visible, infrared, or ultraviolet parts of the spectrum. The power in a continuous beam can range from a fraction of a milliwatt to around 20 kilowatts (kW) in commercial lasers, and up to more than a megawatt in special military lasers. Pulsed lasers can deliver much higher peak powers during a pulse, although the average power levels (including intervals while the laser is off and on) are comparable to those of continuous lasers.

The range of laser devices is broad. The laser medium, or material emitting the laser beam, can be a gas, liquid, glass, crystalline solid, or semiconductor crystal and can range in size from a grain of salt to filling the inside of a moderate-sized building. Not every laser produces a narrow beam of monochromatic, coherent light. Semiconductor diode lasers, for example, produce beams that spread out over an angle of 20 to 40°, hardly a pencil-thin beam. Liquid dye lasers emit at a range of wavelengths broad or narrow depending on the optics used with them. Other types emit at a number of spectral lines, producing light that is neither truly monochromatic nor coherent.

Practically speaking, lasers contain three key elements. One is the laser medium itself, which generates the laser light. A second is the power supply, which delivers energy to the laser medium in the form needed to excite it to emit light. The third is the optical cavity or *resonator,* which concentrates the light to stimulate the emission of laser radiation. All three elements can take various forms, and although they are not always immediately evident in all types of lasers, their functions are essential. Figure 1-1 shows these elements in a ruby and a helium-neon laser; the internal workings of lasers are described in more detail in Chap. 3.

There are several general characteristics which are common to most lasers which new users may not expect. Like most other light sources, lasers are inefficient in converting input energy into light; efficiencies can range from under 0.01 to around 20 percent. These low efficiencies can lead to special cooling requirements and duty-cycle limitations, particularly for high-power lasers. In some cases, special equipment may be needed to produce the right conditions for laser operation, such as cryogenic temperatures for the lead salt semiconductor lasers described in Chap. 19. Operating characteristics of individual lasers depend strongly on structural components such as cavity optics, and in many cases a wide range is possible. Packaging can also have a strong impact on laser characteristics and the use of lasers for certain applications. Thus wide ranges of possible characteristics are specified in many chapters, although single devices will have much more limited ranges of operation.

Differences from Other Light Sources

The basic differences between lasers and other light sources are the characteristics often used to describe a laser: the output beam is narrow, the light is monochromatic, and the emission is coherent. Each of these features is important for certain applications and deserves more explanation.

A typical laser beam has a divergence angle of around a milliradian, meaning that it will spread to one meter in diameter after traveling a kilometer. This figure can vary widely depending on the type of laser and the optics used with it, but in any case it serves to concentrate the output power onto a small area. Thus a modest laser power can produce a high intensity inside the small area of the laser beam; the intensity of light in a 1-mW helium-neon laser beam is comparable to that of sunlight on a clear day, for example. The beams from high-power lasers, delivering tens of watts or more of continuous power or higher peak powers in pulses, can be concentrated to high enough intensities that they can weld, drill, or cut many materials.

The laser beam's concentrated light delivers energy only where it is focused. For example, a tightly focused laser beam can write a spot on a light-sensitive material without exposing the adjacent area, allowing high-resolution printing. Similarly, the beam from a surgical laser can be focused onto a tiny spot for microsurgery, without heating or damaging surrounding tissue. Lenses can focus the parallel rays in a laser beam to a much smaller spot than they can the diverging rays from a point source, a factor which helps compensate for the limited light-production efficiency of lasers.

Most lasers deliver a beam that contains only a narrow range of wavelengths, and thus the beam can be considered monochromatic for all practical purposes. Conventional light sources, in contrast, emit light over much of the visible and infrared spectrum. For most applications, the range of wavelengths emitted by lasers is narrow enough to make life easier for designers by avoiding the need for achromatic optics and simplifying the task of understanding the interaction between laser beam and target. However, for some applications

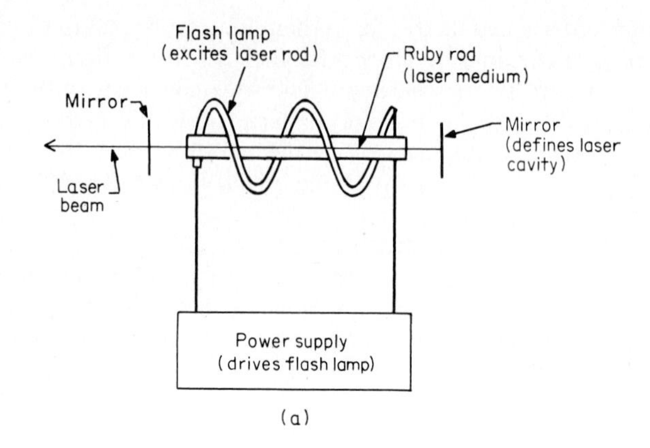

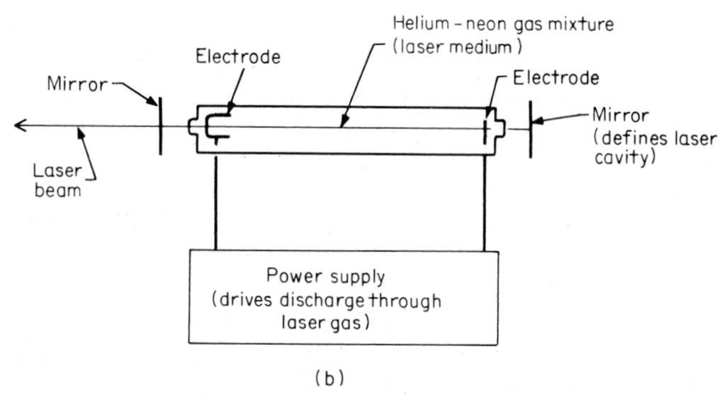

Figure 1-1 Simplified views of two common lasers, *(a)* ruby and *(b)* helium-neon, showing the basic components that make a laser.

in spectroscopy and communications, that range of wavelengths is not narrow enough, and special line-narrowing options may be required.

One of the laser beam's most unique properties is its *coherence,* the property that the light waves it contains are in phase with one another. Strictly speaking, all light sources have a finite coherence length, or distance over which the light they produce is in phase. However, for conventional light sources that distance is essentially zero. For many common lasers, it is a fraction of a meter or more, allowing their use for applications requiring coherent light. The most important of these applications is probably holography, although coherence is useful in some types of spectroscopy, and there is growing interest in communications using coherent light.

Some types of lasers have two other advantages over other light sources: higher power and longer lifetime. In some lasers, such as semiconductor types, lifetime must be traded off against higher power, but in others the life-vs.-power trade-off is minimal. The combination of high power and strong directionality makes certain lasers the logical choice to deliver high light intensities to small areas. For some applications, lasers offer longer lifetimes than other light sources of comparable brightness and cost. In addition, despite their low efficiency, some lasers may be more efficient in converting energy to light than other light sources.

The Laser Industry

Commercial Lasers There is a big difference between the world of laser research and the world of the commercial laser industry. Unfortunately, many text and reference books fail to differentiate between types of lasers that can be built in the laboratory and those which are readily available commercially. That distinction is a crucial one for laser users.

Laser emission has been obtained from hundreds of materials at many thousands of emission lines in laboratories around the world. Extensive tabulations of these laser lines are available (Weber, 1982), and even today researchers are adding more lines to the list. However, most of these laser lines are of purely academic interest. Many are weak lines close to much stronger lines which dominate the emission in practical lasers. Most of the lasers that have been demonstrated in the laboratory have proved to be cumbersome to operate, low in power, inefficient, and/or simply less practical to use than other types.

Only a couple of dozen types of lasers have proved to be commercially viable on any significant scale; these are described in the rest of this book and summarized in Appendix B. Some of these types, notably the ruby and helium-neon lasers, have been around almost since the beginning of the laser era. Others, such as the excimer laser family, are promising newcomers. The family of commercial lasers is expanding slowly, as new types such as alexandrite and copper vapor come on the market, but with the economics of production a factor to be considered, the number of commercially viable lasers will always be limited.

There are many possible reasons why certain lasers do not find their way onto the market. Some require exotic operating conditions or laser media, such as high temperatures or highly reactive metal vapors. Some emit only feeble powers. Others have only limited applications, particularly lasers emitting low powers in the far-infrared, or in parts of the infrared where the atmosphere is opaque. And some simply cannot compete with materials already on the market.

Market Size Sales of lasers per se—including power supply and optics—neared the $500 million mark in the noncommunist world in 1985 (*Lasers & Applications,* 1986). The same survey puts sales of systems containing lasers at nearly $6.4 billion in the same year—a dozen times higher than the sales of "bare" lasers. The difference reflects an important fact: for many applications, the cost of the laser represents only a small part of the system cost.

Because some lasers are much more expensive than others, there are large dichotomies between rankings of laser sales based on number and those based on dollars. Over 2.8 million semiconductor diode lasers were sold in 1985, 90 percent of the total number of lasers. However, diode lasers accounted for less than 25 percent of the dollar volume of sales. Conversely, under 5000 solid-state crystalline lasers (predominantly neodymium) accounted for almost the same dollar volume as diode lasers in 1985.

Types of Organizations Most companies in the laser industry are young, and many are not far removed from humble origins in a garage or basement. The biggest company in the industry, Spectra-Physics Inc., was founded in 1961. The number-two company, Coherent Inc., was established in the mid-1960s. Most companies are even younger, with a fresh round of startups appearing in the early 1980s.

There are some corporate giants in the field. Hughes Aircraft, where the first laser was built in 1960, makes helium-neon and carbon dioxide lasers for the commercial market, and neodymium lasers for the military. RCA and the Japanese giant Hitachi make semiconductor lasers. Companies such as Johnson & Johnson have entered the laser business by buying smaller companies.

Some of the industry giants, such as Spectra-Physics and Coherent, have broad product lines, but many companies specialize in just one or a few types of lasers. While some companies produce related products, others stick to lasers. Likewise, many other companies concentrate on related products such as optics or power supplies and stay away from producing lasers per se. The field is a dynamic one, and readers trying to keep up with who's making what should follow the trade magazines which cover the field: *Lasers & Applications, Laser Focus,* and *Photonics Spectra.*

Market Structure There are few clear-and-fast patterns in the structure of the laser market. The two largest laser makers, Spectra-Physics and Coherent, for many years concentrated on gas and dye lasers, but both now offer broader product lines. Overseas companies, particularly the Japanese, have persistently tried to break into the North American market, but with the exception of semiconductor lasers have not had notable success. Spectra-Physics and Coherent are both truly international corporations, doing significant shares of their business in western Europe. Export restrictions sharply limit sales of American products to eastern Europe and the Soviet Union; the Soviet Union's export organization makes periodic efforts to sell laser products in the United States, but so far without much sign of success.

The biggest laser makers also offer optical components and accessories for single-stop shopping, but many laser makers offer accessories only if they can be used with their own products. There are many small companies (and some large ones) which make laser optics, often custom-made for specific applications. There are several companies which function as catalog supply houses for optics and accessories. All told, about a hundred companies make lasers, and several hundred more provide laser accessories and offer services to the laser industry. The cast of companies changes; for current information consult an industry directory.

An Introduction to Laser Safety

Laser safety has been a controversial topic ever since lasers began appearing in laboratories. There are two major concerns—eye hazards and high voltages within the laser and power supply.

Eye hazards are by far the most controversial and are the subject of laser product performance standards issued by the federal government's National Center for Devices and Radiological Health, formerly known as the Bureau of Radiological Health. Voluntary standards for the safe use of lasers have also been issued by the American National Standards Institute (ANSI). Laser safety is covered extensively in an excellent reference work by David Sliney and Myron Wolbarsht (1980). There are two basic types of eye hazards. High-power lasers can cause permanent damage if a single powerful pulse enters the eye. Lower-power lasers are hazardous for prolonged exposure, that is, if the user stares into the laser beam. The specific hazards depend on the laser power, pulse length, and wavelength. They are described briefly in each chapter of this book, and much more extensively in the Sliney and Wolbarsht book.

High-voltage hazards are less controversial but far more deadly. There is no public record of any person having been killed by a laser beam. However, several people have been killed when they came into contact with high voltages in a laser or power supply. One acquaintance of mine is on that list, electrocuted while servicing a laser. I hope readers of this book will remember to keep their fingers away from the power supply.

Overview of Laser Applications

Laser applications are far too broad and diverse to cover in any detail here. Individual chapters cover the most important applications of each major type. The industrial and commercial applications fall into several basic categories:

- Materials working
- Measurement and inspection
- Reading, writing, and recording of information
- Displays
- Communications
- Holography
- Spectroscopy and analytical chemistry
- Remote sensing
- Surveying, marking, and alignment

There are also several important research applications pursued on various scales by government and industrial organizations and in universities:

- Laser weaponry
- Laser-induced nuclear fusion (sometimes called inertial confinement fusion)
- Isotope enrichment (particularly of uranium and plutonium)
- Spectroscopy and atomic physics

- Measurement
- Plasma diagnostics

Historically, much support for laser research has come from the Department of Defense or quasi-military organizations. Indeed, the principal justification for research in laser fusion has been military interest in simulating the effects of nuclear weapons. The largest military research program now is in efforts to develop laser weapons for use against satellites or for defense against nuclear attack. Laser systems are now in production which serve as range finders, target designators for smart bombs, and scorekeepers for battle-simulation systems.

BIBLIOGRAPHY

ANSI Z-136.1–1980, ANSI Standard for the Safe Use of Lasers, American National Standards Institute, New York, 1980.

Jeff Hecht and Dick Teresi: *Laser: Supertool of the 1980s*, Ticknor & Fields, New York, 1982. Popular-level introduction to lasers and their applications.

Laser Focus Buyers' Guide, PennWell Publishing, Littleton, Mass. Published annually; for subscription information write the publisher at 119 Russell St., Littleton, MA 01460.

Lasers & Applications (staff report), "The Laser Marketplace—1986," *Lasers & Applications* 5(1):45–56 (January 1986).

Lasers & Applications Buying Guide, High Tech Publications, Inc., Torrance, California. Published annually; for subscription information write to the publisher at 23868 Hawthorne Blvd., Torrance, CA 90505.

Optical Industry & Systems Purchasing Directory, Optical Publishing, Pittsfield, Mass. Published annually; for subscription information, write the publisher at Berkshire Common, P.O. Box 1146, Pittsfield, MA 01202.

Performance Standards for Laser Products, National Center for Devices and Radiological Health, publication no. HFX-430, Rockville, MD. Federal standards. To obtain a copy write to 5600 Fishers Lane, Rockville, MD 20857.

David Sliney and Myron Wolbarsht: *Safety with Lasers and Other Optical Sources*, Plenum, New York, 1980.

Marvin J. Weber (ed.): *CRC Handbook of Laser Science and Technology*, 2 vols., CRC Press, Boca Raton, Fla., 1982.

CHAPTER

2

a brief history of the laser

The key concepts of laser physics are far enough removed from the normal realm of classical physics that it took many years to put the conceptual pieces together into a working laser. The basic idea of the stimulated emission of radiation dates back to the early part of this century, but it wasn't until the 1950s that physicists began to work on putting it to practical use. Once the conceptual logjam was broken, the early 1960s saw an amazing and rapid proliferation of laser types, followed by years of experimentation with the new playthings. Although important research advances have also been made in the 1970s and the first part of this decade, there has been extensive work in developing laser applications ranging from fiber-optic communications and information handling to laser fusion and high-energy laser weapons.

This chaper can only touch on some of the high points in the tangled history of the laser. There are really two parallel and largely intertwined stories to be told: the intellectual evolution of the laser concept, and the competition among a number of strong personalities to build the first laser. Readers interested in the intellectual evolution of the field would do well to study *Masers and Lasers, An Historical Approach,* by Mario Bertolotti (1983) or an objective early account by Bela Lengyel (1966). The interplay of individuals has not been treated in as much detail, but I have written historical chapters in books on laser applications (Hecht and Teresi, 1982) and on laser weaponry (Hecht, 1984). I have also written a historical introduction to a collection of interviews with laser pioneers, originally published in *Lasers & Applications* (Hecht, 1985). The Laser History Project, headed by Joan Bromberg, is in the process of collecting oral histories and locating documentary sources on the development of the laser, but that project is not scheduled for completion for several years. It is sponsored by the American Physical Society, the IEEE Quantum Electronics and Applications Society, the Laser Institute of America, and the Optical Society of America.

Stimulated Emission

The idea of stimulated emission of radiation originated with Albert Einstein (1916). Until that time, physicists had believed that a photon could interact with an atom in only two ways: it could be absorbed and raise the atom to a higher energy level or be emitted as the atom dropped to a lower energy level. Einstein proposed a third possibility—that a photon with energy corresponding to that of an energy-level transition could stimulate an atom in the upper level to drop to the lower level, in the process stimulating the emission of another photon with the same energy as the first.

In the normal world, stimulated emission is unlikely because at thermodynamic equilibrium more atoms are in lower energy levels than in higher ones. Thus a photon is much more likely to encounter an atom in a lower level and be absorbed than to encounter one in a higher level and stimulate emission. The first evidence for stimulated emission was not reported until more than a decade after Einstein's prediction (Ladenburg, 1928). For more than two decades thereafter, stimulated emission seemed little more than a laboratory curiosity.

The Coming of the Maser

The first efforts to use stimulated emission were in the microwave region and led to the invention of the maser (from "microwave amplification by the stimulated emission of radiation"). The maser concept evolved nearly simultaneously in the United States and the Soviet Union, but the name usually linked with the invention is that of Charles H. Townes, a physicist then on the faculty of Columbia University and now professor emeritus at the University of California at Berkeley.

Townes says that the maser idea struck him early one morning in the spring of 1951, while he was sitting on a park bench in Washington before going to a meeting of the American Physical Society (Townes, 1978). He had been puzzling over ways to build a source of millimeter waves when he realized that molecules had the right characteristics. He envisioned starting with a molecular beam, which could be split into excited and unexcited portions. Molecules in the excited part could be stimulated to emit microwaves when placed in a special resonant cavity designed to enhance the emission. He outlined the idea to postdoctoral fellow Herbert Zeiger and graduate student James P. Gordon, and by 1953 they had a working maser. Meanwhile, Aleksander M. Prokhorov and Nikolai Basov (1954) of the Lebedev Physics Institute in Moscow had completed detailed calculations of the conditions required for maser action, which were published shortly after Townes's results. The contributions of all three men were recognized in 1964 when they shared the Nobel prize in physics.

Theoretical Groundwork for the Laser

Masers proliferated rapidly in the 1950s. Meanwhile, Townes and other physicists began looking beyond the microwave region to shorter wavelengths. They recognized that at those shorter wavelengths the physical conditions required to produce stimulated emission would be very different than in a maser. Many of the key principles were worked out by Townes and Arthur L. Schawlow, then a Bell Laboratories researcher and now on the faculty at Stanford University. Their results were published in a major paper (Schawlow and Townes, 1958), which pointed out important differences between requirements in the microwave and visible regions, including cavity structure, spontaneous emission ratios, energy differences between energy levels, and excitation mechanisms. The two also filed a patent application before the paper was published, leading to a now-expired U.S. patent no. 2,929,922.

Meanwhile, a Columbia graduate student, Gordon Gould, was working out his own analysis of the conditions required for stimulated emission at visible wavelengths. Gould wrote his proposals in a set of notebooks dated 1957 but did not try to publish his results promptly. Considering himself more an inventor than a researcher, Gould wanted to patent his work. He had his notebooks notarized promptly, but because of bad legal advice he did not file a patent application until April of 1959, about 9 months after the Schawlow-Townes patent application was submitted, and after publication of their paper.

The Schawlow-Townes patent was granted promptly, but Gould's application ran into a lengthy set of interferences. Gould pursued the patent himself for many years, losing some rounds and winning others, but in the process running up massive legal bills. Finally, he turned the litigation and part interest in his patent position over to the Refac Technology Development Corp. of New York. Eventually he was granted two patents (Gould 1977, 1979) based on divisions of his original application. The 1977 patent covers optical techniques for pumping or energizing the laser medium, such as using a flashlamp to drive a dye or neodymium laser. The 1979 patent covers certain laser applications. Efforts have been made to force laser companies to license the technology in these patents, but the validity of the patents remains under litigation, as do requests for issuance of other patents based on Gould's original application. The suits began in 1977, and with the legal system proceeding at a glacial pace, could conceivably outlive the 17-year lifetime of the patents. Ironically, because the laser industry is much larger now than when the Schawlow-Townes patent was in force, the Gould patents could prove much more valuable.

Schawlow and Townes have received many scientific honors for their work, but Gould received little recognition until his patents were issued. The question of who really deserves credit for conceiving the laser may never be definitely resolved. However, while Gould did lose the prestige race, it was he who first coined the word *laser* in his notebooks. Schawlow and Townes described their idea as an "optical maser."

The Race to Build a Laser

Publication of the Schawlow-Townes paper stimulated many efforts to build lasers, and interest spread beyond the narrow scientific community. Gould and executives from TRG Inc., a small Long Island company, got an enthusiastic reception when they presented the Pentagon with a proposal to build a laser. Indeed, the Department of Defense's Advanced Research Projects Agency was so excited it gave TRG a million dollars for the project, rather than the $300,000 the company had requested. Then the military agency proceeded to deny Gould a security clearance to work on the project because of his youthful but long-dormant interest in Marxism (Hecht and Teresi, 1982).

Schawlow, Gould, and most other researchers thought that the best materials for building lasers were gases. Theodore H. Maiman, a young physicist at Hughes Research Laboratories in Malibu, California, quietly disagreed, preferring synthetic ruby crystals, although some theorists insisted that ruby would not work. Maiman, who had studied energy levels in ruby extensively while working on ruby masers, proved that the theorists were wrong. In mid-1960 he proudly demonstrated the world's first laser: a rod of synthetic ruby with reflecting coatings on the ends, surrounded by a helical flashlamp. When the lamp was pulsed, a pulse of red light emerged from one end of the rod, which had a partially transparent coating. The laser era was born.

Maiman's course was not an easy one. Hughes management told him to stop working on the ruby laser, although once it worked the company's public relations officials were delighted to show off the results. The prestigious journal *Physical Review Letters* rejected his report of the ruby laser as "just another maser," a blunder few scientific journals can claim to have matched. A shorter version was published in the British journal *Nature* (Maiman, 1960). Today, however, Maiman is universally recognized as the person who built the first laser and has received a number of honors, although not enough to match Townes and Schawlow. The ruby laser remains in commercial use, although its popularity has declined in recent years.

The Great Laser Explosion

Maiman's demonstration of the ruby laser opened the floodgates. Before the year was out, a second type of solid-state laser was reported, trivalent uranium ions in calcium fluoride, by Peter P. Sorokin and M. J. Stevenson (1960) at the IBM Corp. That laser has never found significant practical use. However, it was followed soon afterward by the demonstration of the helium-neon laser by Ali Javan, W. R. Bennett Jr., and Donald R. Herriott at Bell Telephone Laboratories in Murray Hill, New Jersey (Javan et al., 1961). Their first helium-neon laser (Fig. 2-1) operated at 1.15 micrometers (μm) in the near-infrared; later other researchers found the 632.8-nanometer (nm) red line which has made the helium-neon laser one of the most widespread types (White and Rigden, 1962).

The laser boom really got going in 1961. L. F. Johnson and K. Nassau (1961) demonstrated the first solid-state neodymium laser, in which the neodymium ion was a dopant in calcium tungstate ($CaWO_4$). Elias Snitzer demonstrated the first neodymium-glass laser at American Optical that same year. However, it would be three more years before today's best choice of neodymium host for most commercial applications—yttrium aluminum garnet (YAG)—was demonstrated as a laser material by J. E. Geusic, H. M. Marcos, and L. G. Van Uitert.

The first semiconductor diode lasers were demonstrated nearly simultaneously by three separate groups in the fall of 1962 (Hall, 1976). All three teams—at General Electric Research Laboratories in Schenectady, New York; the IBM Watson Research Center in Yorktown Heights, New York; and MIT's Lincoln Laboratories in Lexington, Massachusetts—demonstrated similar gallium arsenide diodes cooled to the 77 K temperature of liquid nitrogen and pulsed with high-current pulses lasting a few microseconds.

The next few years saw the births of several more of today's most important lasers. William B. Bridges (1964) observed 10 laser transitions in the blue and green parts of the spectrum from singly ionized argon, the basis of today's argon ion laser. C. Kumar N. Patel (1964) obtained a 10.6-μm laser emission from carbon dioxide, today a high-power workhorse in industry. Sorokin and J. R. Lankard (1966) at IBM's Watson Research Center demonstrated the first organic dye laser, today a standard tool of laser spectroscopy. The first chemical laser, hydrogen chloride emitting at 3.7 μm, was demonstrated in 1965 by J. V. V. Kaspar and G. C. Pimentel.

The rate of discovery of new laser types slowed during the 1970s, with the most important discoveries being the excimer and free-electron lasers. The first members of the excimer laser family were rare gas dimers, pairs of rare gas atoms bonded together to form a "molecule" which existed only in the excited state. In the mid-1970s, interest shifted to rare gas halides, which are much more practical light sources and which have become a significant part

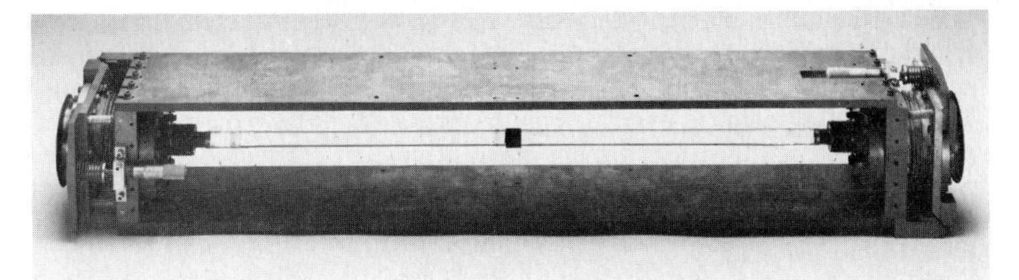

Figure 2-1 The first gas laser, a helium-neon type conceived and developed by Ali Javan, which was demonstrated for the first time on December 12, 1960, at Bell Telephone Laboratories in Murray Hill, New Jersey. The helium-neon gas mixture was contained in an 80-cm-long quartz tube with inner diameter of 1.5 cm; its output was in the near-infrared. *(Photo by Lee Nadel; courtesy of Ali Javan and Laser Science, Inc.)*

of the laser business. In contrast, the free-electron laser, pioneered by John M. J. Madey of Stanford University (Deacon et al., 1977), remains in the laboratory, although there is widespread optimism that it will prove to be of practical importance.

The major research breakthrough of the 1980s may be the x-ray laser. In one set of experiments x rays produced by the explosion of a nuclear bomb stimulate the emission of x rays from suitable materials. However, details of these tests remain shrouded in secrecy because of the potential military applications (Hecht, 1984). More recently, tiny targets illuminated by extremely short and intense laser pulses have produced extreme ultraviolet laser emission at wavelengths as short as 15.5 nm in experiments which have been reported publicly by the Lawrence Livermore National Laboratory.

Evolution of Practical Lasers

It always takes time to translate laboratory developments into practical products, and lasers have been no exception. For a while after Maiman demonstrated the first ruby laser, other laboratories around the world were busy building their own versions. Once they found out that the lasers worked, the researchers set out to see what could be done with them. The answers were not always clear, and one wag labeled the laser "a solution looking for a problem," a description that stuck long after its time had passed.

Maiman was one of the first to try to create a commercial laser industry. Soon after he demonstrated the ruby laser, he resigned from Hughes, and after spending nearly a year at the now-defunct Quantatron, formed Korad, a company which began making ruby and other solid-state lasers. Those early lasers generally went directly to research and development laboratories to see what could be done with them. Testing was sometimes improvised because suitable measurement instruments were not always available. Legend has it that the power of early pulsed lasers was measured in "gillettes"—the number of razor blades that could be pierced by a single pulse. However, it was not long before ruby lasers began finding applications ranging from drilling holes in diamond dies used in drawing wires to finding the ranges to military targets.

It took a while longer to commercialize the now-ubiquitous helium-neon laser. The first operation of that laser was at its 1.15-μm near-infrared line; it was not until 1962 that the now-standard 632.8-nm red line was demonstrated. The helium-neon laser has never been able to produce high powers, but it has proved to be easy to manufacture, and its low-power output at a visible wavelength has found many uses. Prominent early uses included laboratory demonstrations, production of holograms, and construction alignment, but that range of applications has now broadened to include many types of information handling.

The solid-state neodymium laser took more time to perfect because of problems in finding a suitable crystalline host. Although glass was identified early, heat-dissipation problems have always restricted its operation. It took a number of years to settle on YAG as the best crystalline host for neodymium that could be grown in reasonable quality at a reasonable cost. Because of its excellent thermal and optical qualities, Nd–YAG has become the standard solid-state laser material, displacing ruby from many applications.

The carbon dioxide laser also took time to perfect. Much early effort went into finding the gas mixture that gave the optimum output power. New methods of exciting the laser gas have also been developed to enhance the output power and efficiency. Military-sponsored research pushed output powers to many kilowatts, with some demonstration lasers having produced hundreds of kilowatts. In recent years, compact CO_2 lasers have been developed that produce a few watts when the gas passes through a compact waveguide structure—a design that is attractive for many military and civilian uses.

The dye laser has evolved into a powerful tool for spectroscopy and atomic physics. The field has yielded impressive results, aided by dye laser developments pushed by laser pioneer Arthur Schawlow and his Stanford colleague Theodor Hänsch. A number of other laser spectroscopists have played key roles in advancing the field (not always with dye lasers), notably Nicolaas Bloembergen of Harvard University and Vladilen S. Letokhov of the Soviet Institute of Spectroscopy. Schawlow and Bloembergen shared the 1981 Nobel prize in physics for their contributions to laser spectroscopy; the younger Hänsch and Letokhov may be candidates for future awards.

Argon and krypton ion lasers have also become standard commercial products, along with helium-cadmium lasers, rare gas excimers, and a few other types. While the technology for argon, krypton, and helium-cadmium lasers is fairly mature, that for excimer lasers is still developing. Chemical lasers have not become commercially important; although a few small models are on the market, most of the interest in chemical lasers has come from military high-energy laser weapon programs. Laser weapon research remains a driving force behind work on free-electron and x-ray lasers and is also pushing work on high-energy excimer lasers.

The biggest success story of the past few years has been the emergence of the semiconductor diode laser. Although among the first types developed, the diode laser initially had big problems. Early versions required very high drive currents, which quickly heated the devices past their damage thresholds, limiting operation to short pulses at cryogenic temperatures. It was not until the mid-1970s that the first diode lasers capable of continuous operation at room temperature came on the market. Great advances in laser diode structure since then have led to better lasers with longer lifetimes, lower threshold currents, higher powers, a broader range of wavelengths, better beam quality, and more linear light-current curves. These advances and inexpensive mass-production methods have led to widespread applications of diode lasers in fiber-optic communications and information handling.

Current Trends

Laser technology is continuing to advance, but the progress is not spread evenly among the various types. Some types, notably helium-neon, ion, ruby, and neodymium, are generally mature in design; efforts continue to refine performance and improve manufacturing, but dramatic advances seem unlikely. However, major progress continues in other areas.

Excimer laser manufacturers are trying to push their technology from the research laboratory into the industrial world. They face formidable problems because the need to use halogens in rare-gas-halide lasers makes it hard to achieve the highly reliable operation desired by industry. However, they are making significant progress. Government researchers are also making progress in pushing excimer laser output powers to higher levels.

The emergence of the waveguide CO_2 laser, and of new designs for sealed-tube lasers, has breathed new life into the technology of moderate-power carbon dioxide lasers. Several companies are working on compact lasers with output power of 100 W or less that are promising for applications in medicine and materials working.

The trends that have pushed semiconductor laser technology are continuing. New structures are making possible higher powers and higher-speed modulation. New materials and new structures are opening up new wavelength regions. Many observers believe that diode lasers will take over many applications of helium-neon lasers that do not require visible light. The ability to produce visible light efficiently without degrading operating lifetime would open up new frontiers for diode lasers, but practical visible-output diode lasers remain elusive.

The free-electron laser seems destined to find an important place in the family of high-power lasers. Theoreticians spent years puzzling over the nature of the beast, but they now seem to understand it well enough to predict the results of experiments. A planned series of experiments will examine the prospects for reaching high efficiencies and high powers, and there is also work underway on building kilowatt-output free-electron lasers for laboratory research. The prospects for practical uses of x-ray lasers are much less clear, but recent laboratory experiments have shown that energy sources other than a nuclear explosion can power them.

The laser industry per se is remote from military laser weapon programs; only a handful of aerospace contractors both conduct laser weapon research and build commercial lasers. However, a research program spending hundreds of millions of dollars a year on laser-related technology cannot help but impact the laser world. Even if laser weapons never shoot down a single unfriendly missile, some high-power laser technology developed in the laser weapon program has found many peaceful uses.

BIBLIOGRAPHY

N. G. Basov and A. M. Prokharov: "3-level gas oscillator," *Zh. Eksp Teor Fiz (JETP) 27:*431, 1954.

Mario Bertolotti: *Masters & Lasers, An Historical Approach,* Adam Hilger Ltd., Bristol, U.K., 1983.

William B. Bridges: "Laser oscillation in singly ionized argon in the visible spectrum," *Applied Physics Letters 4:*128–130, 1964; erratum, *Applied Physics Letters 5:*39, 1964.

D. A. G. Deacon, et al.: "First operation of a free-electron laser," *Physical Review Letters 38:*892, 1977.

Albert Einstein: *Mitt. Phys. Ges., Zurich 16*(18):47, 1916. English translations appear in

B. L. van der Waerden (ed.), *Sources of Quantum Mechanics,* North-Holland, Amsterdam, 1967; and in D. ter Haar, ed., *The Old Quantum Theory,* Pergamon, Oxford and New York, 1967.
J. E. Geusic, H. M. Marcos, and L. G. Van Uitert: "Laser oscillations in Nd-doped yttrium aluminum, yttrium gallium, and gadolinium garnets," *Applied Physics Letters 4:*182, 1964.
Gordon Gould: U.S. patent 4,053,845, issued October 11, 1977; U.S. patent 4,161,436, issued July 17, 1979.
Robert N. Hall: "Injection lasers," *IEEE Transactions on Electron Devices ED-23*(7):700–704, July 1976.
Jeff Hecht: *Beam Weapons: the Next Arms Race,* Plenum, New York, 1984.
Jeff Hecht: "An introduction to laser history," pp. 1–33 in *Lasers & Applications, LASER PIONEER INTERVIEWS* High Tech Publications, Torrance, Calif., 1985.
Jeff Hecht and Dick Teresi: *Laser: Supertool of the 1980s,* Ticknor & Fields, New York, 1982.
Ali Javan, W. R. Bennett Jr., and D. R. Herriott: "Population inversion and continuous optical maser oscillation in a gas discharge containing a He–Ne mixture," *Physical Review Letters 6:*106, 1961.
L. F. Johnson and K. Nassau: *Proceedings of the Institute of Radio Engineers 49:*1704, 1961.
J. V. V. Kaspar and G. C. Pimentel: "HCl chemical laser," *Physical Review Letters 14:*352, 1965.
Rudolf Ladenburg: "Untersuchungen über die anomale Dispersion angeregter Gase" [Investigation into anomalous dispersion in certain gases], *Zeitschrift für Physik 48:*17–25 (April–May 1928).
Bela Lengyel: "Evolution of masers and lasers," *American Journal of Physics 34*(10):903–913, October, 1966.
Theodore H. Maiman: "Stimulated optical radiation in ruby," *Nature 187:*493, 1960.
C. Kumar N. Patel: "Continuous-wave laser action on vibrational-rotational transitions of CO_2," *Physical Review 136A:*1187, 1964.
Arthur L. Schawlow, "Masers and lasers," *IEEE Transactions on Electron Devices ED-23:*773–779, July 1976.
Arthur L. Schawlow and Charles H. Townes: "Infrared and optical masers," *Physical Review 112:*1940, 1958; also U.S. patent 2,929,922.
Elias Snitzer, "Optical maser action of Nd^{+3} in a barium crown glass," *Physical Review Letters 7:*444, 1961.
Peter P. Sorokin and J. R. Lankard: "Stimulated emission observed from an organic dye, chloroaluminum phtatocyanine," *IBM Journal of Research & Development 10:*162, 1966.
Peter P. Sorokin and M. J. Stevenson, "Stimulated infrared emission from trivalent uranium," *Physical Review Letters 5:*557, 1960.
Charles H. Townes: "The laser's roots: Townes recalls the early days," *Laser Focus 14*(8):52–58, August, 1978.
A. D. White and J. O. Rigden: "Continuous gas maser operation in the visible," *Proceedings of the IRE 50:*1796, 1962.

CHAPTER

3

laser theory and principles

The term *laser* covers a variety of devices, as will be clear from later chapters. While many lasers bear obvious family resemblances to each other, some types are quite distinct. Helium-neon, ion, and carbon dioxide lasers have some important similarities because they all are electrically excited gas lasers, but at first glance they seem to have little in common with semiconductor diode lasers. However, the operation of all lasers relies on certain common physical principles, and it is those principles that will be covered in this chapter.

Laser physics is a discipline that draws upon many other fields, among them quantum mechanics, optics, gas dynamics, semiconductor physics, atomic and molecular physics, and resonator theory. One of the most important factors is the transfer of energy among atoms and molecules, which is closely related to quantum mechanics. Laser characteristics also depend strongly on the optical design of the cavity of the laser medium, and that design, in turn, depends on the physics of the laser medium. This chapter gives an introduction to the fundamentals of lasers; readers seeking a more extensive treatment can consult a number of texts (Svelto, 1982; Tarasov, 1983; Thyagarajan and Ghatak, 1981; Yariv, 1975; O'Shea et al., 1977).

Terminology

The formal name for laser physics is *quantum electronics.* The term dates back to the early days of maser research, and today it is used mainly in connection with academic research. Books with "quantum electronics" in the title tend toward an emphasis on theory rather than practical uses of lasers. The term remains attached to two institutions in the laser field: the biennial International Quantum Electronics Conference, traditionally the most research-oriented of the major meetings in the laser field, and the laser journal published by the Institute of Electrical and Electronics Engineers, the *IEEE Journal of Quantum Electronics.* This book will stay with the more common usage of laser physics.

Much laser physics terminology comes from spectroscopy and the physics of atoms and molecules. One of the most important concepts is the *energy level,* a quantum state of an atom or molecule. The energy of a particular level is measured relative to the ground level, or lowest possible energy level, which is arbitrarily assigned zero energy. The term *species* is used to avoid repeating the phrase "atom or molecule" constantly when discussing energy levels in general terms. The shift of a species from one energy level to another is called a *transition.* A shift to higher energy levels is an excitation to an excited state and requires the absorption of an amount of energy equal to the difference in energy between the two levels. A drop from a higher energy level to a lower one is called a *decay* or *deexcitation* and is accompanied by the release of the transition energy.

Species often absorb or release energy in the form of photons, quanta of electromagnetic radiation. In the laser world, the word "light" is used generically for electromagnetic radiation throughout the infrared and ultraviolet parts of the spectrum, as well as for true visible light. Light itself has a dual nature, as both a particle (a photon) and as a wave. The nature of light can be measured in three ways: as the wavelength of a light wave, as the frequency of oscillation of a wave, or as the energy of a photon. Wavelength is invariably measured in metric units, micrometers (μm), nanometers (nm), or the not-quite-legitimately metric angstrom (Å) (0.1 nm). Frequency and photon energy are actually equivalent measurements, according to the Planck relationship $E = h\nu$, where E is energy, h is Planck's constant, and ν is the frequency. Energy is typically measured in electronvolts (eV; 1 eV is the energy needed to move an electron through a potential of 1 V). Frequency is normally measured in hertz (Hz), the number of oscillations per second. Spectroscopists, particularly those used to working in the infrared, sometimes use the peculiar unit of inverse centimeters (cm^{-1}), which as the name implies is the reciprocal of the wavelength in centimeters; although the units are based on wavelength, they are proportional to frequency or photon energy.

Many other laser terms come from optics or electronics. An extensive listing is given in the glossary in Appendix A at the end of this book.

Energy Levels and Transitions

There are three basic types of energy levels encountered in laser physics, illustrated in Fig. 3-1:

- *Electronic levels* involving the electron configuration in a species. Electronic transitions are the ones portrayed in the Bohr picture of the hydrogen atom, in which the single electron gradually climbs the ladder of energy levels as it moves away from the nucleus, eventually attaining high enough energies to escape from the nucleus. As more electrons are added to the system in more complex atoms and molecules, the picture becomes increasingly complex, and various interactions create a multitude of electronic energy levels. The electronic transitions which normally occur in laser physics involve electrons in the outer shells and typically involve energies on the order of an electron volt or more, corresponding to wavelengths in the near-infrared, the visible, and the ultraviolet. Inner-shell transitions, closer to the nucleus, involve higher energies corresponding to far-ultraviolet or x-ray photons and are less favorable for laser action.
- *Vibrational levels* involving vibration of the atoms in a molecule. Vibrations in various degrees of freedom are quantized, creating a series of levels for each possible vibration pattern. There are also multiple vibration patterns possible, as shown in Fig. 3-2, with the number increasing with the number of atoms in a molecule. Vibrational transitions occur when molecules shift from one vibrational state to another, which can involve a shift from one

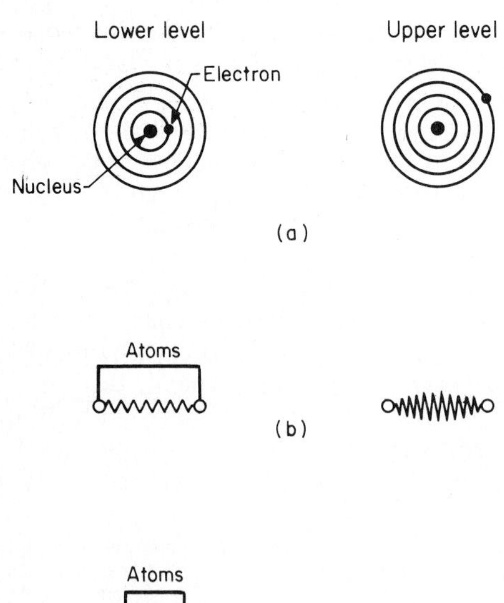

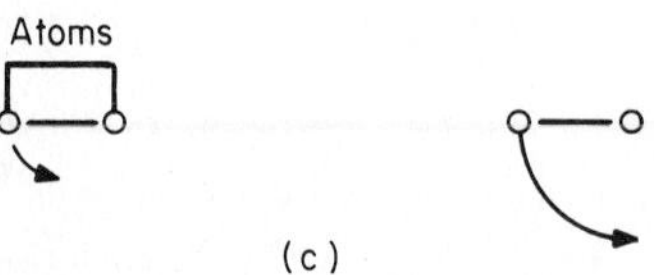

Figure 3-1 Major types of energy-level transitions. *(a)* Electronic transitions occur in atoms and molecules; *(b)* vibrational and *(c)* rotational transitions occur only in molecules.

vibrational pattern to another. Typically vibrational transitions have energies of a fraction of an electron volt and correspond to infrared wavelengths.

- *Rotational levels* involving the quantized rotation of molecules. Energies are typically 0.001 to 0.1 eV and correspond to far-infrared or submillimeter wavelengths.

Complex quantum-mechanical rules determine the likelihood of transitions between different states or energy levels. Some are said to be "forbidden" transitions, although in the world of quantum mechanics forbidden actually means very unlikely, *not* impossible. Others are allowed under the rules of quantum mechanics and thus are likely to occur very quickly. Normally excited states quickly decay to lower-energy states by making one or a series of allowed transitions. However, there are also "metastable" excited levels, states in which a species becomes trapped because there is no readily allowed transition to a lower energy level. These metastable states are important in many lasers because they hold a species in an excited state for a long time on an atomic scale of things.

Although the electronic, vibrational, and rotational states of a species are independent, they can change simultaneously in a single transition. One such case is in the organic dye laser, described in Chap. 17. Electronic transitions of atoms or ions occur independently because such species do not have vibrational or rotational states. However, in molecules, changes in vibrational states are generally accompanied by changes in rotational states.

There is an important difference between transitions involving shifts in only one kind of energy level and those involving simultaneous shifts between two or more types. Transitions between two electronic energy levels of an atom are sharply defined at a specific wavelength. However, if a variety of shifts can occur simultaneously, the transition is spread out over a range of wavelengths, as in dye lasers or carbon dioxide lasers. This may produce a series of discrete lines, as in a CO_2 laser, or a continuum, as in a dye laser.

Excitation

Excitation is the process of raising a species from a lower energy level to a higher one and is a prerequirement for laser action. The transfer of energy into the laser medium can occur by several mechanisms, including:

- Absorption of photons
- Collisions between electrons (or sometimes ions) and species in the active medium

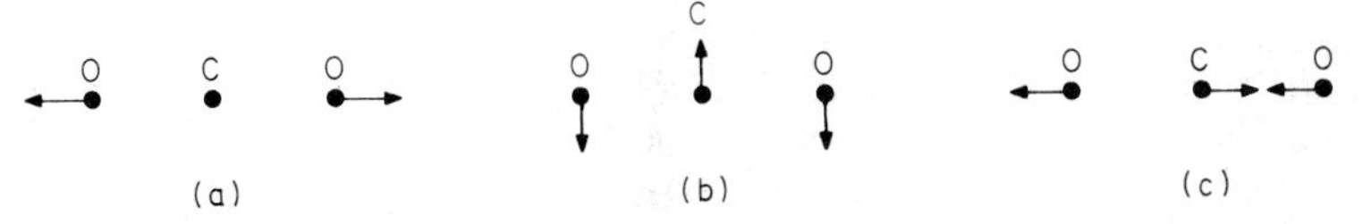

Figure 3-2 Three vibrational modes of the CO_2 molecule. *(a)* Symmetric stretching (v_1 mode); *(b)* bending (v_2 mode); *(c)* asymmetric stretching (v_3 mode). More complex molecules have more complex vibrational modes.

- Collisions among atoms and molecules in the active medium
- Recombination of free electrons with ionized atoms
- Recombination of current carriers in a semiconductor
- Chemical reactions producing excited species
- Acceleration of electrons

The probability of a species absorbing energy by some of these mechanisms is indicated by measurements of *absorption cross section.* Excitation probabilities or cross sections depend on factors such as the wavelength of the illuminating light, the speed of incident electrons, or coincidences in the energy-level structures of species. The excitation of a laser is often a multistep process. For example, in the helium-neon laser electrons passing through the active medium transfer energy to helium atoms, which in turn transfer the absorbed energy to the less-abundant neon atoms, which then emit laser light.

Stimulated and Spontaneous Emission

Excited species can release their excess energy by nonradiative processes, such as collisions with other atoms or molecules, or by emitting a photon. Emission of a photon can be spontaneous or stimulated. As the name implies, spontaneous emission occurs without outside intervention when a species drops to a lower energy level after a natural decay time (analogous to the half-life of radioactive isotopes), typically a tiny fraction of a second.

Emission of a photon and transition to a lower energy level can also be stimulated by the presence of a photon with the same energy as the transition. On first glance, it might seem highly unlikely that a photon of precisely the right energy would wander by in the normally short time before an excited species spontaneously drops to a lower energy level. However, in most cases there is a ready source of such photons: spontaneous emission by the same species. The first few spontaneously emitted photons can trigger stimulated emission of others, leading to a cascade of stimulated emission. Stimulated emission has a couple of special properties. It has the same wavelength as the original photon, and it is in phase (or coherent) with the original light.

Spontaneous emission generates the light we see from the sun, stars, light bulbs, flames, fluorescence, and other normal objects. Stimulated emission produces laser light. Strictly speaking, other light sources can contain small fractions of stimulated emission, but in practice the amount of stimulated emission is undetectable.

Population Inversions

Stimulated emission is rare because of the same thermodynamic fact of life that makes spontaneous emission occur: species tend to drop to the lowest available energy level. Under normal conditions, known as thermodynamic equilibrium, the population of a state tends to decrease as its energy increases, as shown in Fig. 3-3. This means that there should always be a larger population in the lower state of a transition than in the higher state, so a photon is far more likely to be absorbed by a lower-state species than to stimulate emission from one in the higher state. Under these conditions, spontaneous emission dominates.

For stimulated emission to occur, a *population inversion* would be needed—the population of the upper level of a transition would have to be higher than that of the lower level. Then a photon of the transition energy would be more likely to stimulate emission from the excited state than to be absorbed by the lower state. The result is laser gain or amplification, a net increase in the number of photons with the transition energy, corresponding to the difference between stimulated emission and absorption at that wavelength. Typically, gain is measured as percentage increase per pass through the laser medium, or per centimeter of distance through the laser medium. Sometimes gain is measured in the number of added photons generated for each centimeter of laser medium, for example, 4 cm^{-1}, which actually means four per centimeter, not the inverse centimeters of the spectroscopist. Note that a gain of 4 cm^{-1} actually implies that the input power is multiplied by a factor of 5.

Laser gain is proportional to the difference between the chance of stimulated emission and the chance of absorption. This means that the populations of both the upper and lower levels of the laser transition are important. Thus if laser action is to be sustained, the lower level must be depopulated as the upper level is populated, or the population inversion will end. That is indeed what happens in some pulsed lasers.

Laser amplification can occur over a range of wavelengths because no transition is infinitely narrow. The range of wavelengths at which absorption and emission can occur is broadened by molecular motion, vibrational and rotational energy levels, and other factors. Thus laser specialists speak of a

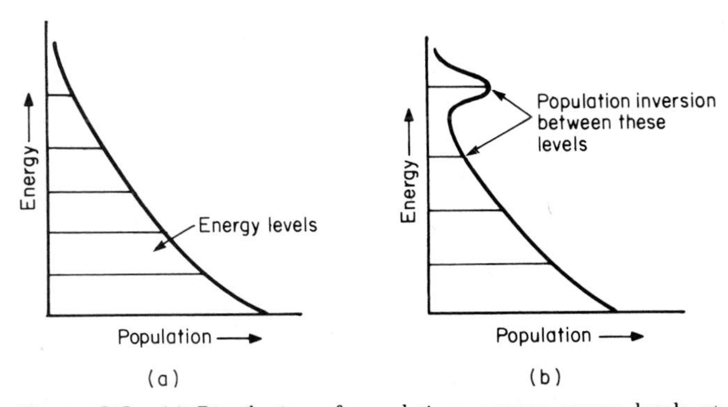

Figure 3-3 *(a)* Distribution of populations among energy levels at equilibrium; and *(b)* during a population inversion. Note that a population inversion does not have to exist for all possible transitions, only for the laser transition. Population inversions can only produce laser emission if they are above the equilibrium energy, the energy for which population is largest in equilibrium.

material having a "gain bandwidth"—a range of wavelengths at which laser gain can occur. The value of the gain factor is highest in the center of the band and lowest at the edges.

Amplification and Oscillation

A population inversion alone is enough to produce "light amplification by the stimulated emission of radiation," but the result is only a coherent, monochromatic light bulb. Most of the useful characteristics of a laser come from oscillation of light within a laser cavity. To understand the difference between laser amplification and laser oscillation, it helps to consider examples of the two phenomena.

The amplification of stimulated emission can occur in nature, although the phenomenon was not discovered until after the invention of the laser. One of the more interesting examples occurs in the upper levels of the Martian atmosphere (Mumma et al., 1981). Solar radiation produces a population inversion in carbon dioxide in the tenuous upper layers of the planet's atmosphere. The gain is low because the gas pressure is low, but the volumes involved are so large that by human standards (although not by cosmic ones) the power emitted on the 10.4-μm CO_2 laser transition is high. However, the energy stored in those CO_2 molecules is extracted inefficiently, and the laser emission is dissipated randomly into space as shown in Fig. 3-4. As a result, the intensity reaching earth is so low that the existence of the laser emission was only discovered using sophisticated spectroscopic instruments.

An ordinary CO_2 laser operates at somewhat higher pressures and uses a much smaller volume of gas—but its real practical advantages are that it extracts energy more efficiently and concentrates it into a narrow beam. This is done by putting a pair of mirrors on either end of a cylindrical tube. If stimulated emission occurs on the axis between the two mirrors, it is reflected back and forth through the tube, stimulating emission again and again from CO_2 molecules. Stimulated emission in other directions is lost out of the laser medium. The result is that the stimulated emission is concentrated in a beam oscillating back and forth between the mirrors.

From an electronic standpoint, the mirrors provide positive feedback. A laser with a pair of totally reflective mirrors, like any positive-feedback system, in theory could amplify itself to infinity, but in practice cavity losses limit the positive feedback and the degree of amplification. In practical lasers one (or sometimes both) of the mirrors lets part of the light escape from the laser cavity, either around its edge, through a hole, or through a partially transparent section. The light that leaks out of the laser cavity forms the laser beam.

The fraction of the beam that is allowed out of the laser cavity depends on the gain of the laser medium. Once a continuous-output laser has started operating, the total gain must equal the sum of the cavity losses plus the fraction of the energy allowed out of the cavity. Thus the lower the gain, the smaller the fraction of light that can be transmitted out of the cavity. If gain is high, transmission can be higher. In some cases, such as some excimer

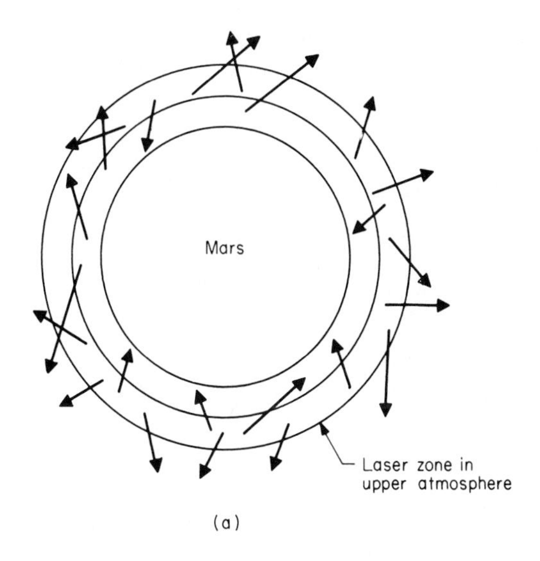

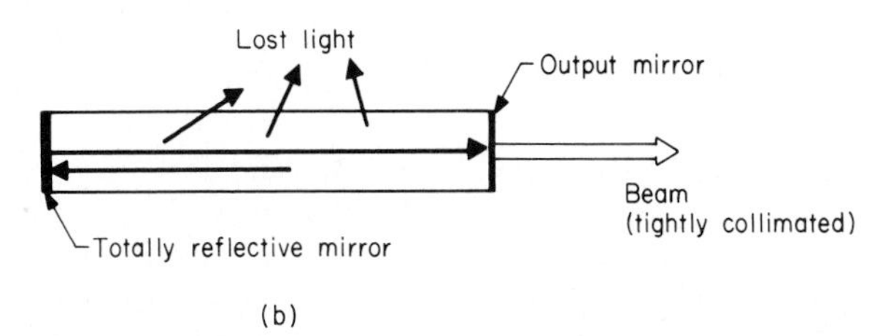

Figure 3-4 The key to production of a laser beam is directivity. *(a)* The upper part of the Martian atmosphere emits "laser light"—stimulated emission—randomly, making it hard to detect from earth. A man-made laser *(b)* includes a resonant cavity that generates a tightly collimated beam.

and nitrogen lasers, the gain is so high that a pair of cavity mirrors is not needed. (Such lasers are said to operate in superradiant mode.)

Another way to look at laser oscillation is as a threshold phenomenon that occurs when gain of the laser medium exceeds the sum of cavity losses and output-mirror transmission. In this case, gain is proportional to operating conditions, including energy input to the laser medium. As input power increases, laser output is zero until the threshold is passed, then increases steeply, as shown in Fig. 3-5. Thresholds differ for different cavity configurations and laser-medium conditions. In practice, different cavity designs and operating conditions are used to produce different output power levels.

The relationship between power inside the laser cavity (intracavity power) and in the laser beam depends on output mirror transparency. For low-gain media, where mirror transparency is low, the power levels inside the cavity

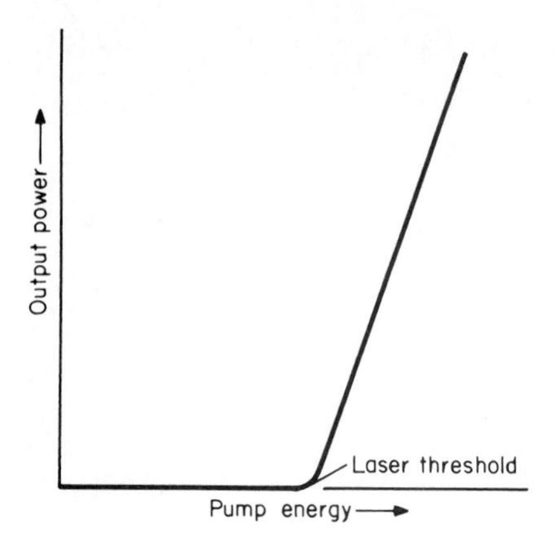

Figure 3-5 Laser threshold phenomenon—a laser does not generate significant optical output until the pump energy passes a threshold. At higher pump energies, the output power increases rapidly. In practice, each laser has limits on output, and eventually the output-input curve bends over.

are much higher than those outside. If high powers are critical, as in harmonic generation or some experiments, it may be desirable to put optical accessories *inside* the laser cavity to take advantage of the higher powers available there. Some lasers are designed with intracavity space to allow placement of optics or samples inside the cavity.

With the exception of high-gain types that operate in superradiant mode, most lasers are oscillators, with cavities defined by a pair of reflectors. However, it is also possible to build laser amplifiers that boost the power produced by separate oscillators. Laser amplifiers have no cavity mirrors and do not generate a beam internally. Instead, they contain an active medium which amplifies the beam from a separate oscillator. Typically, this design is used to produce pulses of high power from laser media with reasonable internal energy storage and fairly high gain. For such lasers it is typically more efficient to attain high powers by using an oscillator and a separate amplifier than by trying to build a larger oscillator. Massive amplifier chains have been built for use in laser fusion experiments (George, 1982).

Resonators, Modes, and Beam Quality

Laser oscillation takes place in a resonant cavity defined by the mirrors at each end. Laser resonator theory draws upon research on radio-frequency resonance, but the large difference in wavelength makes some important changes. Because cavity length is many times laser wavelength, it is possible to oscillate simultaneously on many modes. To keep oscillation modes limited, laser cavities are open with only a pair of small mirrors at opposite ends, unlike radio-wave cavities which can reflect from all sides. (See, for example, Svelto, 1982, pp. 107–144.)

Several configurations are possible for laser resonators, some of which are shown in Fig. 3-6. Conceptually, the simplest is the plane-parallel or Fabry-Perot resonator, in which two flat mirrors are placed at opposite ends of the cavity aligned parallel to each other and perpendicular to the axis of the cavity. Resonances occur when the cavity length D equals an integral number n of half-wavelengths ($\lambda/2$):

$$D = \frac{n\lambda}{2}$$

This expression can also be solved for wavelength, giving the possibility of oscillation at wavelengths

$$\lambda = \frac{2D}{n}$$

Because a typical laser cavity is tens of thousands or hundreds of thousands of wavelengths long, this might seem to permit oscillation at a very large number of wavelengths. However, oscillation is possible only at wavelengths within the gain bandwidth of the laser medium, and special optics can further restrict the oscillation range.

The conceptual simplicity of the plane-parallel resonator is offset by the practical difficulty of aligning the mirrors precisely enough for stable operation of the laser. Similar principles apply to several designs that use one or two concave spherical mirrors to simplify alignment, including

- *Concentric (or spherical) resonator:* Two identically curved spherical mirrors separated by a distance equal to twice their radius of curvature, so the centers of curvature coincide.
- *Confocal resonator:* Two identically curved spherical mirrors separated by twice their focal length, so focal points coincide. (Because focal length of spherical mirrors is half the radius of curvature, this means that the center of curvature of one mirror is at the center of the other.)
- *Hemiconfocal resonator:* A spherical mirror separated from a flat mirror by its focal length.
- *Hemispherical resonator:* A spherical mirror separated from a flat mirror by its radius of curvature.

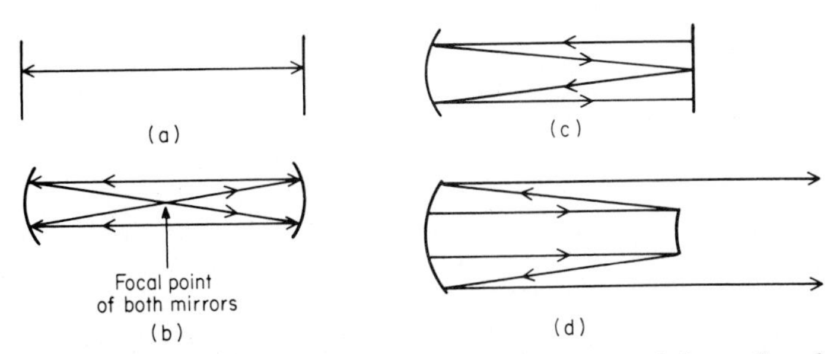

Figure 3-6 A sampling of laser cavity designs, showing how light is reflected between the end mirrors. Note how design of *(d)* the positive-branch confocal unstable resonator leads to production of a beam with doughnutlike cross section. *(a)* Plane-parallel resonator (marginal stability); *(b)* confocal resonator (stable); *(c)* hemiconfocal resonator (stable); *(d)* positive-branch confocal unstable resonator.

These resonators, and some similar designs with other mirror spacings, often are called *stable resonators*. The theory of laser resonators is complex, and strictly speaking, the "stable" label should not be applied to the plane-parallel resonator or to the concentric resonator without further explanation.

The concept of a stable resonator can best be visualized by following the path of a light ray through the cavity. The threshold of stability is reached if a light ray initially parallel to the axis of the laser cavity could be reflected forever back and forth between the two mirrors without escaping from between them, a concept shown in Fig. 3-6. A functional and somewhat stronger condition for stable-resonator operation is that a light ray just slightly out of alignment with the axis should likewise be reflected continually between the resonator mirrors. Another way of expressing that condition is that there has to be some net focusing power within the laser cavity, which in the simplest case is supplied by the mirrors.

The plane-parallel resonator meets the threshold condition for stability, but by itself cannot provide the focusing power needed to meet the stronger condition. Thus, the plane-parallel resonator by itself is not truly a stable resonator, particularly because of the practical problem of aligning the mirrors precisely parallel to one another. However, it can be used if something else in the laser cavity provides the needed focusing power. This is the case in solid-state crystalline and glass lasers, where thermal focusing in the rod itself provides the focusing needed for stable operation.

The concentric resonator is another special case. It meets the theoretical stability criteria, but in practice does not give stable laser operation because it focuses the internal wavefront to a point at its midpoint. In practice, other designs with at least one concave mirror are generally used for low-power gas lasers operating continuously.

Resonators which do not meet the stability criteria are called *unstable resonators*, because the light rays diverge away from the axis. There are many variations on the unstable resonator. One simple example is a convex spherical mirror opposite a flat mirror. Others include concave mirrors of different diameters (so that the light reflected from the larger mirror escapes around the edges of the smaller one), and pairs of convex mirrors (Svelto, 1982).

The two types of resonators have different advantages and different mode patterns. The stable resonator concentrates light along the laser axis, extracting energy efficiently from that region, but not from the outer regions far from the axis. The beam it produces has an intensity peak in the center, and a gaussian drop in intensity with increasing distance from the axis. It is the standard type used with low-gain and continuous-wave lasers, such as helium-neon types.

The unstable resonator tends to spread the light inside the laser cavity over a larger volume. In most cases, the output beam has an annular profile, with peak intensity in a ring around the axis, but a null on axis. This design collects laser energy from more of the volume in the laser cavity, typically leading to higher overall energy conversion efficiency than with a stable resonator, although the efficiency in extracting energy from the region of the laser axis is lower. This works best for high-gain pulsed lasers. One special advantage is the ability to use all-reflective optics, with the output beam escaping around a small mirror, important in working in the ultraviolet where there are few materials able to transmit high powers. Although the doughnut-shaped beam has an intensity null in the near field, it smooths out at greater distances from the laser, giving a more uniform energy distribution. In the far field, some unstable-resonator beams tend to have smaller divergence than beams from stable-resonator lasers.

Laser resonators have two distinct types of modes: transverse and longitudinal. Transverse modes manifest themselves in the cross-sectional profile of the beam, that is, in its intensity pattern. Longitudinal modes correspond to different resonances along the length of the laser cavity which occur at different frequencies or wavelengths within the gain bandwidth of the laser. A single transverse mode laser that oscillates in a single longitudinal mode is oscillating at only a single frequency; one oscillating in two longitudinal modes is simultaneously oscillating at two separate (but usually closely spaced) wavelengths. With the exception of some demanding applications in spectroscopy and communications, controlling the longitudinal modes so that oscillation is only at a single frequency is less important than controlling the transverse modes, which reflect beam quality.

A special terminology has evolved for transverse modes, based on theoretical work done in the early days of laser development (Fox and Li, 1961). Transverse modes are classified according to the number of nulls that appear across the beam cross section in two directions. The lowest-order, or fundamental, mode, where intensity peaks at the center, is known as TEM_{00}. A mode with single null along one axis and no null in the perpendicular direction is TEM_{01} or TEM_{10}, depending on orientation. A sampling of these modes, which are produced by stable resonators, is shown in Fig. 3-7. For most applications, the TEM_{00} mode is considered most desirable, but multimode beams can often deliver more power in a poorer-quality beam, and thus are acceptable for some uses. Somewhat different terminology has evolved for describing the modes of unstable-resonator lasers, but for most practical purposes unstable-resonator lasers can be considered to operate in their lowest-order mode, which produces a beam with annular cross section.

The different transverse-mode patterns described above are visible in the near field, close to the laser's output aperture. As the beam travels through space, the energy distribution becomes more uniform. Far from the laser, in what is called the far field, nonuniformities such as the intensity minimum at the center of an unstable-resonator beam tend to be smoothed out.

For most laser applications, the geometric properties of beam diameter and beam divergence are more important than mode structure. For a gaussian beam, beam diameter is generally measured to the points where intensity drops off to $1/e^2$ the peak intensity. The cavity optics determine beam diameter, which typically is measured at the output mirror. Use of the term *diameter* implies a circular cross section, but laser beams can also have oval or even rectangular profiles.

Beam divergence also depends primarily on cavity optics; it measures the angle at which the beam spreads out after it has left the laser. With the

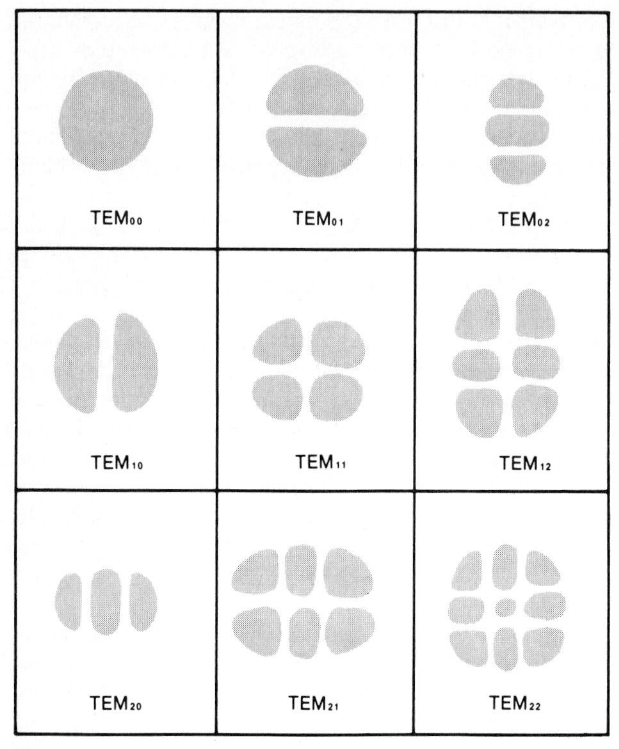

Figure 3-7 Lower-order modes that can be produced by a stable resonator. *(Courtesy of Melles Griot, from* Optics Guide 2.)

exception of semiconductor diode lasers, it is almost invariably less than 10 milliradians (mrad), and for many types it is typically on the order of 1 mrad. The lower limit on beam divergence is set by the diffraction limit, which says that the smallest angular spot which can be produced (measured in radians) is proportional to wavelength λ divided by output aperture D:

$$\text{Divergence} = 1.22 \frac{\lambda}{D}$$

As this relationship indicates, beam divergence can be decreased by using larger output apertures or moving to shorter wavelengths. Lasers with divergence that approximates the diffraction limit are called *diffraction-limited,* a category which includes all lasers producing TEM_{00} beams.

Linewidth and Tuning

By ordinary standards, laser emission is monochromatic, but there are considerable variations in linewidth of the emitted light. The nature of the optical cavity and of the laser medium combine to determine the range of wavelengths emitted by a particular laser.

The simple model of a laser transition occurring between two precisely defined energy states is by its very nature an oversimplification. In some cases, the transition actually involves simultaneous changes in two or more quantum states, leading to a couple of distinct possibilities.

One, seen in carbon dioxide and chemical lasers, is emission on a series of closely spaced transitions. In these molecules, the primary vibrational transition is accompanied by a change in the rotational state. Because the spacing of rotational energy levels is a significant fraction of the energy of the vibrational transition, output is possible on many lines with wavelengths differing by 10 to 20 percent. Commercial carbon dioxide lasers, for example, can emit on over 100 lines near 10 μm.

The second possibility, seen in dye lasers, is a continuum of transitions, a band of wavelengths throughout which the laser can emit light. Pressure-broadening effects, which serve to blur out closely spaced energy levels, cause discrete emission lines to merge into a continuum when the carbon dioxide laser is operated at high pressures.

Lasers can also emit on a family of transitions, which often share some energy levels. Typically this occurs when the excitation process can populate two or more upper laser levels, or when a single laser level can decay in two or more ways that can produce laser emission. Such lasers may be designed to emit on a single line or to produce multiline emission.

Even if only a single laser transition is possible, some line-broadening effects come into play. In gas lasers, motion of the atoms and molecules in the gas causes doppler broadening of the emission lines. Other forms of energy transfer, such as atomic collisions and lattice vibrations in solids, can also cause some line broadening.

The overall result, looking at the laser medium, is a gain curve, a range of wavelengths at which laser gain can occur. If there are many separate lines, they may have separate gain curves, depending on the operating conditions.

The cavity optics set another set of constraints on the wavelengths at which a laser can oscillate. As mentioned earlier, laser resonators can support a number of different longitudinal modes at different wavelengths. Even in a naturally narrow-line laser such as the helium-neon type, several longitudinal cavity modes fall under the laser medium's gain bandwidth. Typically one longitudinal mode will dominate during oscillation, with others present at lower intensities, but the relative intensities at different lines can shift, a phenomenon called *mode hopping*, which can be hard to control in some lasers but is not a problem in others. The spacing of longitudinal modes depends on cavity length—the longer the cavity, the closer the mode spacing. However, except for the tunable semiconductor lasers described in Chap. 19, cavity lengths are generally so many wavelengths long that little can be gained by changing length slightly.

It is possible to insert special optical elements into the laser cavity to restrict oscillation to a narrower range of wavelengths. The usual choice is a Fabry-Perot etalon, a pair of glass plates aligned so they effectively form a resonator. Because the plates are closely spaced, the resonant modes are far apart, and oscillation can readily be restricted to a single longitudinal mode. The spacing can be adjusted by tilting the etalon, by moving one of the plates, or by changing gas pressure inside or outside of the etalon cell. Such adjustments make it possible to change the wavelength selected by the etalon. Oscillation can be limited to a less-narrow range of wavelengths by inserting a narrow-band optical filter that transmits only at those wavelengths into the cavity. Alternatively, the cavity mirrors can be coated to reflect light at only certain wavelengths, preventing oscillation from taking place at other wavelengths.

Where a broad range of emission wavelengths is possible, it is usually desirable to be able to tune or adjust the laser wavelength. Such tuning is usually done by inserting a prism or grating into the optical cavity to serve as a wavelength-selective element. The grating or prism is then tuned (typically by turning) so only light of the desired wavelength would pass back and forth along the optical axis of the laser cavity. Other wavelengths would be directed away from the axis and hence would not oscillate or be amplified. Tuning is commonest with dye lasers but is also used with CO_2 and chemical lasers when the ability to pick a specific wavelength is important.

Coherence and Speckle

The best-known special property of laser light is its coherence. Stimulated emission is in phase with the photon that does the stimulating, and as stimulated emission builds it remains in phase. The coherence is not perfect, but it is far better than that of other light sources.

One way of measuring coherence is as *coherence length,* the distance over which light remains coherent after it leaves the light source. This quantity varies for different types of lasers, depending on the frequency bandwidth of emission. As a rough guideline, coherence length L equals the speed of light c divided by the laser's frequency bandwidth $\Delta\nu$

$$L = \frac{c}{\Delta\nu}$$

(Wilson and Hawkes, 1983). Most estimates of coherence length in this book have been derived using this simple relationship. It is actually based on the concept of temporal coherence, which measures coherence of a light wave emitted over time. The coherence length is equal to the coherence time multiplied by the speed of light.

There is also spatial coherence, which measures the area of a wavefront over which the light is coherent. This concept is independent of temporal coherence; the presence of one type of coherence does not necessarily imply the other.

Theorists have considered coherence in exhausting detail (Born and Wolf, 1975), but most of the theoretical details are of little import to laser users. Coherence is an ordering of light waves that lets them carry large quantities of information. Its major practical importance lies in holography, optical information processing, and development of coherent communication systems.

Coherence also causes a phenomenon called *speckle,* which derives its name from its appearance, as shown in Fig. 3-8. The granular speckle pattern is produced when laser light is scattered from a diffusing surface such as a sheet of paper. Speckle looks like noise but, in the words of Dennis Gabor, who received the 1971 Nobel prize in physics for his invention of holography, "It is not really noise; it is information which we do not want, information on the microscopic unevenness of the paper in which we are not interested" (Gabor, 1971). Speckle can be used in certain measurements, but it also can create problems in some measurement and communication applications, particularly those involving small-area receivers where the speckle pattern could cause fluctuations in received signal. Speckle gives a grainy texture to photographs and holograms recorded with laser light.

Efficiency Limitations

The overall efficiency of lasers is low by electrical standards, although comparable with that of other light sources. Typical commercial lasers convert 0.001 to 20 percent of input electrical energy into energy in the laser beam. For comparison, only a few percent of the electrical energy used by an ordinary incandescent lamp is emitted as visible light, although much of the remainder is emitted in the infrared. The limitations on efficiency are inherent in the nature of a laser.

Energy-level structures of excited species create many losses. Typically the laser transition is only one in a series of transitions that a species undergoes in dropping from the upper excited level to the ground state, and each of those additional transitions leads to losses. For example, the 632.8-nm transition of the helium-neon laser, corresponding to 2 eV of energy, occurs about 20 eV above the ground state, so even in theory only about 10 percent of the excitation energy could emerge in the laser beam.

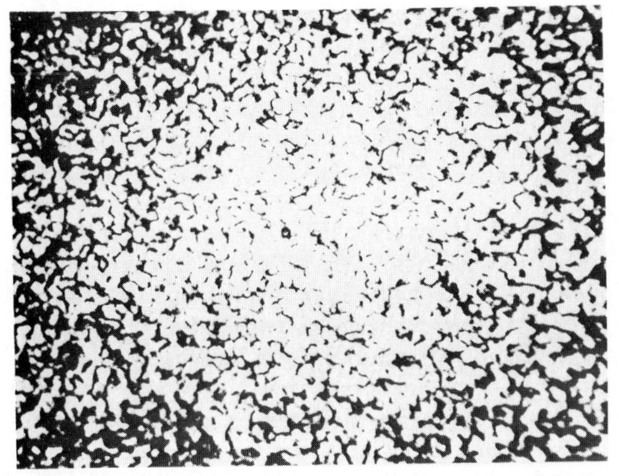

Figure 3-8 Photograph of laser speckle. *(Courtesy of F. P. Chiang.)*

Excitation is also never 100 percent efficient. In the common cases of optical and electrical pumping, much of the energy put into the laser medium either leaks out without exciting anything or simply goes into heating the laser medium. In optical pumping with a flashlamp, for example, some of the wavelengths generated by the lamp may not be absorbed efficiently by the laser medium. Even if the energy is absorbed, it may not be extracted from the laser medium because of the cavity configuration and inherent losses in the optical cavity.

Losses are also inherent in converting the energy from ordinary electrical sources (i.e., wall current) to the form required by the laser, such as high-voltage discharges or flashlamp pulses.

Output Power

Output powers of commercial continuous-wave lasers run from under a milliwatt to tens of kilowatts, with comparable average powers available from pulsed lasers. Typically the range of powers available from any one type is much more limited because of differences in the internal physics. Single-element semiconductor lasers are limited to average or continuous powers well under a watt because of problems with heat dissipation and laser damage, although higher powers have been obtained from monolithic arrays of many semiconductor laser elements. Gas kinetics, heat dissipation, and physical unwieldiness of tubes limit the helium-neon laser to power levels below about 60 mW.

The usual problems are in energy-transfer kinetics. In gas lasers, the kinetic processes needed to sustain the population inversion and stimulated emission can break down for large volumes. In solid-state lasers, heating of the laser medium can cause serious problems. So far the highest continuous or average powers known to have been recorded were produced by chemical lasers used in military laser weapon research, reported to have produced a couple of megawatts (Hecht, 1984). However, few details other than the record power levels have reached the public eye.

Limits on individual types of lasers are described in more detail in the chapters on lasers that follow.

Time Scales of Operation

Lasers can operate continuously for many years, or generate ultrashort pulses lasting only a matter of femtoseconds [(fs) 1 fs = 10^{-15}s]. The differences reflect both the design of the lasers and the internal physics. Users should remember that the time scale of atomic physics is very different from that of the human race, being measured in nanoseconds or microseconds rather than seconds and minutes.

In the laser world, anything operating for a second or more at a time is called "continuous wave." Most commercial continuous-wave lasers can emit for far longer, sometimes many years. But whatever the human perspective, from the standpoint of the physics it is legitimate to consider laser emission that is steady for a second to be continuous-wave. Such short-term continuous-wave operation is most likely for experimental lasers, particularly high-power ones used in military laser weapon research.

Many types of lasers can operate only in pulsed mode, for a variety of reasons. In many solid-state lasers, the key problem is heat dissipation. It takes time for the excess pump energy delivered to the laser rod to make its way out as heat, and continuous-wave operation can cause heat to build up to laser-damaging levels. "Damage" in this case means not just physical damage to the laser rod but also disruption of the laser beam by thermally induced gradients.

Internal physics can also demand pulsed operation. In some cases, the only way to produce a population inversion is by using input-energy pulses with high power that cannot be sustained. In others, the population inversion is inherently short-lived, because of rapid decay of the excited state (as in excimer lasers), or rapid buildup of species in the ground state (as in ruby lasers). In high-pressure gas lasers, pulse length is limited by instabilities inevitable in electric discharges through high-pressure gas.

Pulse repetition rates also vary widely. Ruby and neodymium-glass lasers can generate at most a few pulses per second, while excimers can produce up to a thousand and copper vapor several thousand. Cavity dumping can produce tens of thousands of pulses a second, while modelocking can generate pulses at gigahertz repetition rates. Efforts are under way to increase repetition rates as a way to increase average power from certain lasers, particularly excimers.

Pulse repetition rate, duration, and peak power can be adjusted by a variety of laser accessories, such as Q switches, modelockers, and cavity dumpers, described in Chap. 4. These give laser users added flexibility in matching laser characteristics to their requirements. Details on the pulse characteristics of individual lasers are described in later chapters.

BIBLIOGRAPHY

Max Born and Emil Wolf: *Principles of Optics*, Pergamon Press, Oxford, England, 1975.

J. C. Dainty (ed.): *Laser Speckle and Related Phenomena, 2d ed.*, Springer-Verlag, Berlin and New York, 1984.

A. G. Fox and Tingye Li: "Resonant modes in a maser interferometer," *Bell System Technical Journal 40*:453, 1961.

Dennis Gabor: "Holography 1948–1971," Nobel lecture, December 11, 1971, reprinted in K. Thyagarajan and A. K. Ghatak: *Lasers Theory & Application*, Plenum, New York, 1982, pp. 365–402.

E. Victor George (ed.): *1981 Laser Program Annual Report*, Lawrence Livermore National Laboratory, Livermore, Calif., 1982.

Jeff Hecht: *Beam Weapons: The Next Arms Race*, Plenum, New York, 1984.

C. Breck Hitz: *Understanding Laser Technology*, PennWell Publishing, Tulsa, Okla., 1985.

Michael Mumma et al: "Discovery of natural gain amplification in the 10-micrometer carbon-dioxide laser bands on Mars: a natural laser," *Science 212*:45–50, April 3, 1981.

Donald C. O'Shea, W. Russell Callen, and William T. Rhodes: *An Introduction to Lasers and Their Applications*, Addison-Wesley, Reading, Mass., 1977.

Orazio Svelto, *Principles of Lasers*, 2d ed., Plenum, New York, 1982.

L. V. Tarasov: *Laser Physics*, Mir Publishers, Moscow, 1983. In English.

K. Thyagarajan and A. K. Ghatak: *Lasers: Theory & Application*, Plenum, New York, 1981.
Marvin J. Weber (ed.): *CRC Handbook of Laser Science & Technology*, 2 vols., CRC Press, Boca Raton, Fla. 1982.
J. Wilson and J. F. B. Hawkes: *Optoelectronics: An Introduction*, Prentice-Hall, Englewood Cliffs, N.J., 1983.
Amnon Yariv: *Quantum Electronics, 2d ed.*, Wiley, New York, 1975.

CHAPTER

4

enhancements to laser operation

Laser operation can be enhanced either by modifying the laser device itself or by modifying the beam it generates. The distinction is significant in the world of commercial lasers. Modifications to the laser device typically are made by the manufacturer at the factory, or sometimes at customers' premises. However, it is the user who normally modifies the laser beam.

Factory modifications come in many forms, including substitution of different power supplies, choice of resonator optics, packaging of the device, and selection of control systems. Many data sheets include long lists of options offered by the manufacturers, which give users a considerable range of choices. However, these options are usually specific to a particular model or laser type, and the user has only limited control over them. The major options for each type of laser are described in Chaps. 7 to 25.

This chapter concentrates on modifications the user can make, although often they require hardware purchased from the laser manufacturer. All involve alterations to the laser beam, which can occur within as well as outside the laser cavity. These techniques offer users important flexibility in choice of laser wavelength, pulse length, and (in some cases) output power.

Wavelength Enhancements

Users can control laser wavelength either by altering the optical cavity or by altering the beam that emerges from the laser. Typically changes in the optical cavity are made using equipment supplied by the laser maker, but beam alterations can be made with other hardware. The boundaries are sometimes hazy; harmonic generators, for example, can be put inside or outside the laser cavity, and are offered both as options on certain laser models and as separate products.

Tuning and Wavelength Selection Fixed-wavelength laser emission is fine for many applications, but others would benefit from a choice of wavelengths. With a few lasers, there is no choice. With many, a choice is made by the manufacturer who assembles the laser cavity optics; one example is the helium-neon laser, which could emit in the infrared, but is almost always built with cavity optics that select the 632.8-nanometer (nm) red line. With other types, the user has some degree of selection, either by picking cavity optics or by adjusting a wavelength-selective element inside the laser cavity.

The choice of cavity optics works best for lasers which emit at widely spaced spectral lines. For example, one set of cavity optics can generate a blue beam from a helium-cadmium laser, while a separate set can produce ultraviolet output. In practice, the choice of cavity optics is generally an option.

Insertion of a wavelength-selective element inside the laser cavity allows much finer resolution, either for continuous tuning or for the selection of different laser lines that are not widely separated. The selection normally is made by a wavelength-dispersive element—a prism or diffraction grating—that is placed between the cavity mirrors or serves as a cavity mirror. The dispersive element spreads out a spectrum, and only a narrow range of wavelengths in that spectrum is reflected along the axis of the laser cavity so that it can oscillate. Light at other wavelengths is scattered out of the laser medium. The net effect is to give the optical cavity high losses at all but the desired wavelengths. The wavelength can be changed by turning the dispersive element so different wavelengths are aligned to resonate in the laser cavity. It is used in dye lasers (an example is shown in Fig. 17-3) and carbon dioxide lasers, and to produce single-line output from argon or krypton ion lasers. Such tuning is described in more detail in Chap. 17; it normally comes as an option with lasers that may require it.

Other tuning mechanisms may be used in certain cases. Application of a magnetic field can cause Zeeman shifts of spectral lines, an effect that was used in some early laser-tuning experiments. Internal energy-level structures of materials and refractive indexes of materials are both influenced by temperature and pressure. Shifts in energy levels directly affect transition wavelength. Changes in refractive index of materials in the laser cavity alter the effective length of the cavity and hence the resonant wavelength. However, the only one of these approaches used practically is temperature tuning of the lead salt semiconductor lasers described in Chap. 19.

Line Narrowing and Single-Frequency Operation As indicated in Chap. 3, the doppler-broadened gain profile of even nominally monochromatic lasers normally spans several longitudinal modes, each with slightly different wavelength. This is adequate for most applications, but not for precision measurements. Line-narrowing accessories are also needed to restrict oscillation of dye lasers and other types with closely spaced lines to a limited range of wavelengths.

The basic idea of line narrowing is to insert into the laser cavity optical elements which restrict oscillation to a range of wavelengths so narrow that it includes only a single longitudinal mode. The commonest line-narrowing component is the Fabry-Perot etalon, typically a transparent plate with two reflective surfaces which together form a short resonator that can be inserted within the laser cavity. It can restrict the range of wavelengths, as shown in Fig. 4-1, by requiring that any wavelengths oscillating in the laser cavity meet resonance criteria both of the etalon and of the laser cavity as a whole. Effective spacing of the etalon plates can be changed by rotating the device (i.e., tilting it with respect to the axis of the laser), or by changing internal gas pressure inside a hollow version. In some cases a pair of etalons with different optical characteristics may be used in the same optical cavity to limit operating bandwidth still further.

Frequency bandwidths as low as a megahertz are obtainable from commercial lasers (sometimes called *single-frequency* lasers). Even narrower linewidths have been demonstrated in the laboratory using special (and rather elaborate) techniques. The narrowest linewidths demonstrated are below 100 Hz (Hollberg and Hall, 1983). While laboratory records have been set with custom-built equipment, most users should consider line-narrowing equipment as an option on standard lasers.

Harmonic Generation Nonlinear interactions between light and matter can generate harmonics at multiples of the light-wave frequency, or equivalently

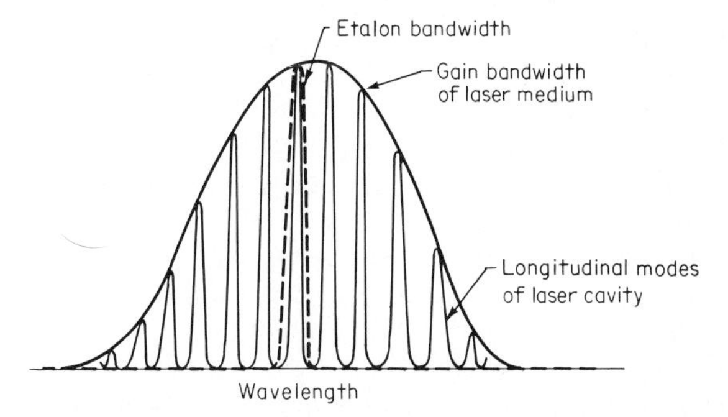

Figure 4-1 Transmission bandwidth of etalon compared with gain bandwidth of laser medium and the longitudinal modes of laser cavity. The etalon restricts oscillation to longitudinal modes of the laser cavity that occur within the etalon bandwidth.

at integral fractions of the wavelength. High-order harmonics have been produced in the laboratory, such as the 28th harmonic of the 1,064-nm output of a neodymium-YAG laser (Reintjes et al., 1977). However, for most applications only the second, third, or fourth harmonics are produced because the difficulty of harmonic-generation techniques, and their inherent energy losses, increases for higher harmonics.

Harmonic generation is a valuable technique because it greatly expands the range of wavelengths at which high laser powers are available, particularly in the ultraviolet and parts of the visible region. The commonest use of harmonic generation is with the 1,064-nm neodymium-YAG laser, producing the 532-nm second harmonic, the 355-nm third harmonic, and the 266-nm fourth harmonic. Dye laser output is also often frequency-doubled to obtain tunable ultraviolet light.

Commercial frequency doublers rely on nonlinear crystals. As a laser beam passes through the crystal, nonlinear interactions between the beam and the material generate an electromagnetic wave at twice the laser frequency. To obtain peak efficiency, the characteristics of the crystal must be matched precisely to the laser wavelength. This is relatively easy when working at a fixed wavelength, such as that of the Nd–YAG laser, but requires adjustment of the beam angle of incidence or the crystal temperature when wavelength is changed, as in a dye laser. Many nonlinear crystals are available, with the commonest including potassium dihydrogen phosphate (KDP, available in normal form or with deuterium substituted for hydrogen-1), lithium iodate ($LiIO_3$), KB_5O_8 (potassium pentaborate, KB5), and ammonium dihydrogen phosphate (ADP); for a more extensive list see Table 12-2 in Yariv and Yeh (1984). The choice of material depends on operating wavelength and power-handling requirements.

The magnitude of the nonlinear effect is proportional to the square of the laser power. This has an important practical effect: high powers are needed to make harmonic generation reasonably efficient. In practice, significant harmonic power is not produced until input fundamental power is on the order of a megawatt per square centimeter (Baldwin, 1974), although this varies considerably with the material. Efficiency of harmonic production increases with the pump power and thus with the intensity of the fundamental-frequency laser beam.

Because of the need for high pump power, harmonic generation is normally used only with lasers that can produce pulses with high peak power, such as Nd–YAG, ruby, and dye types. Because harmonic-generation efficiency depends on peak power, not on pulse energy, the efficiency can be increased by compressing the pulse length, thus increasing the peak power while pulse energy remains the same or is actually reduced.

Normally frequency-doubling crystals are inserted in the laser beam outside the cavity. However, in certain cases they can also be inserted into the cavity to take advantage of the higher powers unavailable in the external beam. Combined with extremely tight focusing, this technique can be used for intracavity generation of the second harmonic of a continuous-wave beam, or of a modest-power pulsed laser.

Generation of third and fourth harmonics using nonlinear crystals usually is a multistep process. The third harmonic is produced by first generating the second harmonic, then mixing it with the fundamental wavelength in another nonlinear crystal so the frequencies add together to produce the third harmonic. To produce the fourth harmonic, the output of a second-harmonic generator is passed through a second frequency doubler. Generation of the third and fourth harmonics requires even higher powers than frequency doubling and normally is done only with high-power pulses from Nd–YAG lasers. A combination of poor efficiency and low ultraviolet transmission makes generation of higher-order YAG harmonics in nonlinear crystals impractical.

Nonlinear interactions in gases or metal vapors, unlike those in crystals, can directly produce odd harmonics (Harris, 1975). As in crystals, the process is more efficient for lower harmonics, particularly the third. Because they produce higher harmonics and are transparent at much shorter wavelengths than crystals, gases or vapors can generate harmonics much further into the ultraviolet. For example, a group at AT&T Bell Laboratories has produced the 35.5-nm seventh harmonic of a krypton fluoride laser's 248-nm output by passing the beam through a pulsed supersonic jet of helium in a vacuum (Bokor et al., 1983); operation in a vacuum was needed to avoid absorption of the extreme-ultraviolet output. The demand for short-wavelength ultraviolet output is limited, however, and gas cells have found few harmonic-generation applications outside the research laboratory.

The chapters describing individual laser types discuss their suitability for use with harmonic generators. Although some nonlinear devices can be made fairly simple to use, the physics underlying their operation is so complex that few users will want to study it in any detail, although references are available that do so (Baldwin, 1974; Yariv and Yeh, 1984).

Sum-and-Difference Frequency Generation Harmonic generation is actually a simplified case of a more general nonlinear process, the generation of sum-and-difference frequencies as two (or perhaps more) light waves pass through a nonlinear material. In general, if the input frequencies are ω_1 and ω_2, the nonlinear process generates additional waves at frequencies $\omega_1 + \omega_2$ and $\omega_1 - \omega_2$. Typically one of these frequencies is selected for use, although the physical interaction generates both.

Sum-and-difference frequency generation can occur in solid or gas phases of suitable nonlinear materials. In practice the process is complex, and like harmonic generation requires high-power beams and careful alignment. Because of its complexity, its use remains much more limited than harmonic generation, although it can be a useful laboratory tool.

Parametric Oscillation Parametric oscillation is another nonlinear phenomenon with complex theoretical underpinnings. It relies on an effect known as parametric amplification, which occurs when an intense "pump" beam at frequency ω_3 is passed through a suitable nonlinear crystal together with a weak "signal" beam at a lower frequency ω_1. Interactions in the crystal transfer energy from the pump frequency to the signal frequency, in effect amplifying the signal frequency. At the same time, an "idler" beam at a frequency $\omega_2 = \omega_3 - \omega_1$ is also generated (Yariv and Yeh, 1984).

In practice, operation of a parametric oscillator differs somewhat from that theoretical picture. A principal difference is that the only optical input required is the pump beam. Parametric noise generated within the nonlinear crystal produces the weak signal beam.

The signal and idler frequencies are determined by the orientation of the crystal and the design of the oscillator cavity in complex ways that are of less relevance to users of parametric amplifiers than to designers of such systems (Svelto, 1982; R. G. Smith, 1976). What the user sees is output—the signal wave—that is continuously tunable in wavelength by adjusting the nonlinear crystal and/or the optical cavity.

The existence of a resonant optical cavity makes parametric oscillators superficially similar to lasers, and like lasers they generate coherent beams. However, there is no stimulated emission within the parametric oscillator cavity, so it is not really a laser.

All parametric oscillator cavities must have end mirrors transparent to the pump beam and reflective at the signal frequency, as shown in Fig. 4-2. There are two basic variations: designs which are doubly resonant and reflect both the signal and idler waves, and those which are singly resonant and reflect only the signal wave. Singly resonant cavities require the high pump power that is available only with pulsed excitation, but they provide more stable output than doubly resonant cavities, even though the latter can be pumped with continuous-wave lasers. Singly resonant cavities are the usual choice in commercial parametric oscillators.

Although parametric oscillation is in theory a general process, in practice it is more limited. Wavelengths of 0.5 to about 14 μm have been produced in the laboratory, but practical devices are limited to the near-infrared and use neodymium lasers as pumps. At one point parametric oscillators were considered promising tools for generating continuously tunable output in the near-infrared. They remain available as accessories for neodymium lasers but have become less popular as color-center lasers and infrared laser dyes have become readily available.

Raman Shifting Another technique gaining popularity as a way of changing laser wavelength is *Raman shifting*. The phenomenon is named for Indian physicist Chandrasekhara V. Raman, who in 1930 received the Nobel prize in physics for discovering it. The Raman shift occurs when molecules scatter light "inelastically," in the process changing the photon energy. The scattering is inelastic because as it occurs, the molecule makes a vibrational and/or rotational transition. If the transition is to a higher energy level, the photon is scattered with a lower energy than it had when it arrived and thus has a longer wavelength than the incident light, a so-called Stokes shift. If the transition is to a lower energy level, the scattered photon carries away the excess energy, and thus has higher energy and longer wavelength than the incident light, an "anti-Stokes shift."

Like harmonic generation, Raman scattering is not likely to occur until

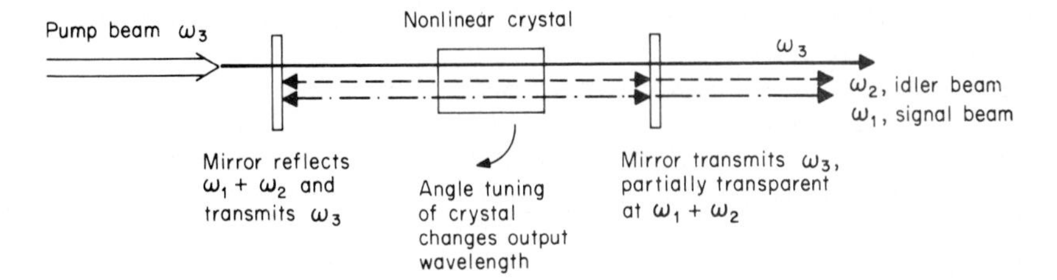

Figure 4-2 Parametric oscillator with a doubly resonant cavity. Mirrors transmit the pump wavelength, but reflect totally or partially the signal and idler beams. In a singly resonant cavity, the idler beam would not be confined to the cavity. The pump beam generates both signal and idler beams as it passes through the nonlinear crystal, which is normally rotated to change the angle of incidence and thereby the output wavelength.

high incident powers are reached. It found only very limited applications until the advent of the laser, which provided the needed source of high-power monochromatic light. Now laser Raman scattering serves as the basis of a type of spectroscopy, which has proved valuable because the scattered light shows the pattern of molecular vibrational transitions, with fine rotational shifts superimposed that are characteristic of particular molecules (Duley, 1983).

Raman scattering can also be used to shift the wavelengths of high-power lasers to regions where they are more useful. Raman shifters have found particular use in downshifting the ultraviolet wavelengths produced by excimer lasers to longer wavelengths in the visible and near-ultraviolet. As shown in Fig. 13-4, this greatly increases the number of wavelengths that can be generated from excimer lasers, although at a sacrifice in power. At very high peak powers the efficiency increases, and military researchers have been studying the prospects for Raman downshifting to generate blue-green pulses for satellite-to-submarine communications (Burnham and Schimitschek, 1981).

Raman shifters have become available commercially. Although they can be used to move to either longer or shorter wavelengths, they are more efficient in generating longer wavelengths, and most current interest centers on "down-shifting" frequency to longer wavelengths.

Changing Pulse Length

It is often desirable to change the "natural" duration of laser pulses. Sometimes the goal is to extend pulse length, but mostly it is to shorten the pulse, either to provide higher peak power or to provide better time resolution.

Some pulse shortening is inherent in nonlinear interactions because of the nature of their dependence on the amplitude of the interacting electromagnetic fields. Laser pulses have finite rise and fall times, and because of the nonlinear dependence of output power on input power, the lower-power parts of the pulses will generate only very weak and often insignificant output. Thus, putting an optical pulse through a nonlinear process, say frequency doubling, tends to shorten it. However, the degree of shortening is limited, and it is of practical importance only in cases where a precise pulse duration is required after a nonlinear process.

Four techniques for producing short pulses are in widespread use. Three of them—*Q* switching, cavity dumping, and modelocking—operate by interacting with light inside the laser cavity, in essence clumping energy together to produce a pulse that is shorter in length and higher in peak power than unaltered laser emission. Such pulse shortening often costs energy, but for many applications that is an acceptable trade-off. The fourth technique is the simple but inelegant approach of switching the beam off and on outside the cavity using some type of shutter—either a mechanical device or an electro-optic or acousto-optic modulator. This does not serve to increase peak power but can produce short pulses; it falls under the heading of external optics and is described in Chap. 5.

Extension of pulse duration without fundamentally altering the structure of the laser is much harder than shortening pulses. Very few applications require such pulse stretching, and it will not be described here.

Q Switching Like any oscillator, a laser cavity has a quality factor *Q* that measures the loss or gain of the cavity. This factor is defined (Thyagarajan and Ghatak, 1981) as

$$Q = \frac{\text{energy stored per pass}}{\text{energy dissipated per pass}}$$

Normally, the *Q* factor of a laser cavity is constant, but modulating the *Q* factor raises interesting possibilities. If the *Q* factor is kept artificially low, say by putting a lossy optical element into the cavity, energy will gradually accumulate in the laser medium because the *Q* factor is too low for laser oscillation to occur and dissipate the energy. If the loss is removed suddenly, the result is a large population inversion in a high-*Q* cavity, producing a high-power burst of light, a few nanoseconds to several hundred nanoseconds long, in which the energy is emitted. This rapid change in cavity *Q* is called *Q switching.*

There are three basic variations on *Q* switching, as shown in Fig. 4-3. One is use of a rotating mirror or prism as the rear cavity mirror. Another is insertion of a modulator into the cavity. The third is insertion of a nonlinear lossy element into the cavity that becomes transparent once intracavity power exceeds a certain level. The first two techniques are called active *Q* switching because they require active user control; the third is considered passive because no special control is needed. *Q* switching of pulsed lasers cannot increase pulse energy, but it can squeeze much of that energy into a pulse a few nanoseconds to a few tens of nanoseconds long, and achieve peak powers in the megawatt range—higher than otherwise possible in most pulsed lasers. Although all types of *Q* switches can operate with pulsed lasers, only active *Q* switches can operate with lasers that are pumped by a continuous source of energy such as a continuous-arc lamp.

Operation of a rotating-mirror *Q* switch relies on the fact that most of the time the mirror is rotating it is not aligned properly with the output mirror, preventing oscillation. Periodically the rotating mirror passes through the point where it is properly aligned, causing cavity Q to increase abruptly and producing an intense Q-switched pulse.

When an optical modulator is used as a Q switch, it normally does not transmit light, blocking off one of the cavity mirrors. When the modulator is switched to transparency, the light in the cavity can reach the mirror, so cavity Q increases to a high level and the laser generates a Q-switched pulse. Normally the modulators used are electro-optic or acousto-optic devices, as described in Chap. 5, although they are packaged and sold as Q switches.

In passive Q switching, a material with nonlinear absorption characteristics (typically a dye in solution housed in a dye cell) is inserted into the laser cavity. At low power levels, the material has high absoprtion, blocking one of the mirrors from the laser medium. When the pump-energy source is pulsed, the amount of light in the laser cavity builds up, eventually reaching a level at which it bleaches out the nonlinear material's absorption. The dye cell thus becomes transparent, cavity Q abruptly increases, and a Q-switched pulse is produced. Typically the bleaching process takes on the order of a

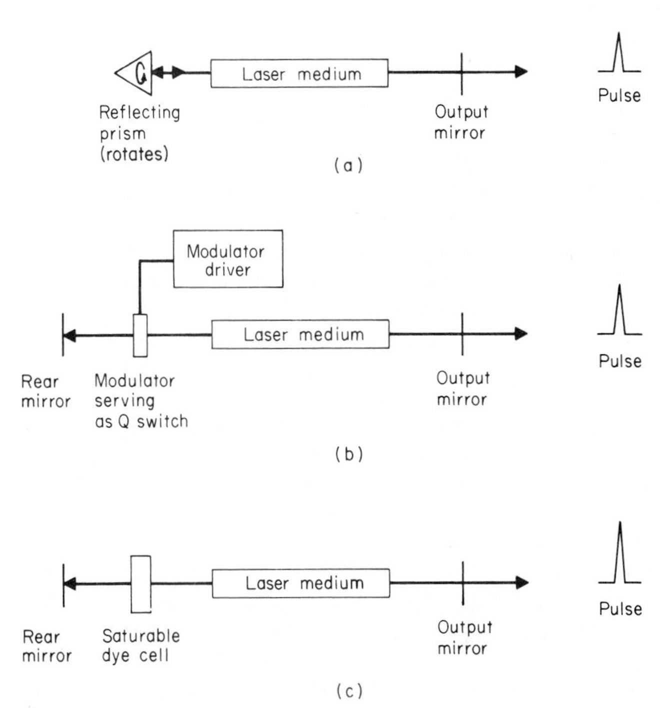

Figure 4-3 Types of Q switching used with lasers: *(a)* rotating mirror or prism; *(b)* active modulator; *(c)* passive. All cases show operation during a pulse. When the laser is not emitting, the rotating prism is turned so none of its faces parallels the output mirror. In other Q switches, the pulse is generated only during the brief interval that the modulator or saturable absorber cell is transparent.

microsecond. Synchronization of passively Q-switched pulses is less precise than that of actively Q-switched pulses, but passive Q switches are simpler and cheaper. Their mode of operation limits passive Q switches to use with pulsed lasers, typically flashlamp-pumped solid-state types such as ruby and neodymium.

The saturation effect that permits passive Q switching is not particularly mysterious. The absorption is due entirely to one transition in the material. As intracavity intensity increases, more and more of the dye molecules absorb light, raising them to an excited state where they cannot absorb any more photons. Although the excited states eventually decay, at some point light intensity reaches a level at which as many species have been raised to the excited state as are in the lower state. At this point, emission is occurring as fast as absorption, so no net absorption can take place, and the dye is said to have been bleached or saturated.

One inherent limitation of Q switching is that it only works for a laser medium capable of storing energy for a time much longer than the Q-switched pulse duration. This requirement comes from the fact that the energy is stored by accumulating the excited species in its upper laser level, and this is done by suppressing the feedback needed for stimulated emission. This is possible only if the spontaneous emission lifetime is relatively long—otherwise the excited species will spontaneously drop down to a lower energy level. Thus Q switching does not work for all types of lasers.

Cavity Dumping Cavity dumping operates on rather different principles than Q switching but produces somewhat similar results. The basic idea of cavity dumping is to couple laser energy directly out of the laser oscillator cavity without having it pass through an output mirror. In this case, both cavity mirrors are totally reflective and sustain a high circulating power within the laser cavity. This circulating power can be dumped out of the cavity by deflecting it. The basic concept can be seen in Fig. 4-4, where a mirror pops up into the laser cavity to deflect the circulating light out. The result is a pulse with length close to the cavity round-trip time.

In practice, cavity dumping is done in different ways. One approach is to put an acousto-optic deflector in the cavity. Normally the deflector transmits light to the cavity mirrors, but in the cavity-dumping mode the deflector switches light out of the laser cavity. Another approach involves an electro-optic modulator and polarizing beamsplitter. When the modulator is off, linearly polarized light is reflected back and forth through the cavity and is not affected by the beamsplitter. However, when the modulator is turned on, light which passes through it is rotated in polarization, and that light is then reflected out of the cavity by the beamsplitter.

Cavity dumping can be used with continuous-wave lasers which do not store energy in their upper excited levels and hence cannot be Q-switched. It also can be combined with Q switching and modelocking to produce special pulse characteristics.

Cavity dumping of a continuous-wave laser can generate pulses of 10 to 50 ns at repetition rates of 0.5 to 5 megahertz (MHz). Typically pulses are

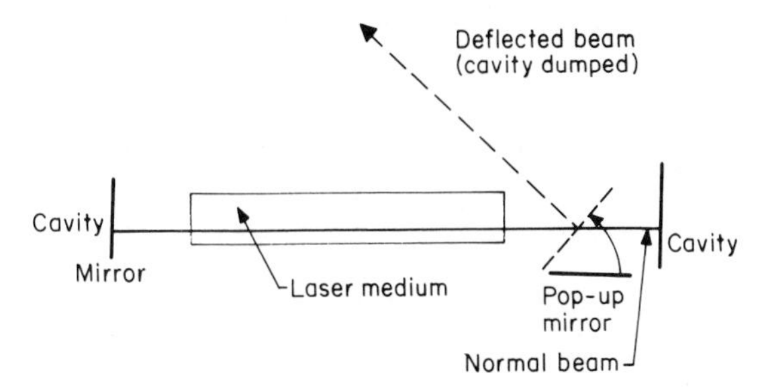

Figure 4-4 The basic concept of cavity dumping is dumping the energy out of the laser cavity by a means such as popping an intracavity deflecting mirror into the beam. Actual cavity dumpers are more complex systems that rely on acousto-optic deflectors or electro-optic modulators.

shorter than *Q*-switched pulses from such lasers, are produced at higher repetition rates, and do not contain as much energy.

Flashlamp-pumped, *Q*-switched solid-state lasers can be cavity-dumped to shorten the normal 3 to 30-ns *Q*-switched pulses to 1 to 3 ns. One *Q*-switched, cavity-dumped pulse is produced per flashlamp pulse.

When used together with modelocking, cavity dumping can reduce the naturally high repetition rates of modelocked lasers to more manageable levels and increase the peak power in the pulses. In this case, the cavity-dumping frequency is synchronized with the generation of modelocked pulses, so periodically one modelocked pulse is switched out of the cavity. Typically, operating frequencies are chosen so the ratio of cavity-dumping frequency to modelocked pulse generation equals an integer plus one-half, say 9½. This can build the peak power of modelocked pulses to as much as a factor of 30 higher than would otherwise be possible. The length of the modelocked pulses is not affected by cavity dumping, which only serves as a gating switch.

Modelocking Pulses in the picosecond regime can be generated by modelocking. The simplest way to visualize modelocked pulses is as a group of photons clumped together and aligned in phase as they oscillate through the laser cavity. Each time they hit the partially transparent output mirror, part of the light escapes as an ultrashort pulse. The clump of photons then makes another round trip through the laser cavity before another pulse is emitted. Thus the pulses are separated by the cavity round-trip time $2L/c$, where L is cavity length and c is the speed of light.

The physics are actually much more complex (see, for example, Svelto, 1982), and the details are well beyond the scope of this book. The basic idea is that modulation of the loss or phase of an optical element in the laser cavity causes many different longitudinal modes to oscillate together in phase. Modelocking occurs when the modulation frequency corresponds to the cavity round-trip time. The modulation can be active, by changing the transmission of a modulator, or passive, by saturation effects similar to those in a passive

Q switch. In either case, interference among the modes produces a series of ultrashort pulses.

Modelocking requires a laser that oscillates in many longitudinal modes. Thus it does not work for many gas lasers with narrow emission lines. However, it can be used with argon or krypton ion lasers, solid-state crystalline lasers such as Nd–YAG, semiconductor lasers, and dye lasers (which have exceptionally wide gain bandwidth). The pulse length is inversely proportional to the laser's oscillating bandwidth, so dye lasers can generate the shortest pulses because of their exceptionally broad gain bandwidths.

Both pulsed and continuous lasers can be modelocked. In either case, modelocking produces a train of pulses separated by the cavity round-trip time. For pulsed lasers, the result is a series of pulses that fall within the envelope defined by the normal pulse length. Single modelocked pulses can be selected by passing the pulse train through gating modulators that allow only a single pulse to pass.

Other Pulse Compression Techniques Modelocking can generate subpicosecond pulses, but the limited gain bandwidth of laser media sets a minimum length on modelocked pulses because of the Fourier transform relationship between pulse duration and spectral bandwidth. The only way to generate shorter pulses is to spread out spectral bandwidth outside the laser. Laboratory researchers have found an ingenious technique for doing so in a way that lets them compress pulse length. An ultrashort modelocked pulse is passed through an optical fiber, a process that spreads out its wavelength spectrum. Spectral dispersion in the glass of the fiber also causes temporal dispersion, but it is an ordered dispersion. Because refractive index is a function of wavelength, the wavelengths for which the index is the smallest travel through the fiber fastest, creating a pulse in which wavelength varies with time. This pulse can then be compressed in time by passing it through a delay-line structure with differential wavelength delays that precisely compensate for the delays in the fiber. That technique (Fujimoto et al., 1984) has been used to generate pulses just 8 femtoseconds (fs) [0.008 picosecond (ps)] long (Shank et al., 1985). However, it is far from routine, even in the laboratory.

Amplification

The output of many lasers can be amplified by passing it through a laser amplifier, which is essentially a laser medium without a resonant cavity. In the simplest configuration, the laser beam makes a single pass through the amplifier, where excitation has produced a population inversion. The laser light stimulates emission from the excited species in the amplifier, and this increases the optical power. In more complex designs, the laser beam can take a zig-zag path through the amplifier to extract energy from a larger volume in the laser medium.

The main value of laser amplifiers is to build up pulses containing high energy with high peak power. There are intrinsic limits on the maximum pulse energy that can be extracted from a particular laser oscillator. This pulse energy can be boosted by passing it through one or more laser amplifiers, because it is possible to extract more energy from a laser medium in amplifier configuration than when it functions as an oscillator.

Oscillator-amplifier configurations are used to extract high-energy pulses from optically pumped solid-state, gas, and dye lasers. Often, the oscillator and amplifier are packaged together in commercial laser systems. Normally commercial laser systems include no more than one or two amplifier stages. The most complex amplifier chains built are the ones assembled from neodymium-glass for laser fusion experiments at the Lawrence Livermore National Laboratory, where the pulse from the master oscillator can pass through up to 19 amplifiers, as shown in Fig. 20-4.

BIBLIOGRAPHY

George C. Baldwin: *An Introduction to Nonlinear Optics*, Plenum/Rosetta, New York, 1974.

J. Bokor, P. H. Bucksbaum, and R. R. Freeman: "Generation of 35.5-nm coherent radiation," *Optics Letters 8*(4):217–219, April 1983.

Ralph Burnham and Erhard J. Schimitschek: "High power blue-green lasers," *Laser Focus 17*(6):54–66, June 1981.

W. W. Duley: *Laser Processing and Analysis of Materials*, Plenum, New York, 1983.

J. G. Fujimoto, A. M. Weiner, and E. P. Ippen: "Generation and measurement of optical pulses as short as 16 fs," *Applied Physics Letters 44*(9):832–834, 1984.

S. E. Harris et al.: "Generation of ultraviolet and vacuum ultraviolet radiation," in S. F. Jacobs et al. (eds.), *Laser Applications to Optics & Spectroscopy*, Addison-Wesley, Reading, Mass., 1975, pp. 181–197.

C. Breck Hitz: *Understanding Laser Technology*, PennWell Publishing, Tulsa, Okla., 1985.

L. Hollberg and J. L. Hall: "Observation of energy-level shifts of Rydberg atoms due to thermal fields," in H. P. Weber and W. Luthy (eds.), *Laser Spectroscopy VI*, Springer-Verlag, Berlin and New York, 1983, p. 229.

J. Reintjes et al.: "Seventh harmonic conversion of modelocked laser pulses to 38.0 nm," *Applied Physics Letters 30*(9):480–482, 1977.

Charles V. Shank et al.: Postdeadline paper at Conference on Lasers and Electro-Optics, Baltimore, May 21–24, 1985.

Peter W. Smith, "Single frequency lasers," in Albert K. Levine and Anthony J. DeMaria (eds.), *Lasers*, vol. 4, Marcel Dekker, New York, 1976, pp. 74–118.

R. G. Smith: "Optical parametric oscillators," in Albert K. Levine and A. J. DeMaria (eds.), *Lasers*, vol. 4, Marcel Dekker, New York, 1976, pp. 190–307.

Orazio Svelto: *Principles of Lasers*, 2d ed., Plenum, New York, 1982.

A. P. Thorne: *Spectrophysics*, Halsted/Wiley, New York, 1974.

K. Thyagarajan and A. K. Ghatak: *Lasers Theory & Applications*, Plenum, New York, 1981.

Amnon Yariv and Poichi Yeh: *Optical Waves in Crystals*, Wiley, New York, 1984.

CHAPTER

5

external optics and their functions

External optics are needed to focus and direct laser beams and to manipulate light in other ways that let it be put to practical use. The field of optics is a science in itself and can legitimately be considered to include laser technology as well as other diverse areas. This chapter cannot hope to tell everything about optics, but it can give an overview of the types of optics often used with laser systems. Readers seeking more detail can select from a number of handbooks (including Department of Defense, 1962; Driscoll, 1978; Melles Griot, 1985; and Wolfe and Zissis, 1978) and textbooks (Ditchburn, 1976; and Jenkins and White, 1976, are examples).

Optical elements perform many tasks. Their simplest functions are the refraction and reflection of light, actions we normally associate with lenses and mirrors. In optical systems, other functions also can be important, such as selectively transmitting certain wavelengths of light, altering the polarization components of light, or breaking light up into its spectral components. These are generally considered "passive" optical functions because the components require no special control or input power to perform them. There are also "active" components which operate under active control or require input energy (other than light) to perform their function; examples include modulators which alter the intensity of a transmitted laser beam and scanners which move a beam.

This chapter concentrates more on passive optics because they are the most widely used. The principles of active optical devices can be extremely complex, so the description here will be limited to an overview of their functions. Although many lasers are used with optical instruments—assemblages of various optical components designed to perform particular functions—that field is far too large to be described properly in this sort of book.

Transmissive Optics

The most familiar types of optical components are transmissive optics, whose action depends on the refraction (bending) of light passing through them. The way in which transmissive optics refract light depends both on the shape of the component and on the refractive index—the crucial measurement of the transmissive characteristics of the material that makes up an optical component.

Refraction and Refractive Index The refractive index was devised as a measurement of how much a material can bend light. Physically, its real meaning is as a measurement of how much a material slows down light traveling through it. The standard is the speed of light c in a vacuum, for which the index of refraction n is defined as equaling 1. For all other materials, the index of refraction n_{mat} is defined as the ratio

$$n_{mat} = \frac{c_{vacuum}}{c_{material}}$$

which normally is greater than 1.

Refractive index depends on the nature of the material and the wavelength of light. As shown in Table 5-1, there are considerable differences among materials. In general, refractive index tends to increase with material density, and for gases there is a definite relationship between gas density and refractive index. The refractive index of a gas at normal density is close to 1, so in practice the indexes of gases are measured relative to that of a vacuum. The index of air is small enough so that in practice the refractive indexes of solids are measured relative to air rather than to a vacuum. This not only simplifies the task of measurement but also gives a more useful value, because virtually all optical systems are used in air, and in such circumstances it is the ratio of refractive indexes of the material and air—rather than the absolute refractive index relative to a vacuum—that is relevant.

From the standpoint of optical design, the importance of the refractive index is its measurement of how strongly light is bent as it passes between two materials, as shown in Fig. 5-1. Note that the frequency of light-wave oscillation remains constant in all materials; the decrease in speed in a denser material translates into a decrease in wavelength. As light enters a material at an acute angle, as in Fig. 5-1, the wavefront is bent by an amount

TABLE 5-1 Refractive Indexes of Selected Materials at Selected Wavelengths*

Material	Wavelength, μm	Refractive index, n
Air (1 atm pressure)	0.59	1.0002765
Pure water	0.59	1.33287
Zinc crown glass	0.59	1.517
Heavy flint glass	0.59	1.650
Vycor optical glass	0.58	1.458
Fused silica	0.59	1.4584
	0.214	1.5337
	0.89	1.4518
	2.325	1.4329
Crystal quartz	0.59	1.54424
As_2S_3	0.6	2.63640
	1.0	2.47773
	2.0	2.42615
	4.0	2.41116
	10.0	2.38155
CaF_2	0.59	1.43384
Germanium	2.06	4.1016
LiF	0.5	1.39430
NaCl	0.59	1.54427
	2.0	1.52670
	10.0	1.49482

* Note that in birefringement materials such as crystal quartz the refractive index depends on polarization of light rays. The values shown are for "ordinary" rays.
Source: Data from Driscoll, 1978; Wolfe and Zissis, 1978; Weast, 1981.

proportional to the differences in the wavelength of light in the two materials. For light going in either direction between materials *a* and *b*, the result is Snell's law of refraction:

$$\frac{n_b}{n_a} = \frac{\sin A}{\sin B}$$

where the n's are the refractive indexes in the two materials, A is the angle from the normal in material *a*, and B is the angle from the normal in material *b*. This can also be expressed in the form:

$$n_b \sin B = n_a \sin A$$

When light travels from a high-index medium to one of lower refractive index, there is no solution to the Snell equation for large angles to the normal because the sine of an angle can be no larger than 1. This is the case of total internal reflection, in which all light incident on the surface between the two materials is totally reflected back into the denser material (for example, back into the prism at the glass-air interface of a prism). The smallest angle from the normal at which this occurs is called the *critical angle*; the concept is shown in Fig. 5-2.

The variation of refractive index with wavelength is sometimes neglected in optical calculations because it is normally small over a small wavelength

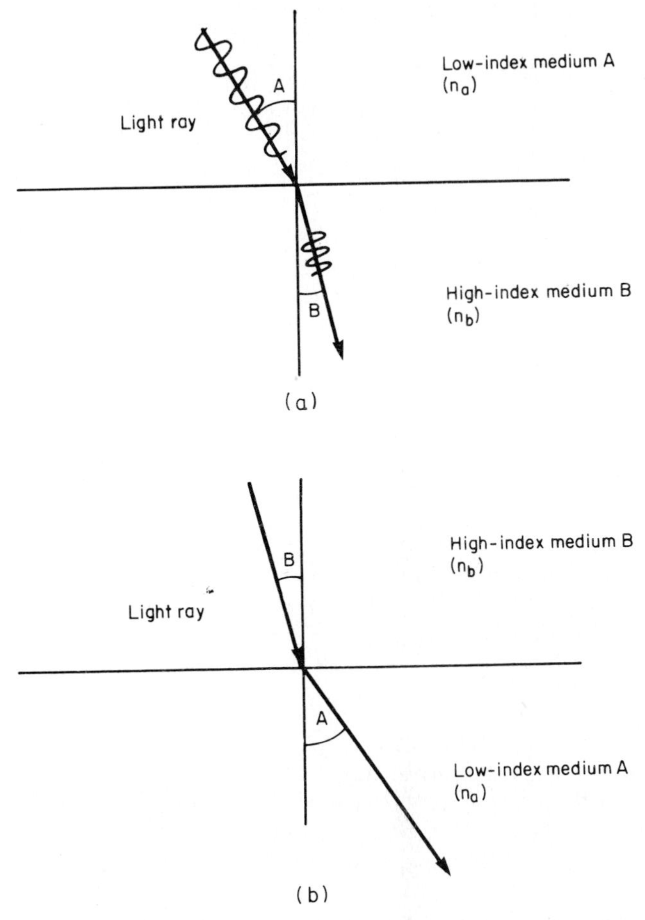

Figure 5-1 Refraction of a light ray passing *(a)* from a low-index medium into a high-index medium and *(b)* of one passing from a high-index medium to a low-index medium. Note that angles are measured from the normal to the surface.

range. However, it can have practical implications, particularly when dealing with a broad range of wavelengths. The variation in refractive index causes light of different wavelengths to be bent at different angles. The most vivid example is in a prism which intentionally spreads out the visible spectrum, but the same sort of spectral dispersion occurs whenever light passes through a medium with refractive index that varies with wavelength. For example, simple lenses show chromatic aberration, because they bring light of different wavelengths to slightly different focal points. And when light pulses covering a range of wavelengths travel long distances through optical fibers, spectral dispersion can spread the pulses out because some wavelengths travel faster than others.

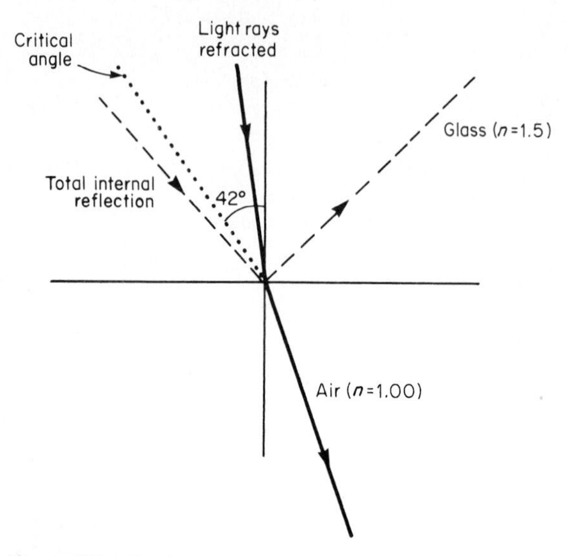

Figure 5-2 Total internal reflection occurs when light in glass strikes the interface with air at an angle from the normal greater than the critical angle, the angle of incidence for which Snell's law gives an angle of refraction of 90°. In the case shown, with glass having $n = 1.5$ and air having $n = 1$, total internal reflection occurs for angles of incidence greater than 42° to the normal.

Lenses Lenses cause a beam of parallel light rays to converge or diverge. There are many different types of lenses, a sampling of which are shown in Fig. 5-3. The commonest types of lenses are radially symmetrical, with one or two surfaces that are sections of a sphere and are generally called spherical lenses. There also are "cylindrical" lenses, in which one or two surfaces are shaped like the sides of a cylinder. While a spherical lens focuses light in two dimensions—bringing a circular beam down to a point—a cylindrical lens focuses only in one dimension, bringing a circular beam down to a line. The surfaces of lenses do not have to be sections of spheres, but that is the commonest choice because of the ease of manufacture. As will be described later, refractive effects of a lens can be obtained by varying the refractive index of glass with distance from the center of an optical element, or by fabricating special types of holograms; however, these approaches are not yet widespread.

Lenses are classified as positive or negative. Positive lenses bend parallel light rays so they converge; negative lenses bend them so they diverge. The focusing power of a lens is normally measured as the focal length. The precise focal length is complex to define (see, for example, Smith, 1978), but a reasonable approximation is to label it the distance from the lens to the focal point where rays initially parallel to the axis are brought to a point. That works for a positive lens because the rays converge. For a negative lens, the

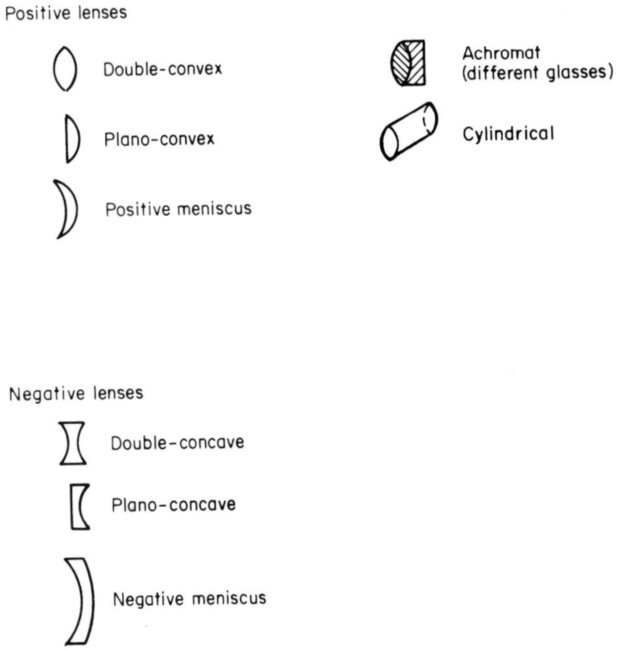

Figure 5-3 A sampling of positive and negative lenses. Achromats, shown only as positive lenses, can also take on other configurations. Cylindrical lenses, which can be positive or negative, are curved only in one dimension; other types shown are curved in two dimensions, but are shown in cross section only for simplicity.

"focal point" is taken as the point behind the lens from which the rays appear to diverge, and the focal length is a negative number. Both cases are shown in Fig. 5-4 for ordinary two-dimensional spherical lenses. Note that strictly speaking the focal length is the distance from the center of the lens to the focal point, *not* from the surface of the lens.

Another difference between positive and negative lenses is in the formation of images. A positive lens can produce a real image—one that can be projected onto a screen—of an object at a suitable position. A negative lens cannot produce a real image, but only a virtual image—one that can be seen by the human eye but not projected onto a screen. (Positive lenses can also produce virtual images in addition to real images.) The difference between real and virtual images is important because only real images can be recorded photographically or by other means.

Focal length of a lens depends on its curvature and on the index of refraction of the lens material. By making some simplifying assumptions that the lens is then compared to its diameter and the size of the optical system, the focal length f can be calculated for a spherical lens using the formula

$$\frac{1}{f} = (n - 1)\ (\frac{1}{r_1} + \frac{1}{r_2})$$

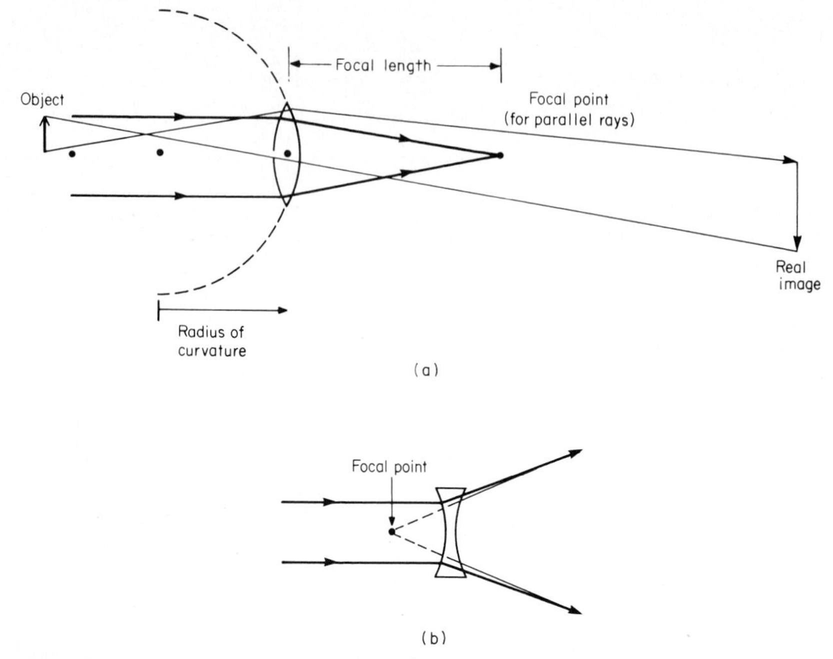

Figure 5-4 *(a)* Refraction by a positive lens bends parallel rays so they meet at a focal point. The real image produced by a positive lens is beyond the focal point. *(b)* For a negative lens, the focal point is the point from which initially parallel rays, diverged by passing through the negative lens, appear to originate. All lens surfaces have a radius of curvature, as shown only for one surface of the positive lens.

where n is the index of refraction, and r_1 and r_2 are the radii of curvature of the lens surfaces.

This focal length, in turn, can be used in other simple calculations of optical characteristics. If an object is at a distance A behind a lens with focal length f, the lens will form an image at a distance B in front of the lens, defined by the relationship

$$\frac{1}{f} = \frac{1}{A} + \frac{1}{B}$$

If two lenses with focal lengths f_1 and f_2 are placed next to each other, the equivalent focal length of the pair of lenses f is defined by

$$\frac{1}{f} = \frac{1}{f_1} + \frac{1}{f_2}$$

As with the calculation of lens focal length, these relationships depend on the assumption that the lens is "thin." Much more complex equations are required if the lens is "thick." More precise formulations of these and other optical relationships can be found in standard optics references (e.g., Smith, 1978).

The focal length enters into calculation of another quantity useful in describing lenses and optical systems, the *focal ratio,* usually written as *f/x,* where *x* is the focal ratio value. It is defined as the ratio of focal length to diameter of the lens or optical system. This is the same focal ratio used in camera lenses and carries the same implications. Low focal ratios mean a short focal length and a working point close to the lens—and require a sharply curved lens. Curvature can be more gradual for a larger focal ratio, with longer focal length and greater working distance. As with cameras, the focal tolerances are greater for larger focal ratios.

Lenses have focusing imperfections known as aberrations, which are present even in optically perfect components. The two most important are spherical aberration and chromatic aberration.

A simple lens with a spherical surface does not bring parallel light rays entering it at different distances from its center (or axis) to precisely the same focal point. This spreading out of the focus is called *spherical aberration* and is most serious for lenses with low focal ratios (e.g., *f*/1). Spherical aberration can be reduced by using two-component doublet lenses, because the positive and negative components have opposite spherical aberrations that nearly cancel out. However, to overcome spherical aberration completely, optics must have nonspherical (aspheric) surfaces designed especially to avoid spherical aberration. These lenses are much harder to make than lenses with spherical surfaces and hence are much more expensive in most cases. Note, however, that spherical aberration becomes less important with increasing focal ratio and can be ignored in most systems with high *f* ratios.

Chromatic aberration is due to the variation of refractive index with wavelength. Because of this variation, a lens brings light of different wavelengths to different focal points. To compensate for chromatic aberration, multielement lenses are assembled from components of different glasses in which refractive index varies in different ways with wavelength. In such achromatic lenses, the spectral dispersion of one element balances out that of the other element or elements. The use of achromats is important in optical systems that cover the entire visible range, such as cameras or binoculars. However, simple lenses usually suffice for laser systems operating at a single wavelength, because the narrow spectral width of the laser makes chromatic aberration insignificant.

One recent variation on the ordinary lens is the gradient (or graded) index lens. In this type, the refractive index of the material changes with distance from the axis. The effect adds focusing power to that supplied by the curvature of the lens. The approach is considered promising for some applications but remains primarily in development.

The functions of a lens also can be mimicked by a suitably designed hologram, known as a holographic optical element. The concept is described briefly at the end of this chapter.

Windows A window is a flat piece of transparent material that should have virtually no effect on the light passing through it. Windows exist in optical systems, as they do in buildings, to separate one physical environment from another. This is typically done when different ambient conditions are required in one part of an optical system than in another; one example is when a

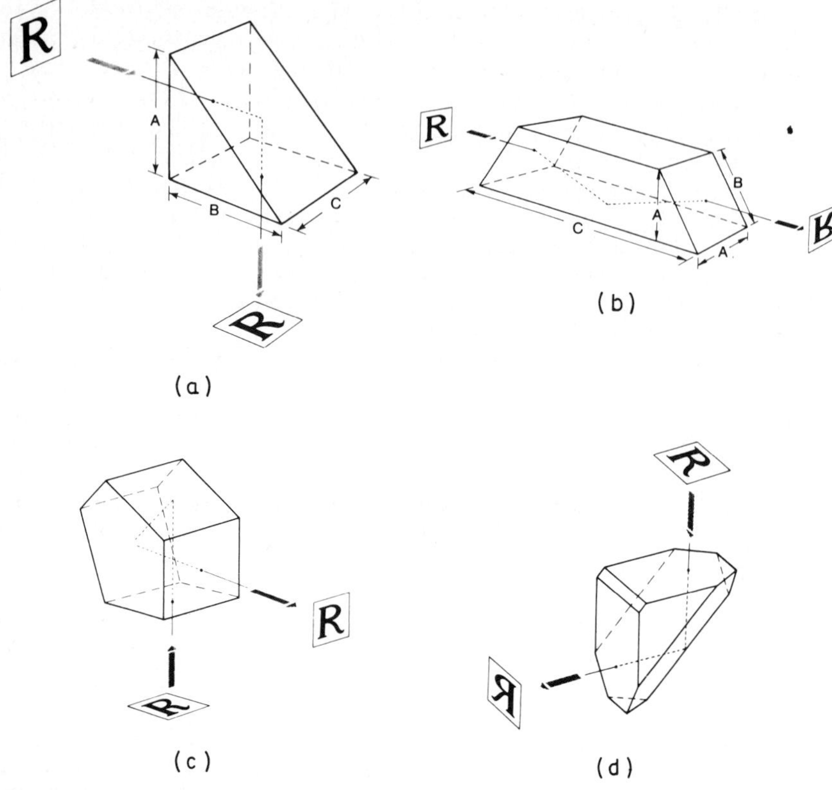

Figure 5-5 A sampling of prism types used for redirecting beams, showing their effect on images passing through them. *(a)* Right-angle prism; *(b)* dove prism; *(c)* penta prism; *(d)* roof ("amici") prism. *(Courtesy of Melles Griot.)*

laser beam is to be passed through a sample of a gas other than air. The window must be transparent at the wavelength used by the optical system, but, as described below, no optical material is transparent at all wavelengths.

Gas-laser tubes often are sealed with windows mounted at *Brewster's angle*, at which light linearly polarized in one direction is transmitted without surface reflective losses, although reflective losses do occur for orthogonally polarized light. Brewster's angle θ_B is defined relative to the normal to the window surface by:

$$\theta_B = \arctan \frac{n_1}{n_0}$$

where n_1 is the window's refractive index, and n_0 is the index of the material (e.g., the laser gas) through which the light travels to reach the window. For glass with index of 1.54, the Brewster angle is 57°. A Brewster angle window does not eliminate light polarized in the orthogonal direction, but it introduces enough losses to make the laser output strongly linearly polarized.

Because all solids absorb some of the light passing through them, solid windows are vulnerable to optical damage when transmitting high-power beams. To avoid this problem, "aerodynamic windows" have been developed, in which gas flows by an opening fast enough to prevent mixing of gases on opposite sides of the opening. Such windows are required only for lasers with very high power output.

Prisms Prisms serve two main functions in optical systems: dispersing light according to wavelength, or redirecting light without changing the size of the beam or the relative alignment of the light rays.

Dispersive prisms rely on the variation in refractive index with wavelength to refract different wavelengths at different angles, spreading out the spectrum. In some optical systems, a slit is used to pick a narrow range of wavelengths out of this spectrum. Alternatively, the prism can be included in a laser resonator aligned so only a certain wavelength will oscillate, as described in Chap. 4.

Beam-redirecting prisms generally rely on total internal reflection to bend light around corners, invert images, reverse images right to left, or perform combinations of these operations. A variety of prisms have been designed for such tasks, such as those shown in Fig. 5-5. Most of these types are used more often in imaging instruments such as binoculars than in laser systems. One exception is the retroreflective prism or "corner cube," designed to return an input beam exactly in the direction from which it came, which is often used with lasers; however, not all retroreflectors are prisms.

The term *prism* can also be used for polygonal blocks which are coated with reflective materials and function as mirrors. These reflect light from their outer faces, so the light never enters the "prism," which in fact need not be made of transparent material. Typically such prisms rotate so the different faces sequentially scan a laser beam, by reflection, across another surface.

Transmissive Materials

The characteristics of lenses, prisms, and windows depend on the transmissive materials from which they are made. The wavelengths at which these materials are transparent depends on the internal energy structure. Strong absorptions occur at wavelengths corresponding to energies at which the material can make a transition from its normal energy state to a higher one. The material is transparent at wavelengths corresponding to energies where there are no possible transitions. Energy-level structures differ, and so do transparencies, but there is no such thing as a material transparent throughout the electromagnetic spectrum. Even air is functionally opaque in much of the ultraviolet and infrared spectral regions.

The amount of light transmitted equals the amount of incident light, minus losses from reflection at the surface and from absorption at the surface and within the material. Reflectivity can vary over a wide range, from nearly 100 percent for many metals in the visible and infrared to near zero for materials such as graphite that look black to the eye. Reflectivity tends to increase with the difference in refractive index between two adjacent materials. Coating the high-index material with a thin layer of material with intermediate refractive index can decrease reflectivity. It is also possible to increase or decrease reflectivity at particular wavelengths by building up multilayer thin-film interference coatings in ways described below.

Absorption occurs both at the surface and within the material. Bulk absorption is cumulative as light passes through the material. This usual measurement is fractional absorption per unit distance traveled through the material, such as 0.1 per centimeter, meaning that 10 percent of the light is absorbed in traveling through 1 cm of material. The same fraction of the light is absorbed in the second centimeter, and so on. Thus the amount of power transmitted through a distance D of material with absorption constant k can be written

$$\text{Transmitted power} = \text{input power} \times e^{-kD}$$

The corresponding absorption law is

$$\text{Absorbed power} = \text{input power} \times (1 - e^{-kD})$$

It is often convenient to write absorption in units of decibels and to measure characteristic absorption in decibels per unit length. In this way absorption (or attenuation) can be obtained simply by multiplying the characteristic value by the distance traveled. This approach is standard when working with fiber optics, where normal units are decibels per kilometer.

The transmissive materials normally used in different spectral regions are described below with two materials often encountered in optics: air and water.

Ultraviolet Optics The ultraviolet spectrum extends from 400 nanometers (nm) to a short-wavelength limit of 1 to 10 nm (definitions vary). The shorter-wavelength end of the ultraviolet spectrum is poorly explored because of experimental problems that increase in severity with shorter wavelength. Laser action also becomes increasingly difficult to produce at shorter wavelengths. A variety of commercially available lasers emit in the 200- to 400-nm range, but only two emit at shorter wavelengths: the argon fluoride excimer at 193 nm, and molecular fluorine at 157 nm. Shorter-wavelength lasers have been demonstrated experimentally but have yet to become practical.

Optical materials used in the near-ultraviolet, wavelengths of 300 to 400 nm, are familiar, because most optical glasses transmit in that region. Special types of quartz can be used at wavelengths shorter than 200 nm, beyond the range of other silicate materials (Driscoll, 1978). For shorter-wavelength operation, the usual choices are magnesium fluoride (transmissive to 110 nm), calcium fluoride (transmissive to 125 nm), or lithium fluoride (transmissive to 104 nm). The first two materials are more durable than LiF (Harshaw, 1982).

Optical damage thresholds tend to be lower in the ultraviolet than in the visible or infrared, so damage is a serious concern. It can occur in the bulk material or at the surface, particularly to coatings applied to enhance or reduce reflections. Nonsilicate ultraviolet windows are subject to some environmental degradation, although the problem is not as severe as for the more delicate materials used in parts of the infrared.

Material problems become increasingly severe at shorter ultraviolet wavelengths. As photon energy increases with decreasing wavelength, it eventually reaches a point sufficient to remove one of the electrons from a completed shell of an atom or molecule. At this threshold, ultraviolet absorption increases abruptly, and the material effectively becomes opaque. The atmosphere becomes opaque at about 200 nm, but rare gases are transparent at shorter wavelengths. Helium, with the highest ionization threshold, is transparent to about 58 nm (Zombeck, 1982). All known solid window materials become opaque by 100 nm. These transmission problems make experimentation in the extreme ultraviolet extremely difficult.

Visible Materials Silicate glasses are the obvious choice for transparent visible-wavelength optics because of their low cost, durability, and high transmission. Standard optical glasses are widely used with lasers. On the other hand, plastic optics, although widely used in consumer optical systems, are rarely used with lasers. Major reasons include concerns about their vulnerability to scratches, and potential scattering from inhomogeneities within the material or from surface scratches. The economies of scale possible with high-volume consumer production are also harder to achieve with the smaller volumes of laser system production.

There is some haziness in the definition of the "visible" region in the laser world. Strictly speaking, the visible region is usually defined as 400 to 700 nm, the wavelengths where the human eye is most sensitive, although the eye can actually detect a somewhat broader range, particularly at longer wavelengths, with very low sensitivity. However, optical engineers may functionally define the visible as the region where common optical glasses and μm plastics can be used—from about 300 nm to 1.5 or 2 micrometers (μm). This reflects the fact that near-infrared optical systems, such as those designed for use with a 1.06-μm neodymium laser, normally will also function in the visible spectrum as well.

Infrared Materials Many materials have been developed for infrared optics, thanks largely to military development of night-vision systems. Development has concentrated in three wavelength bands in which air is transparent: 1 to 2 μm, 3 to 5 μm, and 8 to 12 μm.

The 1- to 2-μm region is usually called the near-infrared, although sometimes that term is used to cover somewhat longer wavelengths, at times stretching to 15 μm. Most silicate glasses are transparent to at least 2 μm, so conventional optical glasses are normally used in this region.

The 3- to 5-μm band, sometimes called the mid-infrared (although that term has also been used for longer wavelengths), is interrupted by some strong atmospheric absorption lines. Chemical and carbon monoxide lasers emit in this region, although some of their lines coincide with strong atmos-

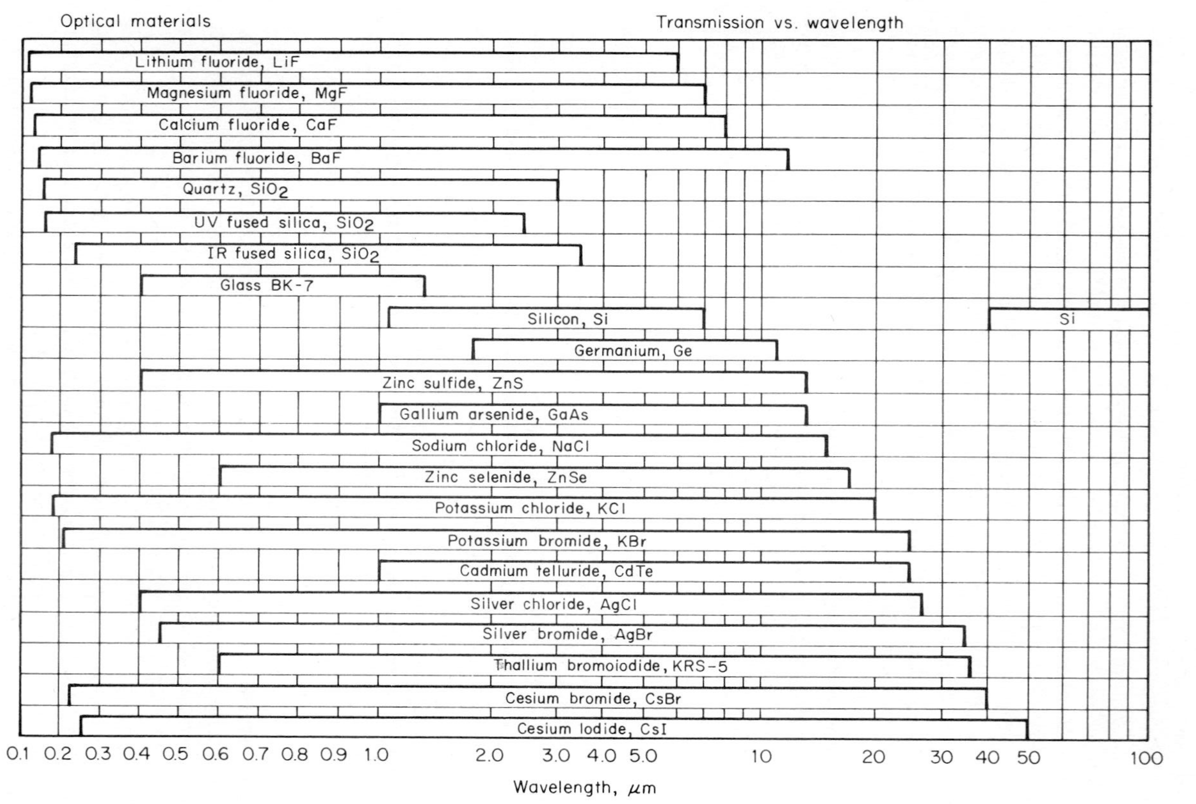

Figure 5-6 Spectral regions at which selected optical materials transmit adequately for optical applications. Note that different forms of silica, including quartz and optical glass, have different transmissions—a consequence of different levels of impurities. *(Courtesy of Janos Technology Inc.)*

pheric absorption bands. Some silicate glasses and quartz are transmissive at the short-wavelength end of this region, but their absorption increases rapidly with longer wavelengths, and they are not usable at the long-wavelength end. Thus the usual choices for such optics are more exotic materials such as those whose transmission ranges are shown in Fig. 5-6.

The 8- to 12-μm region is often called the *thermal* infrared, because those wavelengths correspond to the peak of blackbody emission at room temperature. The region is also known as the far-infrared, a term also widely used for infrared wavelengths of 10 μm and longer. It is important for laser applications because it covers the 10.6-μm wavelength of the carbon dioxide laser. Some important materials used in this wavelength range are shown in Fig. 5-6.

Newcomers to the infrared should be aware of two common features of infrared optical materials. One is that typically they are *not* transparent at visible wavelengths, so an optical system that focuses a CO_2 laser beam may be as clear as mud to the human eye. The other is that many infrared materials are hygroscopic and have such a strong tendency to absorb water from the air that if left unprotected they will slowly dissolve. Users should exercise considerable care in selecting infrared materials, being particularly careful of physical and mechanical characteristics. Important data on infrared materials are available in several references (e.g. Driscoll, 1978; Harshaw Chemical, 1982; Wolfe and Zissis, 1978).

Air Air is almost inevitably a part of any optical system. It can present special problems at any wavelengths where extremely high powers are being transmitted, or where beams must be focused precisely over long distances. Clear air is transparent in the visible region, although dust, haze, smoke, fog, clouds, and precipitation can block visible light. Such problems are rarely significant on a laboratory or industrial scale, but other problems can arise at other wavelengths.

The atmosphere transmits light well at wavelengths as short as about 290 nm, where ozone (O_3) begins absorbing strongly. Fortunately, the main concentration of ozone is in the upper atmosphere, where it blocks short-wavelength solar ultraviolet light. At ground level, air transmits reasonably well to wavelengths as short as 200 nm, where molecular oxygen (O_2) starts absorbing (McCartney, 1983). Shorter wavelengths are usually called the vacuum ultraviolet because they are normally transmitted through a vacuum. However, if vacuum operation is undesirable, rare gas fills can be used at somewhat shorter wavelengths, with operation at wavelengths to about 58 nm possible by using helium.

Strong molecular absorption lines interrupt atmospheric transmission in the infrared and block it altogether at wavelengths longer than 13 μm. Carbon dioxide and trace gases such as methane and ozone make significant contributions to infrared absorption with lines scattered at shorter infrared wavelengths. Water vapor also has strong absorption lines between 1 and 2 μm plus broad absorption bands at 2.5 to 3.5 μm and from 5 to 7 μm, making humidity an important variable. Infrared absorption generally is not as strong

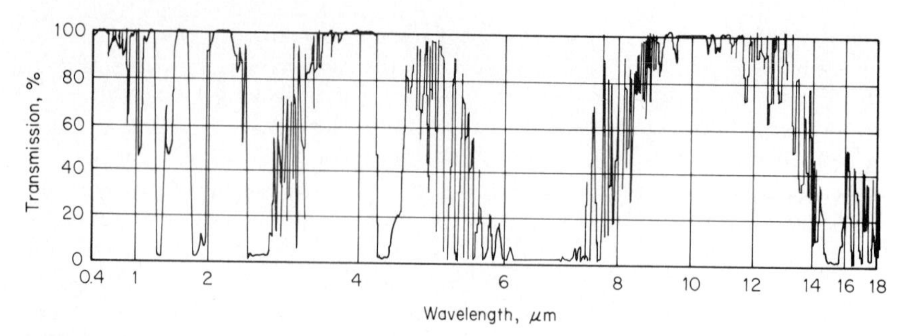

Figure 5-7 Transmission of light in the visible and infrared through a 304-m (1000-ft) horizontal path of sea-level atmosphere, with 5.7 mm of 26°C precipitable water. *(Courtesy of Sanders Associates Inc.)*

as absorption in the vacuum ultraviolet, so it is often possible to transmit beams through short lengths of air even at strong absorption lines. Nonetheless, it is far easier to work in atmospheric windows.

Extensive compilations of data are available which show infrared absorption in air at high resolution (see, for example, Wolfe and Zissis, 1978), and extensive tabulations have been made of transmission data and predictions of atmospheric models. Figure 5-7 gives a view of infrared absorption on a broader scale, without the high resolution needed to show the many narrow lines.

Water Water is opaque to most of the electromagnetic spectrum, with the notable exceptions being in the visible region and extremely low frequency radio waves. The absorption coefficient has a minimum of about 0.0001 per centimeter (cm^{-1}) near 500 nm in the blue-green, rising rapidly at shorter and longer wavelengths. For seawater, absorption reaches 0.01 cm^{-1} near 750 and 300 nm, and has values of 100 to 10,000 cm^{-1} at infrared wavelengths beyond 2 μm (Wolfe and Zissis, 1978). The detailed profile and precise absorption minimum vary depending on the type and purity of water, but in all cases maximum transmission is in the blue or blue-green.

Filters and Coatings

Filters and coatings can alter the transmission characteristics of optics and optical systems. Some filters are discrete optical components which derive their filtering characteristics from the nature of the bulk optical material, but many are actually filtering coatings applied to plain transmissive optical substrates. Modern coatings can be applied in many ways, giving designers great flexibility in selecting properties of an optical system. Coating and filter design are worthy of an in-depth treatment (see, for example, Department of Defense, 1961; Dobrowolski, 1978; Melles Griot, 1985); this section can only provide a brief overview of the field.

Types of Coatings and Their Functions Coatings can be viewed in two ways, by the physical mechanisms of their operation or by their function. The same physical principles can be used for different types of coatings, and a single coating can perform multiple functions. Conversely, the same filtering function (particularly spectral filtering) often can be performed in two or more ways.

The major physical types of filters are

- *Interference coatings:* Types in which alternating layers of two (or sometimes more) different materials, each a fraction of a wavelength of light thick, are deposited on an optical surface. The thicknesses of the layers are carefully controlled. Internal reflections within the multilayer stack cause constructive and destructive interference for certain wavelengths and angles of incidence. Complex but well-quantified design rules let the designer pick the desired coating properties. Because interference effects depend on the distances light has to travel through each layer, the properties of interference coatings change with angle of incidence, and that angle must be known for proper system design. This effect also makes it possible to identify an interference filter, by tilting it and watching for a change in transmission. Also known as *dielectric, dichroic,* or *multilayer* coatings, they are widely used to perform a variety of functions.
- *Simple antireflection coatings:* Types in which a material with refractive index between that of air and an optical element is deposited on the optic to reduce surface-reflection losses.
- *Color-filter coatings:* Dyes applied to the surface of components which transmit light at some wavelengths and absorb it at others.
- *Neutral-density coatings:* Absorptive materials which absorb light uniformly across the range of wavelengths covered by the device or optical system.
- *Metal-film coatings:* Reflective metal layers applied to form mirrors.
- *Metal-dielectric coatings:* Metal-film coatings in which the reflectivity at a particular wavelength is enhanced by applying an overcoat of a transparent dielectric (i.e., nonconductive) material.
- *Protective coatings:* Thin films of materials applied to protect optical components from physical damage or corrosion, such as water-resistant coatings applied to hygroscopic materials, or transparent overcoats deposited on metal to prevent oxidation.

In some cases, notably protective coatings, the physical type also identifies the function, but this is not true for others, particularly interference coatings. From an optical standpoint, there are several different functions:

- *Spectral filters:* Types which selectively transmit and reflect certain wavelengths to separate them, described in more detail below.
- *Antireflection coatings:* Types designed to minimize surface reflection losses. In addition to the simple antireflection coatings described above, multilayer interference coatings can serve to reduce reflection losses at selected wavelengths.

- *Reflecting coatings:* Types designed to provide maximum reflectivity either at a single wavelength (interference coatings) or over a range of wavelengths (interference or metal-film coatings).
- *Polarizing coatings:* Types which selectively transmit light of one polarization. Typically light with one linear polarization is transmitted and that with orthogonal polarization is reflected.
- *Beamsplitting coatings:* Types which transmit and reflect light in a given ratio (say 1:1), thus splitting a single beam into two beams with a desired intensity ratio. Many types are polarizing, splitting light of one polarization in one direction and light of the orthogonal direction in the other direction.
- *Neutral-density coatings:* Types which simply reduce the intensity of light transmitted throughout their working spectrum.
- *Graded coatings:* Attenuating or neutral-density coatings in which the degree of attenuation varies across the aperture of the coating. In certain cases, this can serve an important function in improving beam quality.

Optical components are available either with coatings already applied or without coatings so that user-specified coatings can be custom-applied. Large optics suppliers normally stock components coated for important laser lines, while custom coating may be required for other wavelengths. The application of coatings is something of an art, and design and application must be performed by specialists.

Spectral Filters Diverse applications require selective transmission of certain wavelengths. The spectral filters that do this job, like optical coatings, fall into two sets of categories based on physical mechanisms and optical functions.

For laser applications, the commonest types are interference filters, in which a multilayer interference coating selectively transmits some wavelengths and reflects others. The reflection, rather than absorption, of rejected light is important because laser power density is high enough that absorption could heat absorptive filters enough to cause damage. The design flexibility possible with interference filters also makes it possible to specify a wider range of optical characteristics than for other types. Examples include narrower transmission and rejection bands and sharper changes in transmission with wavelength.

In other types of filters, rejected light is absorbed rather than transmitted, which can cause problems at high power densities. In colored-glass and crystal filters, absorption takes place in the bulk of the material. In gelatin filters, the spectral selection is a thin layer of gelatin containing an organic dye, which is sandwiched between a pair of glass plates. Colored-glass and gelatin filters are in widespread use because of their low cost, but their power-handling capability is limited.

Spectral filters fall into several functional categories. Their characteristics, described briefly below, are illustrated with typical spectral curves in Fig. 5-8:

- High-pass filters transmit wavelengths shorter than a given value and

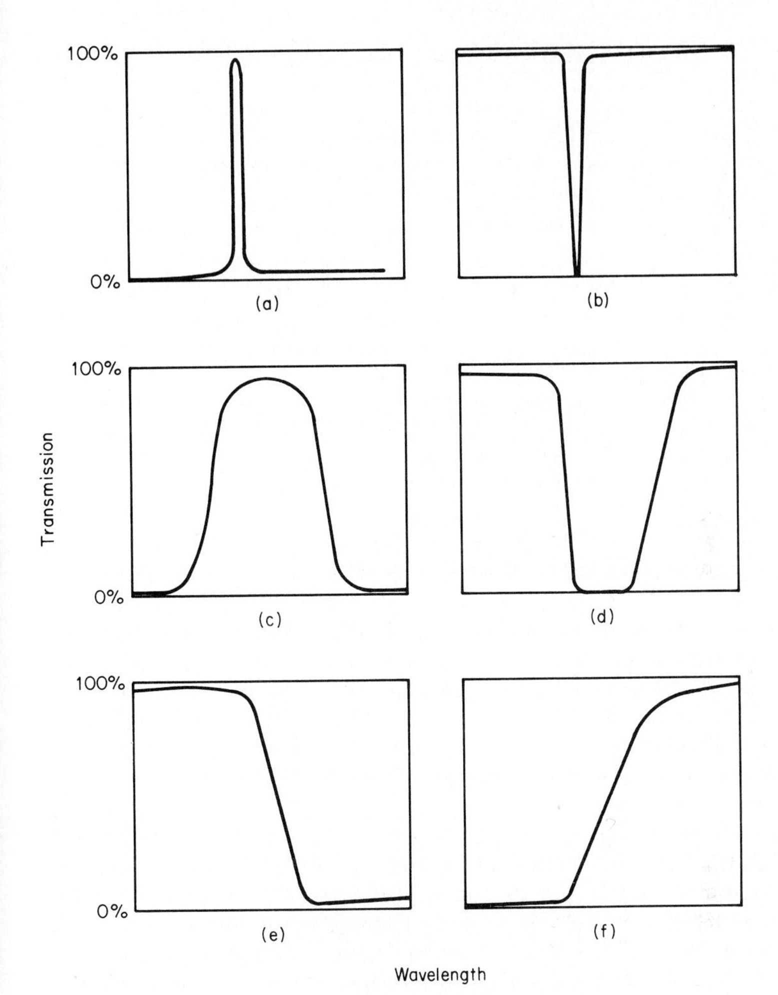

Figure 5-8 Transmission of representative types of filters, shown schematically (these are not actual plots). Wavelength increases to the right on all plots, a common convention in optics. *(a)* Narrow-line filter; *(b)* line-rejection filter; *(c)* bandpass filter; *(d)* band-rejection filter; *(e)* high-pass filter; *(f)* low-pass filter.

reject longer wavelengths. Examples include the "hot mirror" which transmits visible light but reflects in the infrared, and the "heat-absorbing filter," which transmits visible light and absorbs infrared.

- Low-pass filters transmit wavelengths longer than a certain value and reject shorter wavelengths.
- Bandpass filters transmit light within a defined range (e.g., 500 to 600 nm) while rejecting longer and shorter wavelengths. A color filter that transmits only green light would fall under this heading.

- Band-rejection filters specifically reject light within a defined region but transmit other light in that spectral region.
- Edge filters are high- or low-pass filters with a sharp cutoff wavelength.
- Line filters transmit only a very narrow range of wavelengths around a certain value, for example, only light very close to the 632.8-nm helium-neon laser wavelength. The bandpass can be less than 1 nm wide, although peak transmission in such cases is limited.
- Line-rejection filters are the inverse of line filters, selectively rejecting light within a narrow range of wavelengths, such as blocking the helium-neon laser wavelength.

Attenuators or Neutral-Density Filters Neutral-density filters or attenuators are designed to produce attenuation that is uniform regardless of wavelength. Of course, in practice the range over which attenuation is uniform is limited, typically to the visible spectrum. Transmission of a neutral-density filter is measured in units of optical density, defined as:

$$\text{Optical density} = -\log_{10}(\text{transmission})$$

so a filter that transmits 0.001 (0.1 percent) of the incident light has optical density of 3.

Variable attenuators can be useful in many applications. One way to achieve variable attenuation is with a single optical element graded in optical density across its surface. Use of a single element, usually called a *wedge* although not necessarily wedge-shaped, is simple because linear motion or rotation can change optical density. However, the precision of calibration is limited. An alternative is the use of a set of multiple calibrated filters, providing greater accuracy at a cost in convenience.

Spatial Filters In laser applications, it is sometimes necessary to block the outer parts of the laser beam to remove stray light diffracted by dust and lens imperfections, to reduce the effects of lens aberrations, and to improve wavefront quality. This function, sometimes called *aperturing,* is performed by spatial filters, which are essentially holes shaped to block unwanted portions of the beam. Some spatial filters are quite small in diameter and are called *pinholes*—an apt description of their appearance.

An alternative used in some applications is the "apodizing" or "graded" filter, in which optical density varies with the density from the optical axis—increasing as one moves outward. Instead of sharply cutting the beam off at a particular diameter, such a filter gradually reduces power density, an approach desirable in some applications.

Reflective Optics

As mentioned above, many mirrors are thin reflective coatings applied to substrates which have the desired flat or curved shape. Metal coatings are reflective over a broad range of wavelengths, while interference coatings are spectrally selective and reflect a limited range of wavelengths. Both types of coatings can be applied to many substrates, with glass a common choice because of its good optical qualities. Because no light need enter the substrate, glass can be used in any spectral range. Low-expansion materials can be used to avoid thermal distortion.

Mirrors also can be machined from solid metal, with the metal itself, or a reflective coating applied to it, providing the reflective surface. Metal mirrors are highly conductive and can be used with high-power lasers with suitable heat sinks or active cooling (such as water flowing through the substrate). Note, however, that some actively cooled, high-power metal mirrors may fall under export-control regulations.

Metal mirrors are most often used in the infrared, where metal reflectivity is highest. Mirror surfaces can be polished, or machined with a diamond-tipped cutting tool. Diamond turning is particularly valuable for producing nonspherical surfaces, which are difficult or impossible to generate by conventional polishing techniques. Diamond turning does leave small ridges on the surface, but these are not optically significant if they are much smaller than the wavelength. For example, surfaces in which fine machining lines can just be detected by the human eye using visible light work fine for the 10-μm light from a CO_2 laser.

The commonest mirrors are flat or spherical; as for lenses, a spherical mirror is one with a surface that is a section of a sphere. Spherical mirrors can be concave or convex and have a focal length equal to half the radius of curvature. Normally these mirrors are totally reflective (with only a small fraction of the light absorbed and virtually none transmitted), but some flat mirrors are coated with partially reflective coatings that serve as beamsplitters.

Mirrors also can be manufactured with more complex surfaces. Aspherical mirrors are focusing mirrors with surfaces that are neither flat nor spherically curved. One common type is the parabolic mirror, a design used for reflecting telescopes to avoid spherical aberration. Some special applications require off-

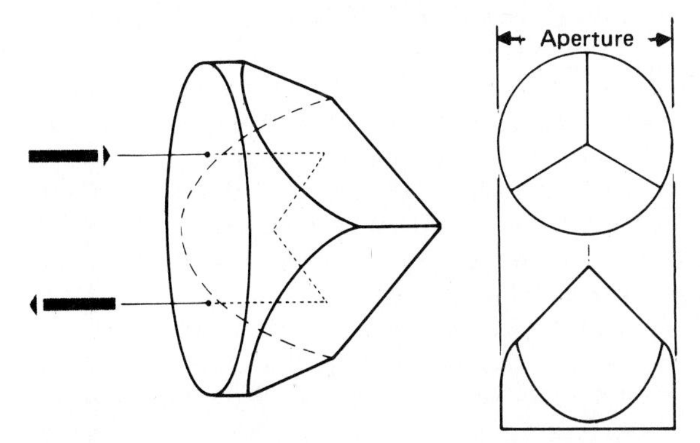

Figure 5-9 A solid-glass retroreflector, in which total internal reflection returns the incident beam to its origin. Corner-cube mirrors are similar, but mirrors aligned at right angles to one another form the retroreflector, rather than a glass block. *(Courtesy of Melles Griot.)*

axis mirrors, in which the curved surface is a section of a three-dimensional surface but does not include the axis of rotation that defined the surface. Parabolic mirrors are fairly easy to produce, but other aspherical surfaces can be expensive to manufacture.

One special type of mirror used in a number of laser applications is the retroreflector or corner cube. This is a prism or a set of three mirrors with faces aligned at right angles with respect to each other, as shown in Fig. 5-9, to reflect any laser beam incident on the device back in the direction from which it came. This function is useful in alignment and measurement systems.

Polygonal mirrors are used in some applications requiring scanning of laser beams across a surface. A multifaceted polygon, typically with 10 or more faces, is rotated about its axis, with each facet successively scanning the beam across the surface. The mirror may be a coated prism or a machined metal block, each with polished reflective surfaces around its perimeter.

Complex reflective surfaces, such as polygons and aspherical mirrors, often are produced by optical replication. This is essentially a mastering process in which a mold is generated from a master and used to form optical-grade surfaces, as described below. The replicated component is then coated with a reflective layer and can serve as a mirror.

Special Components and Techniques

Some specialized components which do not readily fall into the reflective or transmissive categories are used in optical systems. A few of these types have been mentioned in passing above, but the major ones all deserve brief explanations by themselves.

Beamsplitters The function of a beamsplitter is to divide an incident beam into two parts, with one transmitted and the other reflected. Typically it is placed at a 45° angle to the incident beam, so reflected light is deflected at a right angle, while the transmitted light passes straight through. The beam-splitting ratio can be 1:1, or some other value.

Beamsplitters can take several forms, including partly transparent metal films, specially designed prisms, types which reflect one linear polarization and transmit the orthogonal polarization, and multilayer interference coatings which selectively transmit and reflect certain wavelengths. Although polarizing beamsplitters are the only types explicitly intended to affect beam polarization, other types may be sensitive to polarization. For example, certain types may reflect light of one linear polarization more strongly than that of the orthogonal polarization. That can lead to different beamsplitting ratios for differently polarized inputs, a potential source of problems in some optical systems. Users should take care to understand these effects in designing polarization-sensitive systems.

Fiber Optics Optical fibers are best known for their use in signal transmission for communications. However, they were originally developed to serve as optical elements. Single fibers and bundles of randomly arranged fibers can serve as light pipes, conveying light from one point to another along sometimes convoluted paths.

Bundles of fibers in which the fibers maintain their relative alignments at the input and output faces can serve other functions. For example, a simple image inverter can be made by twisting the bundle 180°. A lens can do the same sort of image inversion, but not in such a short linear distance. Aligned or "coherent" fiber bundles can be made rigid or flexible, and both types have been used in many applications, such as in medical endoscopes to view inside regions of the human body not otherwise accessible without surgery. Although most recent writings on fiber optics concentrate on communications, there are some good references on the design and applications of fiber bundles as optical elements (e.g., Siegmund, 1978).

Gratings Diffraction gratings are surfaces covered by closely spaced parallel lines which diffract incident light. Interactions of light waves diffracted by the parallel lines make diffraction angle a function of wavelength, thus like a prism, a diffraction grating spreads out a spectrum. Transmissive gratings can be made, but the normal type is reflective.

Gratings come in three basic types: mechanically ruled, holographic, and replicated. Mechanically ruled gratings are the oldest type, produced by special machines which individually etch each parallel line into a flat plate. The expense of that process stimulated development of methods to replicate many gratings from a single ruled master. The replica gratings are produced by pressing a suitable plastic between a master grating and a flat surface; because of their low cost, they are the commonest type.

Holographic gratings are produced by exposure with laser light in a way that produces a series of interference fringes, which are recorded to form the grating. Lines can be spaced more closely than in ruled gratings, and the process is less expensive than ruling original gratings. Replica gratings can also be produced from holographic originals.

Polarizing Optics The electric and magnetic fields that make up electromagnetic radiation oscillate in directions transverse to each other as the wave travels in a direction orthogonal to both oscillations. Light in which the electric fields are aligned to oscillate in the same direction (and hence the magnetic fields are also aligned in the orthogonal direction) is said to be linearly polarized. Two linear polarizations are possible, orthogonal to each other and to the light ray.

Light can also be polarized circularly or elliptically. For the more general case of elliptical polarization, a stationary observer would see the polarization vector of the light rotating, with its endpoint describing an elliptical pattern. (In this case, direction of the polarization vector indicates polarization direction, while magnitude indicates amplitude of the polarized wave.) Physically, this means that the light ray consists of two linearly polarized components of unequal amplitude, one 90° in phase behind the other. In the special case of circular polarization, the two orthogonal polarization components are equal in amplitude, and the rotating polarization vector describes a circle.

Many polarizing components are available, which either alter the polarization of input light or transmit only light of a given polarization. The simplest are polarizers or polarizing filters, which transmit only light of one linear polarization. Depolarizers scramble the polarization of linearly polarized light, generating randomly polarized light. Polarization rotators rotate the plane of linearly polarized light. Waveplates delay the phase of the part of the input light with one linear polarization by a specified phase with respect to light of the orthogonal polarization. For "quarter-wave" plates, the delay is 90°, while for half-wave plates there is a 180° delay. Note that passing light through a quarter-wave plate polarizes it elliptically. The physics and operation of these and other polarizing optics are covered in more detail in a number of references and standard texts (see, for example, Bennett and Bennett, 1978, for a scholarly discussion, or Melles Griot, 1985, for a practical overview).

Replicated Optics Optical replication, mentioned briefly above, is a process usable for production of many types of reflective optics and a few transmissive components. The attraction of replication is its ability to produce components with unusual surface shapes at lower cost than conventional techniques.

Production of reflective replicas starts with fabrication of a substrate that approximates the shape of the finished piece, but lacks the required precision surface. The substrate is then coated with a plastic-epoxy material and pressed against a master piece to form a replica of the surface, which comes close enough to the quality of the original for many applications. The replica is then reflectively coated for use as an optical element (Weissman, 1978). Transmissive replicas can be made by using transmissive substrates and plastics, but they are used only rarely.

The practical advantages of replication come for quantity production of reflective components with nonspherical surfaces. Such components are hard to produce by conventional techniques, making them expensive singly or in quantity. However, a single original can be used to generate many replicas, at lower cost per component. One notable disadvantage of replicated optics is lower power-handling capability than conventional optics.

Holographic Optical Elements Holograms are best known for their ability to reproduce three-dimensional images. They do so by diffracting light from a two-dimensional interference pattern recorded on a light-sensitive plate or film.

Recently, interest has grown in another application of holograms—the simulation of the function of optical elements. The idea is to create a diffraction pattern which scatters light in the same way that a lens or other optical element would refract or reflect it. Computer synthesis of the holograms (Caulfield, 1983) has proved to be a valuable tool in generating the holographic patterns needed to perform the functions of particular optical elements.

Holographic optical elements reached the market only recently, but they have already found uses in some commercial systems such as laser printers and scanners. Beam scanning is a particularly attractive application because a lightweight holographic optic can replace a heavier and more costly rotating mirror (Kramer, 1981). Holographic optics also can offer otherwise hard-to-get capabilities, such as low-focal-ratio optical systems, at lower cost than conventional optics.

The diffraction angle of the light scattered by a holographic optical element is a function of wavelength, so holographic optics must be designed for operation at a particular wavelength. For the same reason, if holographic optics are to focus light precisely, it must be monochromatic—a requirement met by single-line lasers.

One important issue in holographic optics is diffraction efficiency, because only light diffracted by the hologram is focused to the desired point. Diffraction efficiencies of more than 80 percent have been claimed for commercial products, but such high efficiencies are not always attainable. Low diffraction efficiencies can mean both high losses and high levels of scattered light in the wrong place.

Active Optical Components

The optical components described so far have been passive optics, so-called because their effect on the light they transmit or reflect generally does not change with time. Other types of optics used with laser systems are active, in the sense that their action on light does change with time, typically because of active operator control. A sometimes hazy middle ground is occupied by "nonlinear" optics, which act upon light in a way not linearly dependent on light intensity to perform functions such as harmonic generation (frequency doubling) even without active control.

Some active optical devices properly fall under the heading of beam enhancements and were described in Chap. 4. Those include components designed to change the length of laser pulses, or to alter laser frequency by nonlinear effects. Two other types of active optics widely used with lasers deserve brief mention here: beam intensity modulators and beam deflectors.

Modulators Modulators change the fraction of incident light they transmit in response to external control signals. The modulation may rely on acousto-optic or electro-optic interactions in a suitable crystal, or on mechanical operation of a shutter or aperture.

In acousto-optic modulators, an acoustic wave sets up a pattern of density variations in a suitable material that functions as a diffraction grating. This "effective diffraction grating" diffracts part of the laser light passing through the material at an angle away from the normal beam direction. The fraction of the light energy in the diffracted beam depends on the acoustic-wave strength. Modulation can be observed in both the diffracted and undiffracted beam; commercial devices are designed to use one of those beams as the output.

Commercial models are made mostly for the visible and near-infrared, but some are made for the 10.6-μm CO_2 laser wavelength. These devices have limited beam apertures and require radio-frequency (acoustic-wave) drive powers on the order of a watt. Modulation bandwidths normally are a few megahertz to a few tens of megahertz, but some models have bandwidths as large as 500 MHz. Made of materials including quartz, lead molybdate, and tellurium dioxide, commercial acousto-optic modulators typically sell for a few hundred dollars or up in small quantities.

Electro-optic modulators rely on the effects of electric fields on the refractive index of certain nonlinear materials. Inherent asymmetries within these crystals cause them to be birefringent, meaning that linearly polarized light with electric field vertical experiences a different refractive index than linearly polarized light with orthogonal polarization. The application of an electric field changes this birefringence, in effect rotating the polarization of light. When the input light is linearly polarized, and a linear polarizing filter is used at the output, such a device can function as a modulator; the degree of polarization rotation determines how much light will be transmitted by the output polarizer. The entire assembly (with or without an electronic driver) is sold as an electro-optic modulator.

Prices of electro-optic modulators range from a few hundred dollars to several thousand, with the more costly versions typically including a driver. Typical bandwidths are on the order of 100 MHz, but peak modulation frequencies can exceed 1 GHz. Operating voltages are a few tens of volts to several kilovolts, depending on nonlinear material used. For visible wavelengths, the commonest materials are KDP (potassium dihydrogen phosphate, KH_2PO_4), and related compounds including ADP (ammonium dihydrogen phosphate, $NH_4H_2PO_4$), and some versions with deuterium substituted for hydrogen. A few binary semiconductors are used at longer infrared wavelengths. Certain liquids, such as nitrobenzene, can also be used.

A less elegant, but conceptually simpler, approach is mechanical modulation—interrupting the beam with a shutter. In most cases this is limited to off-on modulation, but it is possible to partly block an illuminated aperture. Several types of shutters are used, including a mechanical plate that slips into a slot, a rotating wheel, and a plate attached to a vibrating tuning fork or galvanometer. Safety regulations require the use of shutters to block output beams on many lasers during certain operating conditions.

The three standard types of modulators all operate on a single beam or aperture at once. Devices called spatial light modulators have been built to simultaneously adjust intensity of multiple points in an aperture. For example, some spatial light modulators can serve as variable transparencies. Such devices, now in developmental stages, are sought for optical computing and signal-processing systems, but remain high in cost and limited in performance.

Beam Deflectors Beam deflectors or scanners are needed for applications in which a laser beam must be moved or scanned across a surface. Examples include printing and reading of bar codes. Both solid-state and mechanical types are available. Solid-state scanners include acousto-optic and electro-optic types, although the latter have found few applications. Mechanical types include rotating mirrors and holographic elements and resonant mirror scanners.

Acousto-optic deflectors, like acousto-optic modulators, rely on the interaction between a laser beam and an effective diffraction grating formed by an acoustic wave in a suitable material. As in a modulator, the grating diffracts a portion of the beam at an angle, while the rest of the beam passes straight through. The deflection angle depends on the grating spacing, which in turn is a function of acoustic-wave frequency, so scanning is accomplished by varying the acoustic drive frequency. Most solid-state scanners on the market are acousto-optic types. They are attractive because of the absence of mechanical parts, but their resolution typically is limited to on the order of a thousand spots per scan (Gottlieb et al., 1983), keeping them from high-resolution applications.

Electro-optic deflectors have also been developed which rely on nonlinear electro-optic interactions. Although they offer better resolution than acousto-optic deflectors, they are more expensive and require high drive voltages, and hence have found few applications.

Rotating-mirror scanners rely on rapid rotation of a polygonal mirror around its axis. The beam arrives from a constant direction, and the reflected light is scanned across a line as rotation changes the angle of incidence on the mirror face. If the mirror faces are uniformly spaced around the polygon, each one scans the laser beam across the same line; thus a single rotation of a 12-facet mirror will scan the beam 12 times. This approach offers high resolution, but it imposes demanding mechanical tolerances on the optics.

Holographic scanners are similar to rotating mirror scanners in that they rely on rapid rotation of scanning optics about their axis. In this case, however, the scanning optics are a set of holographic optical elements mounted around a flat disk. Each holographic optic diffracts most of the incident light at an angle dependent on the angle of incidence, scanning the beam in a line. As with a rotating-mirror scanner, uniformity and symmetry of individual elements are critical for uniform scanning. This is a relatively new approach, and developers claim it offers advantages in light weight, potential low cost, and lower sensitivity to scanning deviations caused by wobble in the rotation (Kramer, 1981).

In resonant-mirror scanners, a mirror is attached to a torsion rod that twists at its characteristic resonance frequency. With the beam incident from a constant direction, twisting of the mirror changes the angle of incidence, scanning the beam. Operating range is limited by the motor, scanning speed, and twisting angle, but such scanners are less expensive than rotating mirrors and do not require as rapid motion.

BIBLIOGRAPHY

Jean M. Bennett and Harold E. Bennett: "Polarization," in Walter J. Driscoll (ed.), *OSA Handbook of Optics*, McGraw-Hill, New York, 1978, sec. 10.

H. John Caulfield: "Computer generated holography," *Lasers & Applications* *2*(5):59–64, May 1983.

Crystal Optics, Harshaw Chemical, Solon, Ohio, 1982. Compilation of data on optical materials available from a major manufacturer, a standard reference in the industry.

Department of Defense: *Military Standardization Handbook—Optical Design*, Mil-Hdbk 141, Defense Supply Agency, Washington, D.C., 1962. This handbook of classical optical design is still available from the Department of Defense and is still valuable as a practical introduction to conventional optics. It is available through the Naval Publications and Forms Center, 5801 Tabor Ave., Philadelphia, PA 19120, (215) 697-3321.

R. W. Ditchburn: *Light*, 3d ed., Academic Press, New York, 1976. Textbook.

J. A. Dobrowolski: "Coatings and filters," in Walter J. Driscoll (ed.); *OSA Handbook of Optics*, McGraw-Hill, New York, 1978, sec. 8.

Walter J. Driscoll (ed.): *OSA Handbook of Optics*, McGraw-Hill, New York, 1978.

Milton Gottlieb, Clive L. M. Ireland, and John Martin Ley: *Electro-Optic and Acousto-Optic Scanning and Deflection*, Marcel Dekker, New York and Basel, 1983.

Francis A. Jenkins and Harvey E. White: *Fundamentals of Optics*, 4th ed., McGraw-Hill, New York, 1976. One of the classic academic optics textbooks.

Charles J. Kramer: "Holographic laser scanners for nonimpact printing," *Laser Focus 17*(6):70–82, June 1981.

Earl J. McCartney: *Absorption & Emission by Atmospheric Gases*, Wiley-Interscience, New York, 1983.

Optics Guide 3, Melles Griot, Irvine, Calif., 1985. A combination handbook of practical optics and catalog issued by one of the country's major suppliers of laser optics.

Walter P. Siegmund: "Fiber optics," in Walter J. Driscoll (ed.): *OSA Handbook of Optics*, McGraw-Hill, New York, 1978, sec. 13.

Warren J. Smith: "Image formation—geometrical and physical optics," in W. J. Driscoll (ed.), *OSA Handbook of Optics*, McGraw-Hill, New York, 1978.

Robert C. Weast (ed.): *CRC Handbook of Chemistry and Physics*, 62d ed., CRC Press, Boca Raton, Fla., 1981.

Harold M. Weissman: "Replication for laser applications," *Laser Focus 14*(12):62–64, December 1978.

William L. Wolfe and George J. Zissis (eds.): *The Infrared Handbook*, Office of Naval Research, Arlington, Va., 1978. An extensive and valuable compilation of information and data on infrared optics and design criteria. Available from the Environmental Research Institute of Michigan, PO Box 8618, Ann Arbor, MI 48107.

Martin V. Zombeck: *Handbook of Space Astronomy & Astrophysics*, Cambridge University Press, Cambridge, England, 1982. Includes some useful optical data.

CHAPTER

6

variations on the laser theme: a classification of major types

Classification of lasers into families of devices is a useful step in moving from a general understanding of lasers to a specific knowledge of how individual lasers work. Several criteria could be used in categorization: type of active medium, excitation method, output wavelengths, output power, pulse characteristics, types of transitions, or spectral bandwidth. The first two criteria are basic to laser operation, and for that reason this chapter will focus on them.

This approach follows the general trend in the laser world, where devices often are grouped as gas, solid-state, semiconductor, and liquid lasers. The latter category is often the fuzziest because in practice it includes only the tunable dye laser described in Chap. 17. For that reason, some classification schemes set up a "tunable" category, but that creates complications because certain gas, solid-state, and semiconductor lasers also have tunable output wavelength.

Because there are so many types of gas lasers, it is useful to break them down by excitation mechanism. For lasers with other types of active media, the nature of the laser medium generally dictates the choice of excitation method.

Gas Lasers

Gas lasers are many and varied. They dominate any list of commercially available types and have produced laser emission on more lines than have been observed in other media (see, for example, Weber, 1982). Researchers have been aided in developing new gas lasers by the ease with which they can be built and tested. The basic elements are easy to assemble: a tube which can be filled with the desired gas mixture, a pair of resonator mirrors for the proper wavelength, and a suitable excitation source. This makes it possible to test new gas mixtures much more readily than new solid-state or semiconductor lasers, which must be custom-grown to the desired specifications.

Developers have also been aided by the flexibility that working with gases allows in finding the ideal mixture for use as a laser medium. The pressure of the laser gas can be changed readily, and extra gases can easily be added to study their effects on energy-transfer kinetics. This is helpful because in practice the "optimum" gas mixture for a particular type of gas laser varies considerably among models with different designs. Users studying specification sheets for lasers which need input gas, such as excimer types, will find that different manufacturers have different specifications—because of differences in operating conditions. Similarly, gas mixtures may differ among sealed-tube lasers of the same type, because of differences in operating conditions. Changes in pressure can lead to significant variations in laser operation. For example, carbon dioxide lasers can produce continuous beams at a small fraction of atmospheric pressure, but discharge instabilities occurring at atmospheric pressure and above limit operation in that regime to short pulses, although with high peak power.

The free movement of species in a gas mixture lets energy be transferred in many ways in a gas. Energy can be transferred by electrons, ions, or photons passing through the medium, and by collisions among species in the gas. Chemical reactions and gas-dynamic processes can produce excited species and population inversions. These mechanisms serve as the basis for the excitation mechanisms described briefly below, and in more detail in the chapters that follow.

Discharge Excitation In discharge excitation, an electric current flows through the laser medium. There are two fundamental variations, shown in Fig. 6-1. In longitudinal excitation, a bias is applied along the laser axis and the current flows the length of the tube. In transverse excitation, the discharge is applied perpendicular to the laser axis. As the electrons pass through the laser medium, they excite species in the gas. In some cases the laser species is excited directly, but more often the electrons will transfer energy to another species, which then loses its excitation energy to the laser species in collisions. In the latter case, the gas that absorbs the energy directly from the electrons is normally at a higher pressure than the species which emits laser light. One example is the helium-neon laser described in Chap. 7.

Discharges can be continuous or pulsed, both of which require an initial electrical breakdown of the gas. In a continuous laser, breakdown is initiated by a voltage spike several times the normal continuous level. After the current starts flowing, the voltage is quickly reduced to a lower level, typically ranging from a kilovolt (kV) to well over 10 kV. Except for lasers operating at high powers, the discharge currents are normally small. Low gas pressures are normally required for a continuous discharge to remain stable.

In pulsed gas lasers, the initial voltage spike—typically peaking in the tens of kilovolts—delivers energy to the laser gas. Pressures well over an atmosphere are possible, with discharge instabilities limiting pulse duration to tens or perhaps hundreds of nanoseconds at such pressures. Fast high-voltage switches are needed for many pulsed gas lasers; such components can cause problems because of high costs and limited lifetimes.

Radio-Frequency Excitation A variation on simple discharge excitation of a gas laser is use of a radio-frequency voltage to maintain a discharge passing through the gas. The main application of this technique in commercial lasers has been in waveguide carbon dioxide types.

Electron-Beam and Ion-Beam Excitation Excitation with an electron beam is essentially a variation on the pulsed-discharge approach. Electrons are

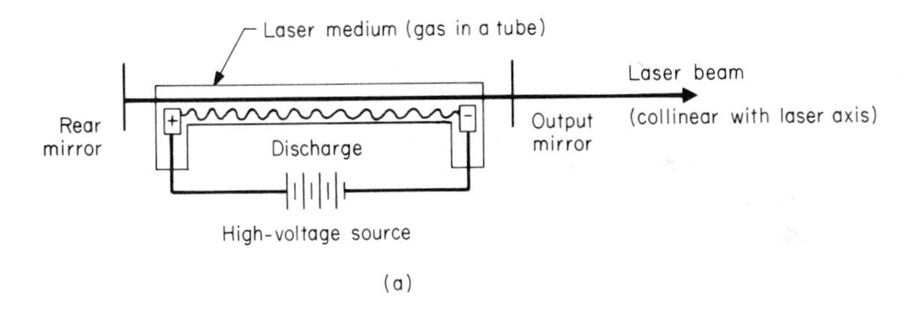

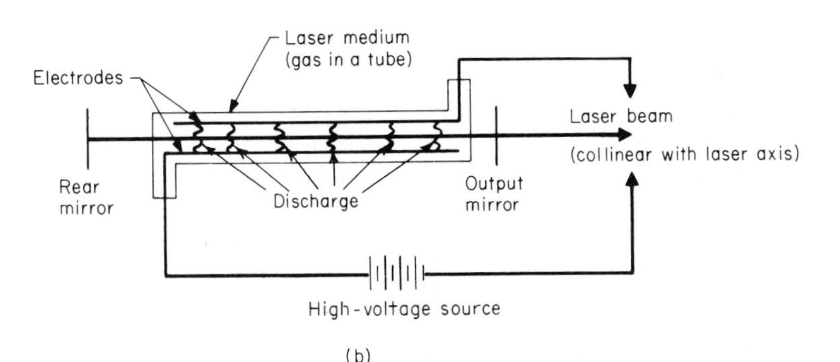

Figure 6-1 An electrical discharge exciting a gas laser can be applied *(a)* longitudinally, along the laser axis, or *(b)* transversely, perpendicular to the laser axis.

deposited in the laser gas not by a discharge between a pair of electrodes, but by a pulse of electrons from an accelerator (an electron-beam generator). Electron-beam excitation has the advantage of quickly transferring plenty of energy to the laser gas. However, the need for a bulky and expensive electron accelerator limits its use to large lasers and research laboratories.

Pulsed ion beams can be used to excite laser gas mixtures in much the same way as electron beams. Excitation with proton beams has been demonstrated in the laboratory (Golden et al., 1978), but such ion-beam excitation is even more cumbersome than electron-beam excitation and is not in common use.

Chemical Excitation Energy from an exothermic chemical reaction in a gas can excite a laser. In the commonest example, two species react to produce a third species, which carries at least part of the exothermic energy of reaction as vibrational energy. If the reaction is rapid, the result can be a population inversion of the excited species produced by the reaction. The most important such chemical lasers are the family of hydrogen halides, particularly hydrogen fluoride.

A chemical laser is an unusual device by laser standards. Gas flow is crucial; the fuels must be fed into the laser continually, mix smoothly, and flow rapidly out an exhaust so reaction products do not accumulate in the ground state and end the population inversion. Both pulsed and continuous operation are possible. Gas pressure in the laser chamber is normally quite low, which along with the toxicity of the hydrogen halides imposes some practical operational constraints on gas-flow techniques.

Although chemical lasers nominally draw their operating energy from a chemical reaction, an electrical "spark" may be needed to start the chemical reaction or to prepare one or both gas components for the reaction. The need for this modest electrical input is not a serious operational problem, but it can lead to quotes of electrical efficiency greater than 100 percent. In those cases, electrical efficiency is thermodynamically meaningless because the bulk of the energy in the laser was generated by the chemical reaction.

Gas-dynamic Lasers Another way to produce a population inversion is to expand a hot, high-pressure gas into a vacuum. The gas-dynamic expansion cools the gas, but enough of the molecules remain in the same energy level they occupied in the hotter, higher-pressure gas to produce a population inversion. The inversion lasts only until the gas can start approaching equilibrium, so, as in the chemical laser, rapid gas flow is vital. Within a gas-dynamic laser, the population inversion exists only for a short distance downstream of the expansion nozzles, and resonator mirrors must be placed there, as shown in Fig. 10-3.

The gas-dynamic laser principle is a general one, but because it relies on thermal energy distribution, its operation is limited to the infrared. In practice, gas-dynamic lasers are assumed to be carbon dioxide types, although a gas-dynamic carbon monoxide laser has been demonstrated (Murthy, 1976). The main use of gas-dynamic lasers is to produce powers of tens or hundreds of kilowatts for studies of high-energy laser effects; they are not available commercially.

Optical Excitation A conceptually simple approach to laser excitation is optical pumping. In this approach, photons passing through the laser medium excite species to high energy levels, leading to a population inversion and thus to laser action.

Although some gas lasers are optically pumped, practical applications of the technique are limited by its low overall efficiency. The fundamental problem is that two energy-conversion steps are needed: one to produce the pump light and one to convert the energy in the pump light into laser output. As a result, the overall efficiency is the product of the efficiencies of the two steps, and thus is low, generally lower than possible when directly exciting the laser medium with an electrical discharge.

There are some exceptions. The major ones are the far-infrared gas lasers described in Chap. 15. Optical pumping is the best approach for these lasers because it offers the strongly state-selective excitation required to produce a population inversion on the closely spaced vibrational-rotational and rotational transitions (Coleman, 1982). Discharge excitation produces a more general excitation of the laser gas, populating many higher-energy states. Similarly, optical pumping is a valuable tool in laboratory demonstrations of laser action because it can excite specific states, which may be needed to produce laser action at specific wavelengths. However, even when pumping with such efficient light sources as the carbon dioxide laser, the general result is a laser with overall efficiency so low it is used only when its specific output wavelengths are required.

Solid-State Lasers

In the laser world, *solid-state* has a different meaning than in electronics. A solid-state laser is one in which the active medium is a nonconductive solid, a crystalline material, or glass doped with a species that can emit laser light. Semiconductor lasers are considered functionally different types, even though they are made of solid-state materials, because of fundamental differences in their operation.

In a crystalline or glass solid-state laser, the active species is an ion embedded in a matrix of another material, generally called the host. A variety of crystals and glasses can serve as hosts, with the main requirements being transparency at wavelengths used for optical pumping, ease of growth, and good heat-transfer characteristics. Useful crystals include synthetic garnets and ruby. Silicate, phosphate, fluorophosphate, and other types of glasses can also serve as hosts. In commercial lasers, the active species functionally is an impurity introduced into the host during fabrication so that it makes up on the order of 1 percent of the finished material. Lasers have been demonstrated in which the active species is a stoichiometric component of the host crystal—for example neodymium in NdP_5O_{14} (Chinn and Zwicker, 1977)—but the active species is still at low concentration, and such lasers remain in the laboratory.

The active species in solid-state lasers are locked within the matrix of an insulator, making it impossible to excite it by passing a discharge through the laser medium. Optical pumping is the only practical approach, dictating the use of crystals which are transparent at the absorption wavelengths of the laser species. The efficiency of optical pumping is enhanced by broadening of absorption lines of the active species, caused by the host matrix, making excitation with the broadband emission of a flashlamp reasonably efficient. Nonetheless, the maximum overall efficiency of optically pumped solid-state lasers remains well below the peak efficiency possible with the best gas lasers.

Lasers are rarely used for direct optical excitation of other lasers because of the low efficiency of converting electrical input to laser output, particularly at wavelengths short enough to excite solid-state lasers. One exception is the color-center laser, described in Chap. 23, where the resulting disadvantage of low efficiency is offset by its broad tunability. There is also growing interest in the use of semiconductor diode lasers for optical pumping of neodymium lasers because diode lasers can efficiently generate light in the neodymium pump band.

Semiconductor Lasers

Practical semiconductor lasers are excited when current carriers in a semiconductor recombine at the junction of regions doped with *n*- and *p*-type donor materials. This occurs when current is flowing through a forward-biased diode made from certain semiconductor materials. At low current densities, recombination at the diode junction generates excited states which spontaneously emit light, and the device operates as an incoherent light-emitting diode (LED). If the current density is high, and if the semiconductor device includes reflective facets to provide optical feedback, the diode can operate as a laser as described in Chap. 18.

Two other types of semiconductor lasers have been demonstrated in the laboratory, although neither has proved practical. In one, an electron beam pumps an intrinsic semiconductor material that does not contain a junction layer. Such devices have been packaged in a manner similar to a cathode ray tube, with the electron beam striking a sheet of semiconductor material from the rear, stimulating the emission of a laser beam from the surface of the tube. The 3M Co. attempted to commercialize the technology in the mid-1970s (*Laser Focus*, 1974), but eventually abandoned the effort. Similar devices have been studied in the Soviet Union.

Optically pumped semiconductor lasers have also been demonstrated in the laboratory. Their main use is in research on semiconductor laser physics. The principal attraction is the ability to excite materials in which electrical excitation is impractical—generally because the current levels required to produce laser action would destroy the device either before or shortly after laser threshold was reached.

Liquid Lasers

Discussions of liquid lasers almost invariably start and end with the tunable dye laser, described in Chap. 17. The active medium in dye lasers is a fluorescent organic dye, dissolved in a liquid solvent. As in solid-state lasers, the only reasonable excitation technique is optical pumping, with a flashlamp or (more often) with an external laser. The low efficiency of laser pumping is tolerable because it offers better-quality output. The main attractions of the dye laser, its tunable output wavelength and ability to produce ultrashort pulses or ultranarrow linewidth, are so important for many applications that its low overall efficiency is entirely acceptable. Laser action has been demonstrated from dyes in the vapor phase, or embedded in a solid host, but such lasers have not proved practical.

It is possible to produce laser action by optically pumping rare earth ions in solution (Samelson, 1982). However, such lasers have not found practical applications.

Laboratory Developments

Most lasers that have been demonstrated in the laboratory fall into the categories described above, but there have been a handful of exceptions. The most important of these are the free-electron laser, the nuclear-pumped laser, and the x-ray laser.

Free-Electron Lasers The free-electron laser is a device which extracts light energy from a beam of free electrons passing through a spatially periodic magnetic field, as shown in Fig. 6-2. It is unlike other lasers because it relies on light emitted by electrons that are not bound to atoms. The underlying theory is complex (see, for example, Jacobs et al., 1982; Marshall, 1985), and while it has some unique characteristics, certain aspects of its operation resemble those of certain microwave devices.

Free-electron laser research is widespread and advancing on many fronts. The impetus comes from some unique advantages the device offers over other types of lasers:

- Potential for producing very high powers by extending technology developed for existing electron accelerators.
- Prospects for efficiency of 20 percent or more at high powers.
- Potential for tunable operation from the millimeter-wave region to the extreme-ultraviolet or soft x-ray region, although no single device would operate over such a broad range. (The broad tuning range comes from the fact that the electrons are not bound when they emit light; it would be achieved by varying the electron energy and the spacing and strength of the magnetic field.)
- Avoidance of medium-inhomogeneity problems that have limited power obtainable from other high-power lasers, because the electron beam travels through a vacuum.

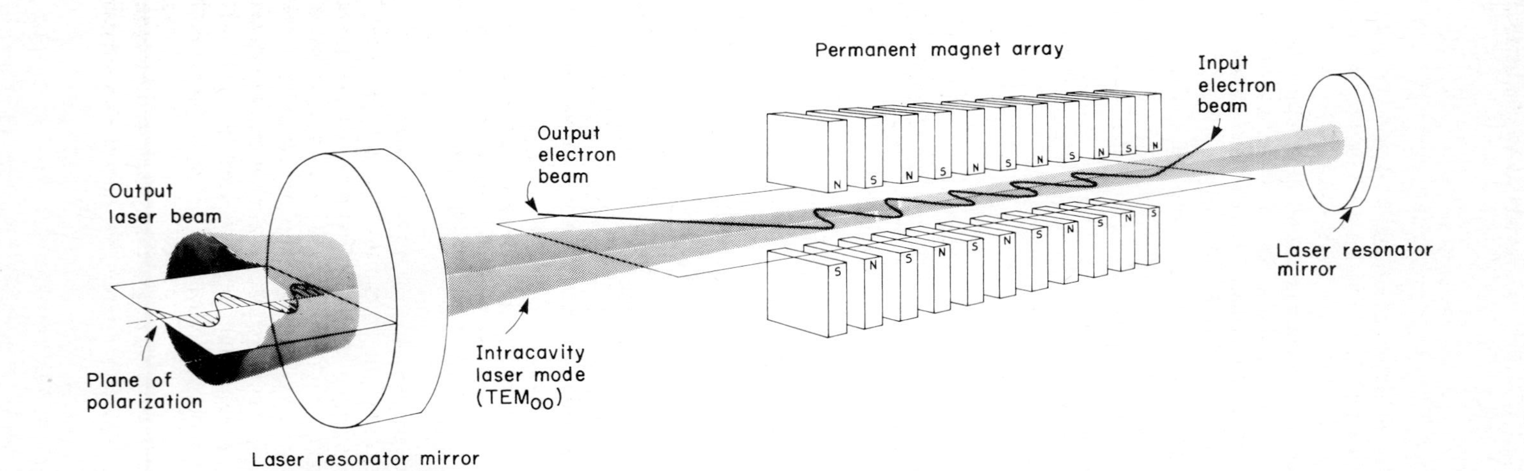

Figure 6-2 Basic components of a free-electron laser. *(Courtesy of Luis Elias, University of California at Santa Barbara Quantum Institute.)*

Research on free-electron lasers remains at an early stage, with only a few devices demonstrated. Much of the support comes from the Department of Defense's high-energy laser weapon program, so much research is aimed at high-power versions. However, there is also increasing interest in developing free-electron lasers that would generate comparatively modest powers, probably no more than a few kilowatts and more likely under a kilowatt, at wavelengths in the infrared and visible where laser emission has so far been unavailable (*Lasers & Applications,* 1982). Medical researchers have expressed strong interest in such devices, and it is possible that some early devices could be on the market in a few years.

Nuclear-Pumped Gas Lasers Gas lasers in which the excitation energy is transferred from products of a nuclear reaction to the excited species have been demonstrated in the laboratory. In the laboratory demonstrations, intense pulses of neutrons have caused fission of isotopes including helium 3, boron 10, and uranium 235. The ions produced by the fission then transfer their energy to the laser medium and laser action occurs (Marcum, 1978). Early interest in the concept came from hopes of building a gas-core nuclear reactor that would allow energy to be extracted in the form of a high-power laser beam. Output power of about a kilowatt has been demonstrated (DeYoung, 1981), but practical devices seem far off, research has slowed greatly, and little work is under way now.

X-Ray Lasers A very different type of nuclear pumping has reportedly been used to produce laser action in the x-ray region (Robinson, 1981; Hecht, 1984). Unofficial reports of highly classified experiments performed by physicists from the Lawrence Livermore National Laboratory indicate that a burst of x rays from the explosion of a nuclear bomb produced an intense, collimated pulse of x rays with the characteristics of laser emission. (No adequate mirror materials exist for x-ray wavelengths, so the stimulated emission would be amplified but could not resonate in an optical cavity.) Details have never been discussed publicly by those involved in the program, and some outside observers are skeptical about some of the published claims. The heavy secrecy reflects the fact that the main interest in the concept is for applications in ballistic missile defense above the atmosphere; the nature of the energy source makes it unlikely that such x-ray lasers would find many peaceful applications.

Other Livermore researchers recently produced stimulated emission at 15 to 20 nm by focusing extremely high power laser pulses onto thin metal foils (Matthews, 1984). The intense laser pulses vaporized the foil, and recombination of electrons with highly ionized metal atoms produced the excited species and stimulated emission. Gain was much higher than in earlier experiments where laser vaporization of carbon fibers apparently produced a population inversion on an 18.2-nm transition (Jacoby et al., 1981).

BIBLIOGRAPHY

S. R. Chinn and W. K. Zwicker: "Flash-lamp excited NdP_5O_{14} laser," *Applied Physics Letters 31* 3, 178 (1977)

Paul D. Coleman: "Far-infrared lasers, introduction," in Marvin L. Weber (ed.), *CRC Handbook of Laser Science & Technology,* vol. 2, CRC Press, Boca Raton, Fla., 1982, pp. 411–419.

R. J. DeYoung: "Kilowatt multiple-path ^3He–Ar nuclear pumped laser," *Applied Physics Letters 38*(5):297, 1981.

"Free-electron lasers suggested for medicine," *Lasers & Applications 1*(2):26, October 1982.

J. Golden et al.: "Intense proton-beam pumped Ar-N_2 laser," *Applied Physics Letters 33*(2):143, 1978.

Jeff Hecht: *Beam Weapons, The Next Arms Race,* Plenum, New York, 1984.

IEEE Journal of Quantum Electronics, March 1983. Special issue on free-electron lasers.

S. F. Jacobs et al. (eds.): *Free-Electron Generators of Coherent Radiation,* 2 vols., Addison-Wesley, Reading, Mass., 1982.

D. Jacoby et al.: "Observation of gain in a possible extreme ultraviolet lasing system," *Optics Communications 37:*193, 1981.

"3M seeks applications for a laser 'operationally analogous to a crt,' " *Laser Focus 10*(12):22–24, December 1974.

S. Douglas Marcum: "Slow progress in nuclear pumping," *Laser Focus 14*(10):12–24, October 1978.

Thomas C. Marshall: *Free Electron Lasers,* Macmillan, New York, 1985.

Dennis L. Matthews: Paper WWI presented at Optical Society of America Annual Meeting, San Diego, Calif., October 30 to November 2, 1984.

S. N. B. Murthy: "Gas-dynamic and chemical lasers, gas dynamics," in E.R. Pike (ed.): *High-Power Gas Lasers 1975,* Institute of Physics, Bristol and London, 1976, pp. 222–242.

Clarence Robinson Jr.: "Advance made on high-energy laser," *Aviation Week & Space Technology,* February 23, 1981, pp. 25–27.

Harold Samelson: "Inorganic liquid lasers," in Marvin L. Weber (ed.): *CRC Handbook of Laser Science & Technology,* 2 vols., CRC Press, Boca Raton, Fla., 1982.

CHAPTER

7

helium-neon lasers

The helium-neon laser is the commonest and most economical visible laser, and the least-expensive gas laser. It has found applications ranging from construction alignment to laboratory research because of its ability to produce powers from a fraction of a milliwatt to a few tens of milliwatts at 632.8 nanometers (nm) in the red. The low cost and visible output of the red "He–Ne" has made it the standard choice for school and museum demonstrations—making it the one laser virtually every reader of this book is sure to have seen in action. Less well known are other helium-neon transitions which serve as the basis for commercial infrared lasers, and a weak green line which recently became available in commercial lasers (Eerkens and Lee, 1985).

The helium-neon laser was one of the first types developed, initially operating at 1153 nm in the infrared (Javan et al., 1961). Others soon found that the same gas mixture could lase in the red (White and Rigden, 1962), and because the visible line generally is more useful, commercial development centered on that line. The red line is not ideal for all applications, but the helium-neon laser has become successful because it is easier and less costly to manufacture than other visible lasers. Annual production of He–Ne lasers has passed the 200,000 mark (Hitz, 1985), exceeding all other types except semiconductor diode lasers. Diode lasers are displacing He–Ne's from many applications where visible output is not crucial, but the He–Ne seems to have a secure place in the laser world unless cheap, long-lived diode lasers with visible output and good beam quality come on the market. That technology has yet to materialize (Hecht, 1984). The recent introduction of commercial green He–Ne lasers may open up new applications, but the field has not yet had time to work with that technology.

Internal Workings

Basic Physics The active medium in a helium-neon laser is a mixture of helium and neon and total pressures of a fraction of a torr to several torr, with best working pressure depending on discharge-tube diameter. Power is highest when gas pressure (in torr) times discharge diameter (in millimeters) is 3.6 to 4 (Svelto, 1982). Typically the gas mixture contains above five times more helium than neon for 632.8-nm operation, with somewhat different ratios used for infrared operation. (Many details of green laser operation are considered proprietary.)

The energy in a helium-neon laser comes from an electric discharge, which passes a few milliamperes through the laser tube at a couple of thousand volts when the laser is in steady operation. (An ignition voltage of about 10 kV is needed to start laser operation.) Electrons passing through the active medium collide with both helium and neon atoms, raising them to excited levels. The more abundant helium atoms collect most of the energy, then transfer that energy readily to neon atoms, which have excited states at about the same energy above their ground states. The neon atoms then lose their excitation energy and drop to lower energy levels via several transitions, as shown in Fig. 7-1. Emission is possible on several transitions indicated in the figure. The wavelength an individual laser emits depends on the choice of optics and operating conditions of the tube.

Internal Structure A helium-neon laser is a gas-filled tube, with internal electrodes exciting the gas to emit light. Mirrors on each end of the tube define the laser cavity. Early helium-neon lasers were quite simple, but over the years manufacturers have incorporated a number of refinements to improve performance.

Figure 7-2 shows the internal structure of a typical modern helium-neon laser with output in the milliwatt range. The discharge passes from the cathode at one end of the tube to the anode at the other, going through a capillary bore one to a few millimeters in diameter. The capillary structure concentrates the discharge, thus improving overall efficiency. The small diameter of the discharge bore also helps control laser beam diameter, mode, and beam divergence. Much effort goes into selecting electrode shapes to make the discharge uniform.

In early helium-neon lasers, windows were epoxied to the ends of the laser tube, and the cavity mirrors mounted externally. That approach presented problems because helium could diffuse out of the tube through the epoxy (the

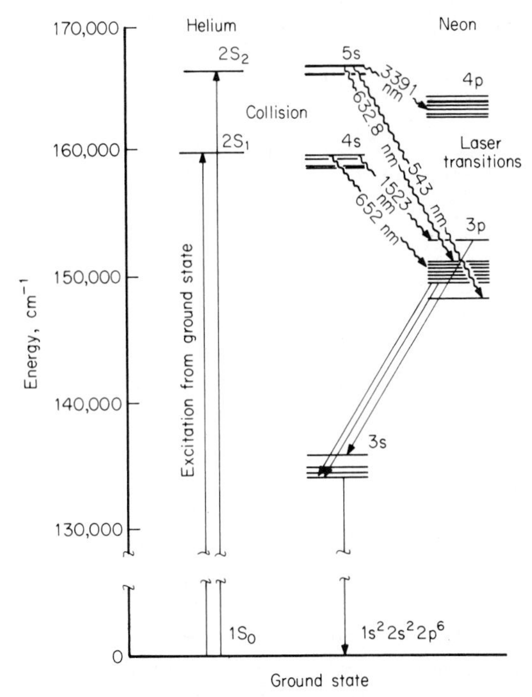

Figure 7-1 Energy levels in a helium-neon laser, with major laser transitions indicated. Collisions transfer energy from helium atoms to neon atoms, which drop through laser transitions on their way to the ground state. *(Courtesy of Melles Griot.)*

gas could also diffuse through thin tube walls of soft glass). Epoxy seals also let water vapor seep slowly into the tube, leading to degradation of performance after a few thousand hours of operation.

Now virtually all He–Ne lasers are made with hard seals, in which the glass is bonded directly to metal at high temperatures. Typically the laser cavity mirrors are bonded to metal end plates, which in turn are bonded to the glass laser tube, without the use of epoxies. Hard seals reduce the helium leakage rate to under 0.01 torr (1.33 Pa) per year, and make contamination from water vapor and other materials insignificant (Palecki, 1982). This can extend operating life of He–Ne tubes to 20,000 hours or more.

In today's mass-produced He–Ne lasers, cavity mirrors typically are bonded directly to the tube, through the metal end plate, a simpler structure than possible if the mirrors are separate from the laser tube output windows. This approach exposes the mirror coating directly to the discharge inside the laser tube, but current hard coatings can withstand such conditions. Some He–Ne lasers are still produced with Brewster angle windows and external cavity mirrors, but these are typically special-purpose models produced in small quantities, often specifically for laboratory applications.

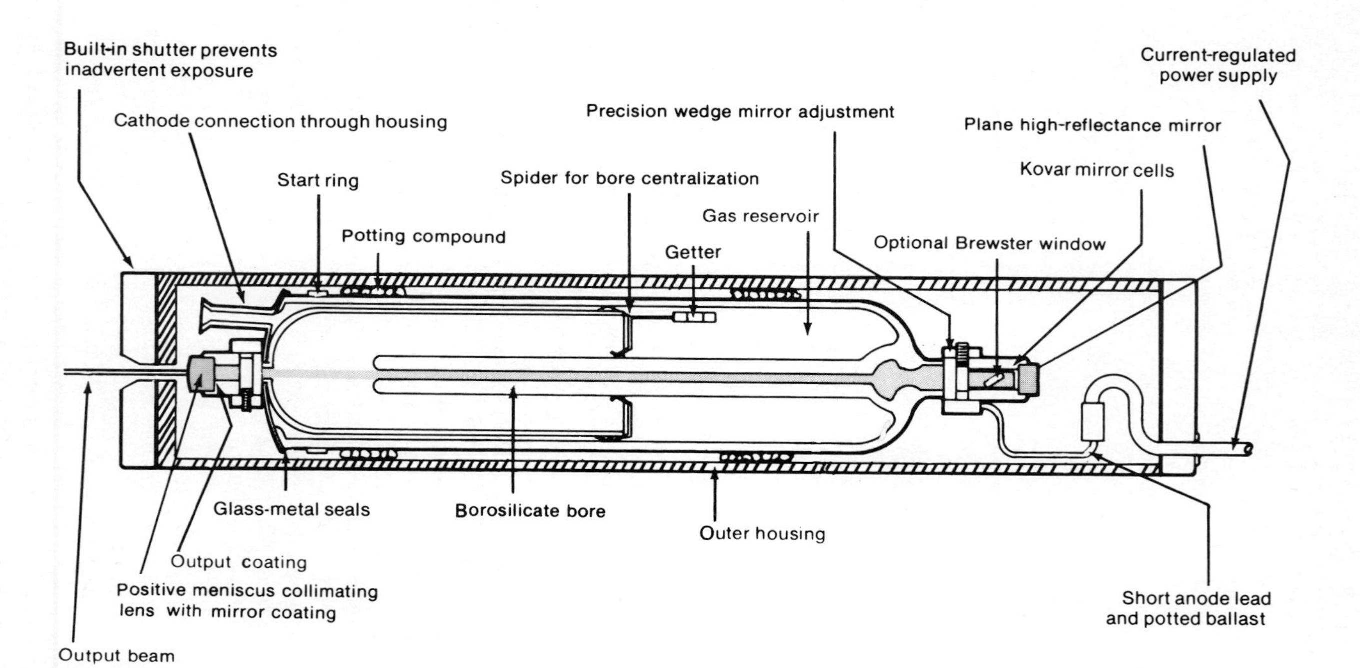

Figure 7-2 Internal design of a modern hard-sealed helium-neon laser. *(Courtesy of Melles Griot.)*

In standard helium-neon lasers, one mirror is totally reflective, while the one at the opposite end of the cavity transmits the fraction of the light that becomes the external laser beam. Cavity lengths range from around 10 cm for low-power models to a couple of meters for lasers with output in the 50-mW range.

Inherent Trade-offs The most obvious trade-off with helium-neon lasers is the increase in bulk and price of the laser when moving to higher powers. While the helium-neon laser can readily produce a few milliwatts, the internal physics make it much harder to produce higher powers. Output above 60 mW has been demonstrated in the laboratory but is not practical and is not available from commercial lasers.

In general, infrared and green He–Ne lasers are more expensive per milliwatt of output than red models. A major reason is the much smaller production runs. A contributing factor for the green laser is the weakness of the line, together with the newness of the commercial version.

Variations and Types Covered Most helium-neon lasers are linear tubes that generate 0.5 to 10 mW in the red, although a few models produce somewhat higher power. There are three significant variations on this basic theme: green emission, infrared emission, and ring lasers.

Although a rich spectrum of atomic neon laser lines has been seen in the laboratory (Davis, 1982), traditionally only two infrared lines were available other than the usual 632.8-nm red line. Those two wavelengths are 1.15 and 3.39 micrometers (μm), both used mostly for research. In 1985, commercial lasers emitting on the 543-nm green line, 594-nm yellow line, 612-nm orange line, and the 1523-nm infrared line were introduced. Those new models use hard-seal tube technology similar to that in commercial red He–Ne's. However, some 1.15- and 3.39-μm lasers are based on general-purpose tube designs with Brewster angle windows, with wavelength-selecting optics added outside the cavity.

The ring laser is a variation in which three (or sometimes four) mirrors define a "ring" path inside the laser resonator. This requires a triangular or square laser tube, such as is shown in Fig. 7-3. Ring lasers are used as gyroscopes to detect rotation about their central axis (Hecht, 1982; Aronowitz, 1971). These lasers are specially built for use in military and commercial systems and are not sold separately as components on the open market. Because of their specialized nature, ring lasers are not covered in this chapter.

Beam Characteristics

Wavelength and Output Power The standard wavelength of helium-neon lasers is 632.8 nm in the red. Commercial models emit continuous beams from a few tenths of a milliwatt to 60 mW, with most in the 0.5- to 7-mW range.

The other visible lines are weaker, with initial models providing continuous-wave output of 0.5 mW at 543 nm, 0.2 mW at 594 nm, and 1.0 mW at 612 nm.

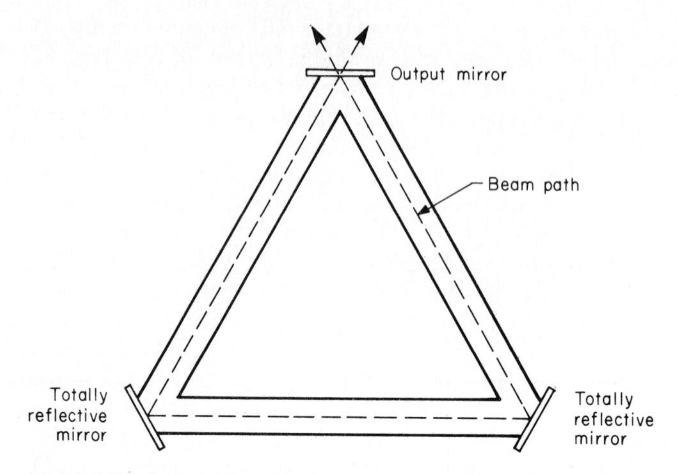

Figure 7-3 Outline of a triangular ring laser used as a rotation sensor. Light propagates in both directions around the ring, and subtle rotation-induced changes in wavelength are detected by observing interference effects at the output mirror.

Three infrared helium-neon lines are available in commercial lasers: 1.15, 1.523, and 3.39 μm. Beams are continuous, with powers typically in the 1- to 10-mW range. A few models have been designed to emit on more than one line, with the choice among lines made by selecting optics; these generally emit lower powers in the infrared than in the red.

Because the other visible and infrared lasers are uncommon, the rest of this chapter will concentrate on the characteristics of the red helium-neon laser, except where noted otherwise. Generally, characteristics of the other visible and infrared versions are similar except for power and wavelength.

Efficiency Overall efficiency of a helium-neon laser is low, typically in the range of 0.01 to 0.1 percent.

Temporal Characteristics of Output Normally helium-neon lasers produce continuous beams. The continuous output can be modulated at rates to about 1 megahertz (MHz) by varying drive current in models designed for that purpose. He–Ne lasers are normally not operated in pulsed mode.

Spectral Bandwidth and Frequency Stability The doppler-broadened gain curve of mass-produced helium-neon lasers spans around 1.4 GHz, the equivalent of 1.9×10^{-3} nm at the laser's red wavelength as shown in Fig. 7-4. The doppler gain curve encompasses several cavity modes, each much narrower than the broad doppler curve. The number of cavity modes depends on mode spacing—roughly 500 MHz on the average, but larger for smaller cavities and smaller for larger cavities. The shorter the laser cavity, the fewer cavity modes falling under the gain curve.

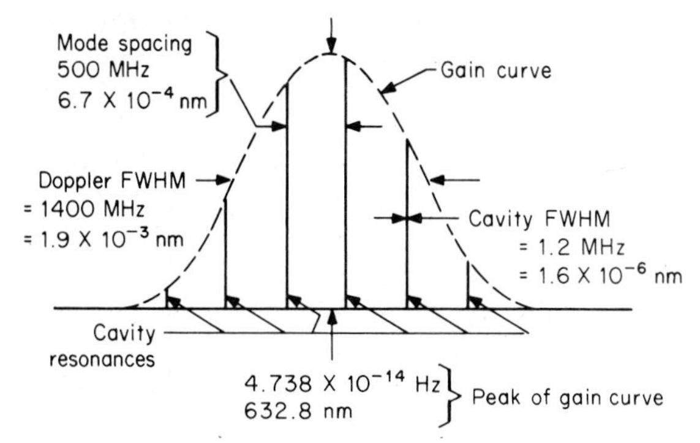

Figure 7-4 Doppler-broadened gain curve of a helium-neon laser, showing the much narrower cavity resonances that lie within that curve. FWHM = full width at half maximum. *(Courtesy of Melles Griot.)*

For most practical purposes, the linewidth of 1 part in 300,000 is adequately narrow, and minor mode shifts do not cause noticeable effects. When narrower bandwidth is required, as in certain scientific applications, special optics can be added to restrict laser oscillation to one of the cavity modes within the gain curve. Frequency stability of such single-frequency lasers is on the order of ±1-MHz drift (0.002 part per million) during a 5-minute interval when operated at constant temperature. However, temperature-induced variations in the cavity length can cause the fundamental frequency to vary by around 5 MHz per degree Celsius change in ambient temperature, so temperature stability is a must for stable frequency output. Normally warm-up times of half an hour to an hour are needed to obtain stable single-frequency output, longer than needed to get stable output from a laser with a normally broad gain curve.

Amplitude Noise Root-mean-square (rms) amplitude noise at important frequencies, between about 30 Hz and 10 MHz, typically is rated at 1 percent or less in mass-produced He–Ne lasers. Noise ratings of 2 percent are possible for lasers at the upper end of the commercial He–Ne power range. Long-term drift is typically 2.5 or 5 percent over an 8-hour interval.

Beam Quality, Polarization, and Modes Most helium-neon lasers produce a good-quality TEM_{00} beam with a classic gaussian intensity distribution, although some inexpensive models emit in multiple transverse modes. The beam may be linearly polarized or nominally unpolarized, but the unpolarized beams are generally *not* truly randomly polarized and may contain shifting proportions of light of different linear polarizations. Although this is acceptable for many applications, users who require an even distribution of polarizations or true random polarization may wish to add external depolarizers. Linear polarization is typically 500 to 1.

Coherence Length Mass-produced He–Ne lasers have coherence lengths around 20 or 30 cm, adequate for holography of small objects if care is taken that the object and reference beams travel similar distances. Single-frequency He–Ne lasers have much narrower spectral bandwidths, and hence much longer coherence lengths, but are considerably more expensive.

Beam Diameter and Divergence Beam diameters of mass-produced helium-neon lasers with TEM_{00} output in the milliwatt range are around a millimeter and tend to increase with output power. Divergence is on the order of 1 milliradian (mrad), dropping with increasing beam diameter because these lasers normally operate near the diffraction limit. Short, low-power models may have small-diameter beams and larger divergences—to about 0.34 mm and 2.4 mrad. For longer, high-power models, beam diameters can reach a couple of millimeters, with divergence about 0.5 mrad.

Diameter and divergence of multimode beams from some lasers do not come close to the diffraction limit. For example, one 5-mW multimode laser has 2-mm beam diameter and 8-mrad divergence, far above the diffraction-limited beam quality of TEM_{00} lasers.

Stability of Beam Direction Although many data sheets do not specify beam-direction stability, typical values for mass-produced He–Ne lasers are around ±0.2 mrad for operation from a cold start, and around ±0.02 to 0.03 mrad after a 15- to 20-minute warm-up period.

Suitability for Use With Laser Accessories The low powers of helium-neon lasers effectively prevent their use with accessories requiring high laser powers, such as harmonic generators. Although modelocking and *Q* switching are possible, they are extremely rare in practice; other lasers are much better choices for modelocking or *Q* switching because they offer higher powers. The continuous beams from helium-neon lasers can be modulated by external modulators. Note, however, that these accessories can easily cost more than a typical He–Ne laser.

Operating Requirements

Input Power Most packaged helium-neon lasers operate from ordinary 115-V wall current, with 230-V operation a standard option. Power consumption runs around 20 W for a 1-mW laser to over 400 W for a 50-mW version. Low-power versions are also available for battery operation.

Laser heads and tubes are available which require a few milliamperes of direct current at 1500 to 2000 V. The tubes themselves require a ballast resistance in the range of 45 to 75 kΩ for proper operation. Peak voltages of around 10,000 V are needed to initiate the discharge in the laser tube.

Cooling Low-power helium-neon lasers rely on passive air cooling, while higher-power models have fans that provide forced-air cooling. Even the highest-power He–Ne's do not produce a tremendous amount of waste heat, no more than a few hundred watts.

Consumables Practically speaking, there are no consumables required for He-Ne lasers.

Required Accessories A complete packaged helium-neon laser requires no special accessories other than those needed for a particular application. Laser "heads" packaged without internal power supplies do require a separate module to convert wall current to the high-voltage direct current needed to power the laser tube. Bare laser tubes are also sold; these require both a high-voltage direct-current source and mounting accessories.

Operating Conditions and Temperatures Although most helium-neon lasers are intended for a normal laboratory or industrial environment, they generally can withstand a broader range of conditions. Standard mass-produced He–Ne lasers are rated for operating temperatures of 0 to 40°C to −20 to +50°C. Storage-temperature extremes can be as broad as −40 to +80°C. The lasers are rated to operate at altitudes to about 3000 m (10,000 ft), and at humidities from 0 to 100 percent. Because helium-neon lasers are mass-produced and used in a wide range of conditions, specification sheets often provide much more detail on operating ranges than do data sheets for other types of lasers.

Some helium-neon lasers are ruggedized for use in hostile environments, such as mines, agriculture, and construction sites. These lasers generally are sealed to keep dust out of the laser head and to minimize high-voltage hazards.

Typically, a helium-neon laser begins producing light within a few seconds after it is turned on. However, it may take 15 to 60 minutes to warm up enough to give rated performance. Adequate warm-up times are particularly critical for single-frequency lasers, which require stable temperature for stable operation.

Mechanical Considerations In recent years, some very compact helium-neon lasers have come on the market, with low-power tubes as short as about 10 cm (4 in) and as thin as 1.6 cm (⅝ in). Packaged laser heads are only slightly larger, although many users concerned with small size prefer to do their own packaging of the laser tube. Lasers producing a few milliwatts are typically a foot (0.3 m) long, while the most powerful models can run a couple of meters long. A bare tube can weigh as little as 0.1 kg, although weights of 0.25 to 0.5 kg are more common; packaged laser heads weigh about twice that much. Addition of a power supply can bring total weight to a kilogram or two for low-power lasers, while a complete laser capable of delivering a few milliwatts would weigh a few kilograms. The highest-power He–Ne lasers can weigh as much as 100 kg complete with laser head and power supply.

Safety The basic rules for safe use of helium-neon lasers are to avoid staring into the laser beam and contacting the high voltages that power the discharge.

Laser powers of a few milliwatts or less are far from deadly and may not even be felt as a warm spot on your skin. However, the intensity of a helium-neon laser beam is comparable to that of sunlight (although it is spread over a much smaller area). Because both sunlight and a laser beam are made of parallel rays, the eye can focus both onto very small spots on the retina. If the eye stares at either the sun or a laser beam long enough, the intensity at the retina can be high enough to cause permanent damage. There are no hard-and-fast rules for exactly how much exposure will cause what type of damage, so caution should be the watchword. However, an inadvertent momentary glance into a low-power He–Ne beam is no more likely to cause instantaneous blindness than an accidental glance at the sun. The hazards increase with laser power, so particular care should be taken with higher-power He–Ne's.

Many applications require that helium-neon beams be scanned across surfaces to read information. Careful design can limit powers to levels low enough and maintain scanning speeds high enough that the beam can be scanned through regions accessible by people without presenting any danger. An excellent example is the use of laser scanning systems in automated supermarket checkout; although millions of people pass by such systems daily, stringent federal standards do not even require a "Caution" sign on the laser reader.

Reliability and Maintenance

Lifetime Hard-sealed helium-neon laser tubes typically have rated operating lifetimes of 10,000 to 20,000 hours, and shelf lives of 10 years, although shelf lives can be lower for higher-power models. Tubes without hard seals typically have operating lifetimes of a few thousand hours.

Maintenance and Adjustments Needed Helium-neon lasers are largely maintenance-free, and most should require no adjustments under normal operating conditions. However, laser heads which are dropped may require realignment, while those operated in particularly dirty environments may require periodic cleaning.

If extremely stable operation is required, it may be necessary to leave the laser running continually. Complete He–Ne lasers come with beam-blocking shutters which allow the laser to run continually without having the beam emerging from the head.

Mechanical Durability He–Ne lasers generally are built to withstand moderately rough handling, and some are ruggedized specifically for use in mines, construction, agriculture, and surveying. Normal mass-produced He–Ne heads and tubes are reasonably rugged, although because they are made of glass, breakage is always possible if they are dropped.

Some manufacturers specify shock resistance for laser heads and tubes. One company specifies shock resistance of 15 times the force of gravity (15*g*'s) for 11 milliseconds (ms) for laser heads, and 50*g*'s for 1 ms for laser tubes. The different specifications reflect the different damage vulnerabilities. Laser heads can be knocked out of alignment, but the main damage to tubes is breakage.

Failure Modes and Causes Loss of helium, poisoning of the helium-neon mixture by such contaminants as water vapor, or damage to the internal optics will cause gradual degradation in output power. Malfunction of the high-voltage system and shock damage to the laser tube or optics are other potential problems. Slow degradation of output power level is inevitable even with hard-sealed tubes, but users should realize that the rated lifetimes of helium-neon lasers are long compared with those of other light sources. Also note that reduced output level does not always prevent operation of a system or experiment.

Possible Repairs Helium-neon laser tubes can be cleaned and refilled with a fresh gas mixture, but this costs more than buying a new mass-produced laser tube. (It may, however, be practical for high-power models.) Mass-produced tubes are cheap, and the simplest course is usually substituting a new one. Power supplies are also normally replaceable.

Commercial Devices

Standard Configurations Helium-neon lasers are produced in large quantity and offered in several forms. The three commonest offerings, shown in Fig. 7-5, are

- *Laser tube:* A sealed glass tube containing the laser gas and electrodes, typically with integral cavity mirrors sealed to the ends of the tube.
- *Laser head:* A laser tube packaged into a housing, typically a cylindrical tube. The laser head includes optics and safety features as well as the protective housing, but it does not include a power supply to generate the high voltages needed to drive the tube.
- *Complete laser system:* The laser head plus power supply, packaged together with safety interlocks and all other equipment needed to permit operation. The laser is ready to run once it is plugged into a power outlet and switched on. The complete laser may be housed in one integral box or may consist of a head attached to a power supply and control box.

These different versions are intended for users with different needs. Complete laser systems are intended for those who want a ready-to-use laser, such as a demonstration laser for a physics laboratory or a light source for holography. Laser heads and laser tubes are intended for incorporation into other equipment, such as bar-code readers or laser printers. Tubes are often used as replacement components. Power supplies, which convert a low-voltage source of direct or alternating current to the high voltages needed for powering the laser tube are also sold as separate modules which users can connect to laser heads or tubes.

Options Many options for helium-neon lasers concern packaging. In addition to those listed above, they include ruggedization, hermetic seals, and battery operation to permit use in the field or under hostile conditions. There are also other options that deal more with the laser's operating mode, including

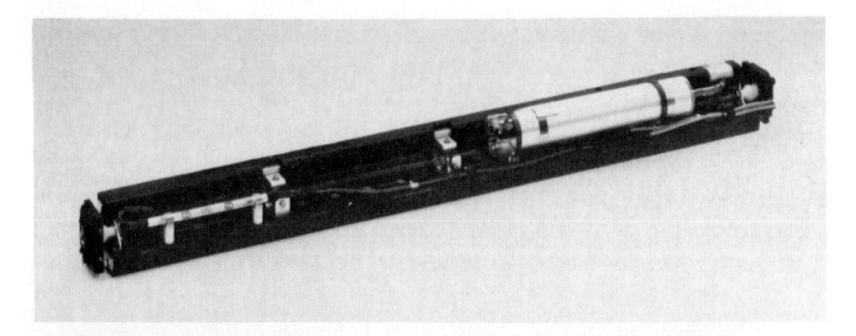

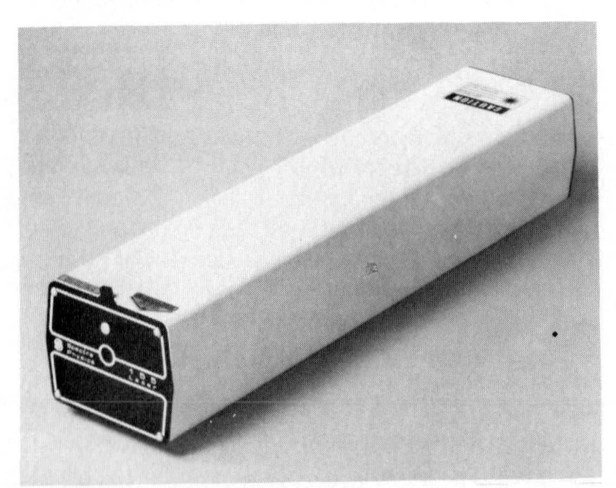

Figure 7-5 Standard configurations of helium-neon lasers, showing a laser tube, laser head, and complete laser. *(Courtesy of Spectra-Physics Inc.)*

- Choice of polarized or unpolarized output
- Single-frequency output
- Type of power supply
- Operation at infrared wavelengths
- Intracavity space for insertion of laser accessories or experimental samples

(This option is available only with laboratory lasers that lack integral optics and offers higher power than is available in the beam that emerges from the laser cavity.)

Special notes He-Ne lasers with output of a few milliwatts or less are generally mass-produced, but those with powers of 10 mW or more are made in smaller quantities and often packaged primarily for laboratory applications.

Pricing As of this writing, single-unit prices of red helium-neon laser tubes run from the $100 range for small models delivering around a milliwatt of "randomly polarized" light to well over $1000 for tubes producing over 10 mW. Complete laser systems with packaged laser heads and power supplies typically are priced at about twice as much as bare laser tubes capable of the same output power. The most powerful He–Ne lasers, with output around 50 mW, can run over $10,000 complete. Options such as special packaging can raise the price well above the normal level. Single-frequency output is also expensive, with such lasers generally selling for a few thousand dollars each, despite delivering only a few milliwatts of power.

Prices can drop steeply as the quantity purchased increases, particularly for laser tubes and heads. There have been reports that prices of low-power He–Ne tubes have dropped to as low as around $20 each when purchased in massive quantities, but users of tubes in moderate quantities should not expect to come close to such levels.

Prices are significantly higher for green and infrared models. The first green He–Ne head on the market carried an $850 price tag and required a $200 power supply to drive it. Infrared models are even more expensive, typically running from $2000 to over $10,000 for the most expensive versions.

Suppliers There are many suppliers of helium-neon lasers and tubes in the United States, Japan, and Europe. A number of companies specialize in mass production of lasers and tubes, while others concentrate on producing specialized products in smaller quantity. The market has been an intensely competitive one. A prime reason for that competition has been overcapacity for mass production of low-power tubes, particularly in Japan and Europe, built up in anticipation of a large videodisk market that never materialized. However, increasing demand for tubes could reduce that overcapacity.

Only a few companies supply infrared He–Ne lasers, and as of late 1985 only two offered green models.

Applications

For many years, the helium-neon laser has been the standard choice for applications which require an inexpensive laser producing power in the milliwatt range. It is the least-expensive visible laser by about an order of magnitude (more if large quantities are purchased), and until recent improvements in semiconductor diode lasers was the least expensive laser capable of continuous output. As a result, helium-neon lasers have found a broad range of applications that can be broken down into several broad areas:

- *Alignment and positioning:* The use of a laser beam to draw a straight line for optical or electronic detection to aid in aligning or positioning. Construction workers use instruments with scanning He–Ne beams to define a plane or line while they are building walls and hanging ceilings. He–Ne lasers draw straight lines to keep sewer pipes and tunnels on straight courses. He–Ne lasers also define straight paths for machine tools and help medical personnel position patients in x-ray imaging systems. He–Ne beams are used to align grading equipment used in heavy construction and agriculture. Diode lasers have begun to replace He–Ne lasers in systems which use electronic sensors, but they cannot do the job when visible light is required.
- *Reading and scanning:* The tightly focused beam from a helium-neon laser can be scanned across a surface to read bar codes, special characters, or other symbols. The beam repetitively scans a well-defined pattern, and variations in the scattered light are detected optically. The resulting electronic signal can be decoded to indicate the pattern that is being read. The most widespread such application is reading bar codes for supermarket checkout, but bar-code technology is spreading rapidly to other applications. Helium-neon laser scanners have also been used for optical character recognition.
- *Writing and recording data:* A modulated laser beam can record information on a light-sensitive surface if the beam is scanned across the surface as it is modulated. In laser printers, the laser beam scans a photoconductive drum, discharging the electrostatic charge held by the surface at points where the beam is "on." The resulting pattern is then transferred to paper by a photocopier-like toner process, producing a printed page. Lasers can also encode data as a series of dots on a disk for use in computer data storage systems. Although diode lasers are claiming an increasing share of this market, helium-neon lasers remain in use in many systems.
- *Videodisk playback:* Helium-neon lasers can play back video programs that are prerecorded as dense dot patterns on videodisks. Originally developed for home entertainment, the technology has found growing use in video games, simulations, and other commercial applications. Here, too, diode lasers are increasingly supplanting helium-neon lasers. He–Ne lasers have never been used in the digital audio disk players which use diode lasers for similar storage and playback techniques.
- *Holography:* The good coherence, low cost, and visible output of helium-neon lasers make them the standard choice for recording holograms of stationary objects.
- *Medicine:* Helium-neon lasers are used in several types of medical therapy, primarily unconventional treatments such as laser acupuncture, biostimulation, stimulation of wound healing, and alleviation of pain. Helium-

neon lasers are used by the more orthodox medical establishment in instruments which count cells by measuring light scattering and which measure other quantities of medical interest. They also help align patients in x-ray imaging systems.

- *Measurement:* Many measurement instruments use helium-neon lasers, including interferometers which measure surface contours, light-scattering instruments, and equipment to measure physical dimensions. Ring lasers are used as rotation sensors or "laser gyroscopes" for navigation; their first commercial uses are in the Boeing 757 and 767 aircraft.
- *Demonstrations and displays:* Helium-neon lasers are the standard choice for demonstrations of how lasers work, and for many types of displays needing only modest amounts of visible light.
- *Research and development:* Like most other lasers, helium-neon types find many laboratory applications as simple, inexpensive sources of visible coherent light in a tightly focused beam.
- *Sensing:* Helium-neon lasers are often used as light sources for various types of sensors, including those which rely on changes in the optical properties of optical fibers and bulk materials.
- *Communications:* Red helium-neon lasers are used in some through-the-air communication systems which require narrow beams, but the numbers used are small. Most types of laser communications rely on diode lasers, particularly fiber-optic systems. The 1.523-μm He–Ne laser is being marketed for use with fiber optics, because glass fibers have their lowest attenuation at that wavelength and 1.5-μm diode lasers are hard to make. However, the gas laser is more likely to be used in measurement systems than in communication systems.

BIBLIOGRAPHY

Frederick Aronowitz: "The laser gyro," in Monte Ross (ed.), *Laser Applications*, vol. 1, Academic Press, New York, 1971, pp. 134–200.

G. Bouwhuis and J. J. M. Braat: "Recording and reading of information on optical disks," in Robert R. Shannon and James C. Wyant (eds.), *Applied Optics and Optical Engineering*, vol. IX, Academic Press, New York, 1983, pp. 77–110.

Christopher C. Davis: "Neutral gas lasers," in Marvin J. Weber (ed.), *CRC Handbook of Laser Science and Technology*, vol. 2, CRC Press, Boca Raton, Fla., 1982, pp. 3–68.

Jeff W. Eerkens and William M. Lee: "New He-Ne lasers with green (543 nm) and fiber-optimum IR (1523 nm) outputs," paper WM16 at Conference on Lasers and Electro-Optics, Baltimore, May 21-24, 1985.

Jeff Hecht: "Laser gyros, the guiding light," *High Technology* 2(3):24–28, May/June 1982.

Jeff Hecht: "Outlook brightens for semiconductor lasers," *High Technology* 4(1):43–50, January 1984.

C. Breck Hitz: "The Laser Marketplace—1985," *Lasers & Applications* 4(1):47–56, January 1985.

A. Javan, W. R. Bennett Jr., and D. R. Herriott: "Population inversion and continuous optical maser oscillation in a gas discharge containing a He–Ne Mixture," *Physical Review Letters* 6:106, 1961.

Optics Guide 3, Melles Griot, Irvine, Calif., 1985, pp. 333–368.

Gerald S. Palecki: "Helium-neon lasers, reliability & lifetime," *Lasers & Applications* 1(1):73–75, September 1982.

D. L. Perry: "CW Laser oscillation at 5433 Å in neon," *IEEE Journal of Quantum Electronics* 7:107, 1971.
David Sliney and Myron Wolbarsht: *Safety with Lasers and Other Optical Sources*, Plenum, New York, 1980.
Orazio Svelto, *Principles of Lasers*, 2d ed., Plenum, New York, 1982, p. 208.
A. D. White and J. D. Rigden: "Continuous gas maser operation in the visible," *Proceedings of the IRE 50*:1796, 1962.

CHAPTER 8

noble gas ion lasers

The label *ion laser* is a vague one that the laser world applies specifically to types in which the active medium is an ionized rare gas. Argon, with strong lines in the blue-green and weaker lines in the ultraviolet and near-infrared, is the most important type commercially. Krypton ion lasers are also marketed, despite their weaker output, because they offer a wider range of visible wavelengths. Laser action has also been demonstrated on lines of ionized neon and xenon, but those types are not marketed except for the pulsed xenon laser described briefly in Chap. 16. (The helium-neon laser emits on *neutral* neon lines.)

The prime attraction of argon and krypton lasers is their ability to produce continuous-wave output of a few milliwatts to, from argon, tens of watts in the visible, and up to a few watts in the ultraviolet. The technology is not easy to master. Ion lasers carry price tags of a few thousand to a few tens of thousands of dollars, have tubes with operating lifetimes limited to 1000 to 10,000 hours, and tend to be delicate. But users who need visible or near-ultraviolet output in that power range often have little choice, and today's ion lasers are marked improvements over earlier versions. Ion lasers may not meet the system designer's ideal specifications, but they can offer cost-effective solutions to problems in building systems for medical, printing, entertainment, and data-storage applications, as well as maintaining a place as a laboratory laser.

Introduction and Description

Active Medium In an ion laser, the active medium is a rare gas from which one or more electrons have been stripped to form a positive ion. In argon and krypton lasers, pure gas is used, with normal operating pressure slightly under 1 torr (133 Pa). The two gases can also be mixed in what is called a *mixed-gas* laser, which emits on spectral lines of both ions. The laser transitions are between upper energy levels of the ions, with singly ionized species emitting in the visible and near-infrared, and doubly ionized species emitting in the near-ultraviolet.

The commonest member of the ion-laser family is argon, which emits on the lines shown in Fig. 8-1. The strongest line is the 514.5-nm line of singly ionized argon (Ar II in the spectroscopist's notation), followed by 488.0 nm. Several other weaker lines appear in untuned blue-green emission from argon, or when the laser is tuned specifically to those wavelengths. There are also near-ultraviolet lines of Ar III (Ar^{2+}) and a near-infrared line of Ar^{+} (Ar II).

Krypton is a less-efficient laser gas and hence is used less often. However, as shown in Fig. 8-1, it has lines in parts of the spectrum where argon does not emit. By far the strongest line is the 647.1-nm Kr^{+} line in the red, but there are also important lines in the green and yellow, unavailable from other

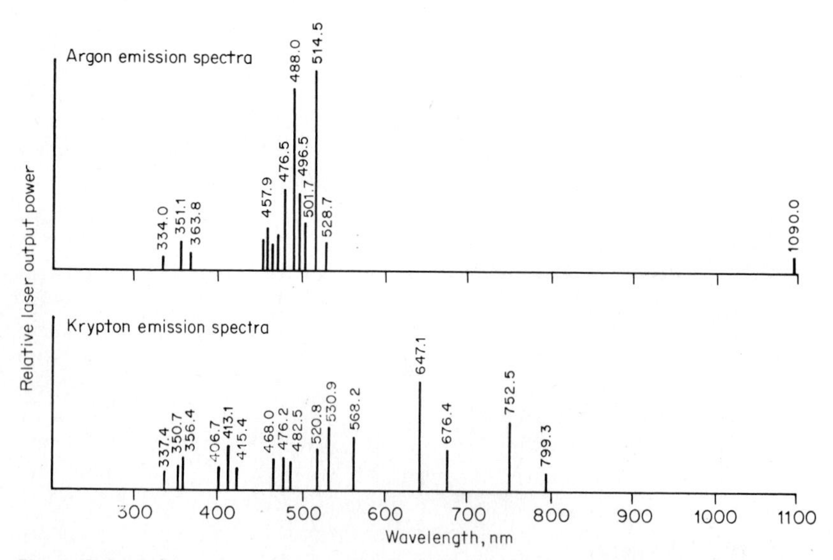

Figure 8-1 Relative intensities of major lines of argon and krypton for one laser model. Strengths of various lines may differ somewhat between models. *(Courtesy of Spectra-Physics Inc.)*

sources, as well as in the red and blue. Kr^+ also has two lines in the 700- to 800-nm region that is properly considered near-infrared, although it is weakly visible to the human eye. Kr^{2+} has two near-ultraviolet lines as well.

Energy Transfer Ion lasers are excited by a high-current discharge that passes along the length of the laser tube and is concentrated in the small-diameter bore or center of the tube. An initial spike of a few thousand volts breaks down the gas, then voltage drops to between about 90 and 400 V and discharge current jumps to 10 to 70 A. The high current densities in the center of the tube both ionize the gas and provide the energy that excites the ions to the upper laser levels. Stimulated emission brings the excited species to the lower laser level, which quickly decays to the ground-state ion, emitting extreme-ultraviolet light.

A simplified view of Ar^+ kinetics is given in Fig. 8-2; the details are complex and far beyond the scope of this discussion (see, for example, Bridges, 1982). Three major mechanisms raise Ar^+ to the upper laser level. One is energy transfer from an electron to a ground-state argon ion. Another is collisional transfer of energy from an electron to Ar^+ in an excited metastable state. The third is decay of higher levels produced by electron excitation. The contributions of the various processes appear to depend on operating conditions and are difficult to separate because all three lead to the observed variation of output with the square of current density in the tube.

Even this simplified view of the Ar^+ energy-level structure indicates some of the problems in building ion lasers. The upper laser levels lie nearly 20 eV above the ground state of Ar^+, and nearly 36 eV above the ground state of the argon atom. This means that powerful excitation is needed.

Another problem, not obvious from the diagram, is the very short lifetime of the lower laser level, which drops to the ground state of Ar^+ spontaneously with emission of a 74-nm extreme-ultraviolet photon. This short lifetime helps sustain a population inversion despite the relatively inefficient ways in which the upper laser levels are populated. However, it also means that production of a laser photon with roughly 2 eV of energy is accompanied by the loss of about 18 eV in spontaneous emission—limiting efficiency severely. The short-wavelength ultraviolet light also can damage many optical materials.

As might be expected, the ionized plasma is hot—about 3000 K, according to spectral measurements of argon laser lines. Together with the high current, this creates extremely hostile conditions inside the laser tube, causing sputtering of material from tube walls and electrodes even when high-temperature materials are used. These and other factors combine to limit ion-laser efficiency and to make design of ion-laser tubes a very difficult problem.

The energy-level structure of argon is more complex than is shown in Fig. 8-2, because there are many possible upper laser levels, some of which share the same lower level as the 488.0- and 514.5-nm transitions. There is also a separate set of energy levels for Ar^{2+}, in which the upper laser levels are 25 to 30 eV above the Ar^{2+} ground state, which itself is 43 eV above the atomic ground state. Thus even higher current densities are needed for laser

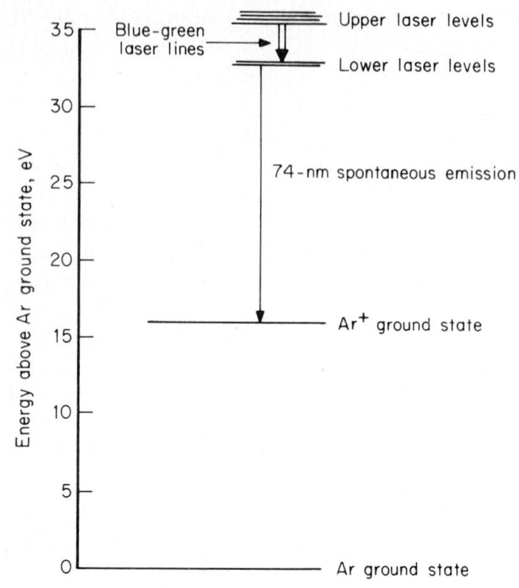

Figure 8-2 Energy levels of singly ionized argon, showing only the blue-green laser lines. Actual energy-level structure is considerably more complex, as indicated by the many possible emission lines.

emission on the ultraviolet lines of Ar^{2+} than on the blue-green Ar^+ lines, and efficiencies and operating lifetimes are lower.

The picture is qualitatively similar for krypton. However, energy levels for the heavier atom are somewhat lower and more closely spaced. Most upper laser levels of Kr^+ are 16 to 19 eV above the ion's ground state, and the wavelengths of laser transitions are somewhat longer than those of the corresponding argon transitions. Krypton is a less-efficient laser medium than argon because its transitions have lower gain, but the spacing of energy levels gives a broader range of laser wavelengths. The higher mass of the krypton ions contributes to one other problem: higher sputtering and gas-depletion rates than in argon because the accelerated krypton ions have more energy than the lower-mass argon ions.

Internal Structure

The two decades since the first argon laser was demonstrated by William B. Bridges at Hughes Research Laboratories (Bridges, 1964) have seen a gradual evolution of ion-laser design. Designers have settled on ceramic and metal-ceramic tubes with rated lifetimes ranging up to 10,000 hours. Although different manufacturers have taken different approaches, common elements are present in many designs. This reflects the need to overcome the same technological obstacles.

To attain the high current densities needed for excitation, the discharge applied along the length of the laser tube must be confined to a narrow "bore" a millimeter or two in diameter in the tube's center. The high current densities create a hot plasma in the laser tube and lead to sputtering of the bore or tube materials by the plasma, which can erode elements in the tube, introduce contaminants into the laser gas, and deplete the gas supply by trapping some atoms. The strong electron flow in the discharge tends to push neutral atoms toward the positively charged anode, while ions migrate toward the negative cathode, creating a need for gas circulation in the tube. A supply of extra laser gas is needed to replenish gas depleted during operation. In addition, the low efficiency of laser operation requires some way of getting rid of waste heat. These factors have led to several common design elements: use of high-temperature ceramics and metals resistant to sputtering in the tube, a gas-flow path separate from the central laser bore, a gas reservoir attached to the tube, and active cooling with flowing water or forced air.

A common design for low- and moderate-power argon lasers is shown in the photograph of Fig. 8-3. The discharge current passes from the cathode in the fat part of the tube along the narrow bore to be collected at the anode. The "necks" at each end of the tube are optical connections ending in Brewster angle windows, which allow linearly polarized light to escape without losses and serve as the connections with an external resonator. The bore is typically made of beryllia (BeO), which is chosen for its ability to withstand high temperatures, low sputtering, good insulation, high thermal conductivity, good mechanical strength, and low porosity. Cooling is by water flowing along the ceramic outer surface of the bore, or by air forced to pass over it. A magnet which creates a magnetic field parallel to the bore axis may be used to help confine the discharge current to the center of the bore.

Builders of high-power argon lasers are increasingly turning to an alternative (Mefferd, 1983) shown in Fig. 8-4. This approach relies on conductive metal disks brazed to a large-diameter ceramic tube. The bore is defined by central holes in tungsten disks, which are brazed to copper cups, which in turn are brazed directly to the ceramic tube envelope. The tungsten is chosen for its resistivity to sputtering and high temperature, and the copper is used because of its high thermal conductivity, essential for transferring heat to the envelope and thus to the cooling water flowing around it. Holes in the copper cups allow the required return gas flow within the laser tube. The use of conductive water cooling is a key element in making this approach practical.

The design of krypton tubes is similar to that of argon tubes, but there are a few differences worth noting. Because krypton is depleted faster than argon, the tube must contain much more reserve gas. Krypton lines other than the dominant 647.1-nm wavelength vary strongly in intensity with pressure, so active pressure control is important in krypton lasers designed to operate at those lines. The lower efficiency of krypton also requires a higher drive current density to achieve the same output power as argon. For similar reasons, most krypton lasers are water-cooled, although that is not an absolute must.

Ion lasers designed for ultraviolet operation are similar to krypton in having low efficiency and generally needing water cooling. Production of the ultraviolet

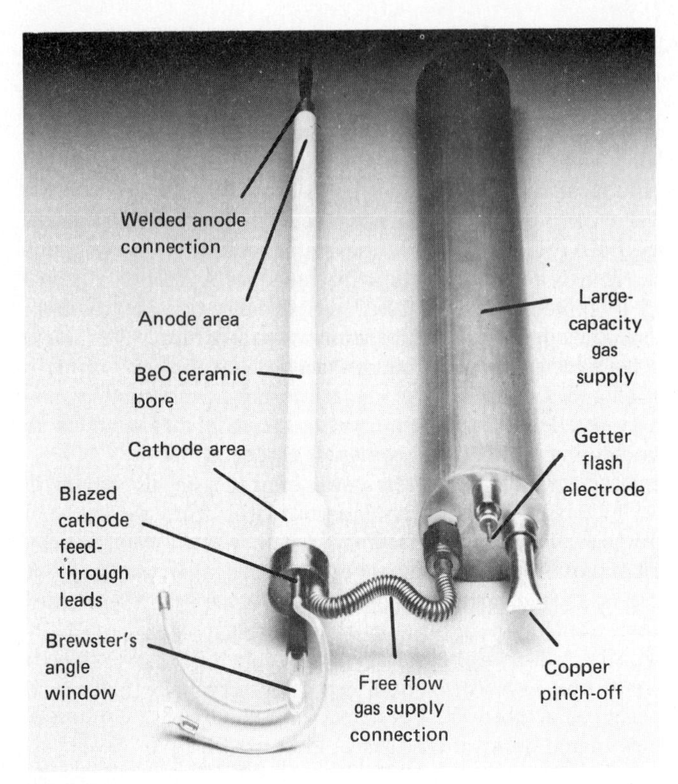

Figure 8-3 Structural components of a ceramic-bore ion laser, showing the characteristic long narrow discharge bore and fatter cathode region. The tube at right is a gas reservoir. *(Courtesy of Cooper LaserSonics.)*

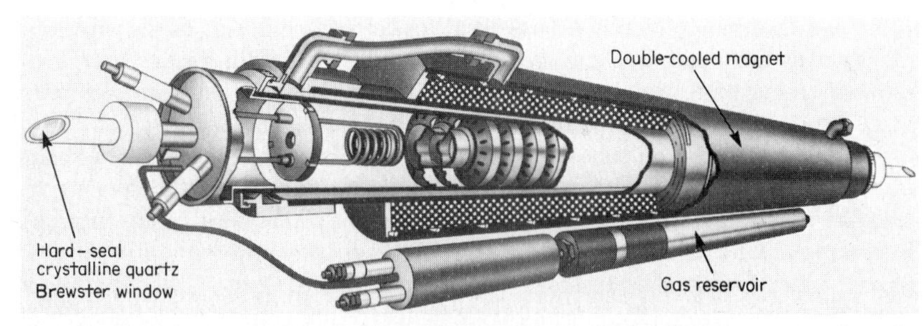

Figure 8-4 Hybrid metal-disk ceramic ion laser, in which the discharge passes through the bore defined by the central holes in the internal metal disks. Return flow is through holes outside of the central region. Heat is conducted from the central tungsten disks through copper cups to the ceramic material and thus to the cooling water flowing over the ceramic. *(Courtesy of Coherent Inc.)*

lines of the doubly charged ions also requires a higher discharge current density than the longer-wavelength lines of singly charged ions, leading to a greater reliance on current-confining magnetic fields. In addition, optics resistant to ultraviolet damage are very necessary in lasers emitting on the ultraviolet lines. Ultraviolet output also requires higher internal magnetic fields than visible output, as shown in Fig. 8-5.

Optics Most ion-laser tubes have "stems" at each end, terminating in Brewster angle windows which transmit linearly polarized light to external cavity optics. In early designs, the Brewster windows were sealed to the tube with epoxy, but now longer-lived hard seals are the usual choice. In some low-power ion lasers, fused silica is used as the Brewster window material for operation at visible lines. However, short-wavelength ultraviolet light produced in the laser plasma causes formation of color centers in fused silica, so crystalline quartz windows are used in higher-power lasers or those intended for ultraviolet operation.

Generally cavity optics are separate from ion-laser tubes, but in some low-power models they are assembled in an integral unit with the tube. Standard cavity optics are reflective over much of the visible spectrum, allowing multiline oscillation in the blue-green with argon, or multiline oscillation in much or all of the visible with krypton. Selection of a single oscillation wavelength requires insertion of a wavelength-dispersive element—normally a prism—between the tube and one of the cavity mirrors. Special reflective optics are needed for operation on the ultraviolet lines of argon or krypton. Because of the low gain of ion-laser media, the output mirror transmits only a small fraction of the light inside the laser cavity.

Cavity Length and Configuration Compact air-cooled ion lasers can have cavities as short as a quarter meter, with high-power models having cavities to about 2 m long. Models intended for industrial use generally have little extra room inside the cavity, but those intended for laboratory applications typically leave some space between the laser tube and the cavity optics for insertion of a mode-selecting etalon. Modelockers directly replace the tuning-

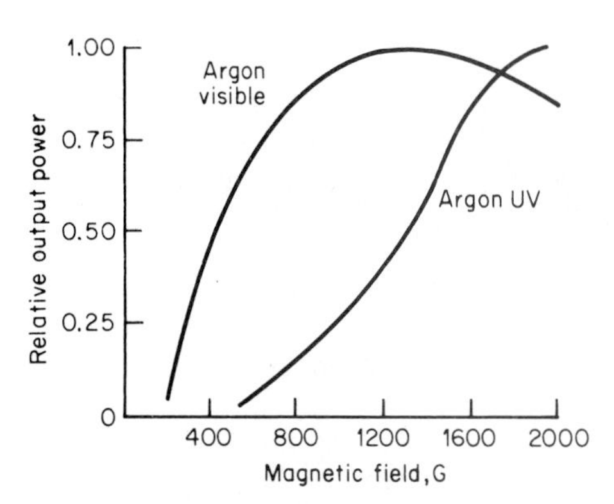

Figure 8-5 Effects of current-confining magnetic field on gain on singly ionized argon lines in the visible and on doubly ionized ultraviolet lines. *(Courtesy of Spectra-Physics Inc.)*

prism assembly. For cavity dumping, an external module is added on the laser's output coupler end.

Stable resonators are the rule for continuous-wave, low-gain ion lasers. Normally the cavity mirrors have weak focusing power because the gain region is the long, narrow bore in the center of the tube through which the discharge current flows. This design matches the resonator and gain volumes, but it leaves the laser vulnerable to misalignment if mechanical or thermal stresses are present. To avoid such problems, many manufacturers use stabilized resonator structures made of low-expansion materials.

Inherent Trade-offs

Ion lasers confront both users and designers with several inherent trade-offs:

- Spectral purity of single-wavelength oscillation vs. higher power of multiline operation.
- Greater range of visible wavelengths from krypton vs. higher power and longer life of argon.
- An increase in output power is generally accompanied by increases in price and complexity of the laser and a decrease in tube lifetime.
- The simplicity of forced-air cooling vs. the greater effectiveness of heat removal with convective flowing water cooling.
- Ultraviolet output, more effective for some applications, comes only at higher cost, shorter tube life, and lower output power and tube efficiency than visible-wavelength output.
- The potential versatility of mixed-gas lasers emitting on both argon and krypton lines is offset by the faster depletion of krypton and accompanying loss of the krypton lines, making the mixed-gas tubes short-lived.

Variations and Types Covered

Argon is by far the most important ion laser commercially, accounting for most of the units sold, with krypton in second place, accounting for about 10 percent of the market. Sales of mixed-gas argon-krypton lasers are very much smaller, but that type is included here because of its close relationship to argon and krypton lasers. No other rare gas ion laser is commercially significant.

This chapter focuses on argon and krypton lasers in their various forms, including ultraviolet and visible versions, air- and water-cooled models, and types emitting on one or multiple lines.

Beam Characteristics

Wavelength and Output Power Multiline visible output of commercial argon lasers ranges from a few milliwatts to about 20 W. The laboratory record is 500 W (Alferov et al., 1973), but such high powers have not proved practical. The less-efficient krypton laser is limited to lower powers; commercial versions can produce several watts of multiline visible output, with much of the power on the 647-nm red line. Commercial lasers can produce up to a few watts on the ultraviolet lines combined. The specifications are given for TEM_{00} output, which is standard for high-power ion lasers.

Both argon and krypton can emit on individual lines or closely spaced groups of lines, as well as oscillating simultaneously on many lines. The major lines from commercial lasers are shown graphically in Fig. 8-1, with relative intensities shown for one high-power commercial laser. The wavelengths available from any particular laser depend on cavity optics, tube design, and the adjustment of any wavelength-dispersive tuning element used in the laser cavity.

Efficiency Ion lasers are inefficient. The best of them, argon, has typical overall (or "wall-plug") efficiency of 0.01 to 0.001 percent for multiline visible output. Values are lower for single-line output, for krypton lasers, or for ultraviolet operation of either gas.

Temporal Characteristics Commercial ion lasers produce continuous-wave output normally. Modelocking or cavity dumping can produce short pulses of light from an ion laser in which the discharge is operating continuously. Because of the time needed to build up a stable high-current discharge, it is not wise to pulse the laser by turning its electrical power on and off. If modelocking and/or cavity dumping cannot provide the desired pulse modulation, the best course is use of an external modulator.

Spectral Bandwidth and Frequency Stability The spectral bandwidth of ion lasers is a function of their mode of operation. Multiline emission from argon lasers is concentrated at six wavelengths between 457.9 and 514.5 nm; multiline krypton emission can span an even broader wavelength range. Spectral bandwidth of a single line is about 5 GHz, or roughly 0.004 nm, about 1 part in 10^5 at the 6×10^{14} Hz frequency of the blue-green argon lines. The 5-GHz linewidth, in turn, is the doppler bandwidth and contains multiple longitudinal modes, each with much narrower linewidth. Linewidth of a single longitudinal mode of an ion laser is about 3 MHz, which can be obtained by using an etalon.

Frequency stability becomes an important concern with ion lasers only for operation in a single longitudinal mode. In a well-designed ion laser, temperature-induced variations in the length of the etalon or the laser cavity can cause frequency shifts, with changes in etalon cavity size normally dominant. Temperature stabilization of the etalon can keep frequency stable to within several megahertz over a 1-s interval, and to within several tens of megahertz over 10 hours. In most cases, such stabilization is good enough to prevent hopping between longitudinal modes, normally spaced 80 to 250 MHz apart, which could cause rapid frequency shifts.

Amplitude Noise Control of the current flowing through the laser tube can limit root-mean-square (rms) amplitude noise level to from 1 percent to a few percent. Peak-to-peak variations, specified by some manufacturers, are inevitably larger, and can be specified as high as 15 percent.

Much better control of amplitude noise is possible by monitoring the optical output and using that information to control input current. Some ion-laser makers claim rms noise values of 0.1 percent using light control, although typical specifications are somewhat higher. Makers who specify peak-to-peak values for noise specify values as high as 5 percent with light control, which is necessarily higher than the rms value for the same quantity.

Specifications of amplitude noise do not differ greatly for single- or multiline operation of ion lasers. All assume that the laser has been warmed up before measurement.

Beam Quality, Polarization, and Modes Brewster angle windows on ion-laser tubes lead to linearly polarized ouput; the typical design calls for vertical polarization, specified at a ratio of 100 to 1 or more. TEM_{00} is the commonest transverse mode, especially on scientific lasers. Because somewhat higher powers are possible with multiple transverse modes—albeit at a sacrifice in beam quality—some models seeking to deliver maximum power for minimum cost produce multimode output. A few models emit in the doughnut-shaped TEM_{01}^{*} mode.

Coherence Length Coherence length of an ion laser emitting on multiple lines is very small. For single-line oscillation with typical 6-GHz bandwidth, coherence length is about 50 mm. Use of an etalon to limit oscillation to a single longitudinal mode can extend coherence length to tens of meters or more.

Beam Diameter and Divergence Typical values of beam diameter and divergence of ion lasers are 0.6 to 2 mm and 0.4 to 1.2 mrad, respectively, varying somewhat among models, with the product of the two quantities typically being about 0.9. Both beam diameter and divergence are slightly larger for krypton lasers emitting in the red than for blue-green argon lasers.

Stability of Beam Direction The directional stability of an ion-laser beam is affected by random fluctuations within the laser tube and by variations in the refractive index of wavelength-selecting prisms caused by changes in temperature. Variations caused by the prism are larger, with one manufacturer reporting a change of 11 arcseconds (0.05 mrad) per degree Celsius, and saying that a temperature change as small as 10°C can completely detune the laser. This problem can be avoided by temperature compensation of prism wavelength selectors. Because the wavelength-selecting optics are the source of the strongest variations, multiline lasers can be stabler in beam direction. Although most makers do not specify the stability of ion-laser beam direction, one specification sheet does list stability of 0.1 mrad for an air-cooled, multiline argon laser.

Suitability for Use with Accessories The most important accessories used with ion lasers are

- Modelockers, to produce trains of picosecond pulses from lasers emitting in multiple longitudinal modes. With the typical 5-GHz linewidth of ion-laser lines, the resulting pulses are about 90 to 200 picoseconds (ps) long, produced at repetition rates in the 75- to 150-MHz range.
- Cavity dumpers, which can produce high peak power pulses in the 5- to 20-ns range, or which can be used with a modelocker to slow down the repetition rate and increase the amplitude of modelocked pulses.
- Etalons, used to narrow spectral width to a single longitudinal mode.

Wavelength conversion is possible, generally by using the ion laser to pump a dye laser, which can be tuned continuously from the near-infrared to the near-ultraviolet, as described in Chap. 17. Harmonic generation is possible by tight focusing of a high-power beam in a suitable material, but it is generally a second choice and not often used in practice, one reason being the difficulty in producing pulses with high peak power from ion lasers. The short upper-state lifetimes prevent energy storage in the media, making Q switching impractical.

Operating Requirements

Input Power Input electrical requirements for ion lasers range from 10 or 15 A single-phase current at 120 V to 60 or 70 A of three-phase current at 460 V, with input requirements increasing with output power. An argon laser with multiline output of 3 to 5 W typically would require electrical input of 8 to 26 kW. Similar electrical input would generate about 0.75 W of multiline krypton output. Up to 26 kW of electrical input is needed to generate 0.4 W in the ultraviolet from argon or 0.15 W of ultraviolet from krypton.

The power supply converts the electrical power from the line supply into a form suitable for driving the laser. The discharge is triggered by a peak voltage of a few thousand volts. The current then rapidly builds to 10 to 70 A, concentrated in the bore, while the voltage drops to a steady value of 90 to 400 V, depending on the length of the tube. Light output is roughly proportional to the square of current density, and hence increases faster than the input electrical power.

Cooling Ion lasers require either forced air or water cooling, with water cooling needed for higher-power models. Specified tap-water flow rates for laser heads are 5.6 to 22.7 L/min, and higher-power models require water cooling for their power-supply modules. Water cooling is used for some argon lasers delivering as little as 100 mW of multiline output. Generally, water flows along the outside of the laser tube, as shown in Fig. 8-6, conducting away excess heat.

Forced-air cooling can be used for argon-laser tubes delivering multiline outputs as high as a few watts, although at that power level both gas flow rate and exhaust temperature must be high. One 4-W multiline argon laser requires 250 ft^3 (7 m^3) of air per minute and can heat the exhaust to 180°F (80°C); specified tube lifetime is 10,000 hours. Air cooling is the rule for very low power argon lasers.

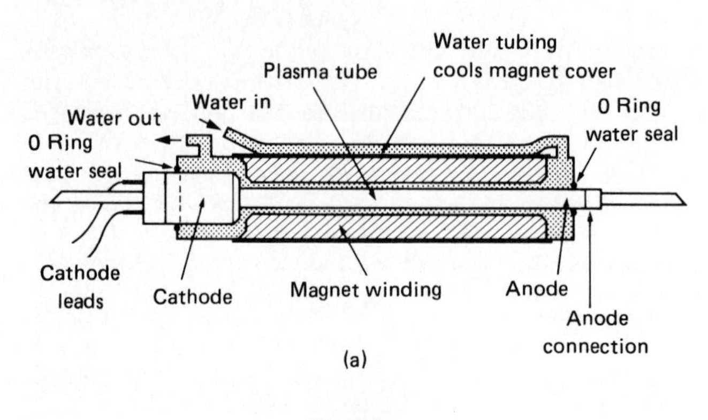

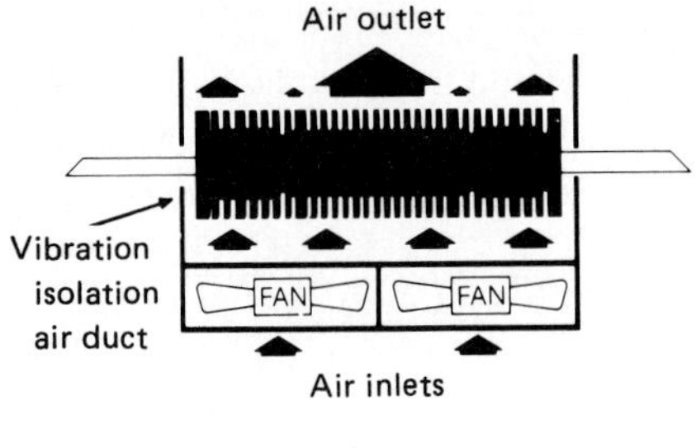

Figure 8-6 In water-cooled ion laser (top) the water flows around the outside of the laser tube to remove heat. In air-cooled types (bottom) fans force air past the tube, and the air removes excess heat from cooling fins attached to the tube. *(Courtesy of Cooper LaserSonics.)*

Consumables Outside of cooling water, there are no true consumables associated with ion lasers, with the possible exception of the tube itself. Tubes have lifetimes rated in the thousands of hours and account for about a third of the price of the complete laser. As described below, refurbishment is an alternative to replacement with a new tube.

Required Accessories Pulsed operation of an ion laser requires a cavity dumper, modelocker, and/or external modulator. Single-frequency operation requires a temperature-stabilized etalon. Applications demanding particular stability in frequency, beam direction, or output power may require accessories for temperature stabilization of the laser or components, or light control of the laser output. Special optics are generally needed to obtain output on the less-common laser lines and are a must for ultraviolet operation of either argon or krypton.

Operating Conditions and Temperature Warm-up of ion lasers is needed to establish a stable current discharge within the tube and to reach proper temperatures for stable output power, wavelength, and beam pointing. Models intended for demanding scientific applications give specifications for stability of operation measured after as much as 2 hours of warm-up time, although other models intended for the same uses call for shorter warm-up intervals. However, models intended for industrial applications have warm-up times measured in seconds or minutes.

Water-cooled ion lasers require tap-water input and drainage for their flow-through cooling systems. Air-cooled models require adequate ventilation for exhaust fans. Otherwise, ion lasers are designed to operate under normal laboratory conditions or under the temperature-controlled conditions typically found in an office or computer room.

Mechanical Considerations Ion-laser sizes vary widely, increasing with output power. The heads of the largest types weigh about 65 kg, are about 2 m long, and are about 20 by 20 cm in cross section. They require separate power supplies, often including water connections and controls, that measure up to about 70 by 65 by 60 cm and weigh up to about 220 kg.

On the other end of the scale, a compact air-cooled version delivering a few milliwatts at 488 nm has a 6-kg head 41.5 cm long that measures 17.5 by 13.2 cm in cross section. That model comes with a 27-kg power supply that measures 51.5 by 40 by 16.3 cm, but another model with slightly larger laser head comes with a power supply that weighs just 8 kg and measures 34 by 23 by 14 cm.

Safety Argon lasers fall under Class IIIb or Class IV of the federal laser safety code, and users should take care to avoid direct exposure to the beam, particularly eye exposure. Standard safety goggles are available which strongly attenuate argon's blue-green lines, and the laser does not produce the single pulses of high peak power that have proven particularly dangerous with other types of laser. However, the unattenuated, narrow-divergence beam from an argon laser can penetrate the eyeball and permanently affect the retina.

Krypton lasers, although generally lower in power, fall into the same safety classes. Multiline visible output from krypton is a particular problem because the laser wavelengths are scattered across so much of the spectrum that goggles made to block the laser lines would also block most visible light. This makes it vital for users of multiline krypton lasers to exercise special care in avoiding eye exposure.

The ultraviolet lines of both argon and krypton are strongly absorbed by the lens of the eye, and only a small fraction of such light reaches the retina. There are few continuous-wave lasers emitting at these wavelengths, and eye hazards are not well quantified, so the use of ultraviolet-absorbing goggles would be a prudent precaution. Care should be taken with the krypton lines at 752.5, 793.1, and 799.3 nm in the near-infrared; those lines can be weakly perceived by the human eye, but users should not make the mistake of judging their intensity by eye.

Like other discharge-driven lasers, ion lasers need high voltages, but only to initiate the discharge. After the discharge is established, voltage drops to the 90- to 400-V level, with currents of 10 A or more flowing through the tube. The hazards presented by such levels of current capacity are different than those at higher voltages, resembling the dangers of an open wall socket.

Special Considerations Ion-laser alignment is delicate and can be affected by changes in temperature and mechanical stresses. Manufacturers design their resonators to stay rigid in the face of reasonable mechanical and thermal stresses, but they cannot be expected to take countermeasures to protect against abuses such as improper mounting, incorrect adjustment, or placement next to a hot-air vent.

Reliability and Maintenance

Operating lifetime of an ion laser is limited by the tube. Typical rated lifetimes range from 1000 to 7000 hours for water-cooled argon lasers and from 3000 to 10,000 hours for air-cooled argon, both for operation at visible lines, but not all these ratings are backed by manufacturer warranties. Lifetimes are shorter for krypton lasers, or for operation on ultraviolet lines, with krypton lifetime limited by gas depletion and ultraviolet lifetime limited by the high operating current used to produce ultraviolet emission. The operating life, defined as how long the tube meets specifications, is shortest for argon-krypton mixed-gas lasers, because the krypton is depleted rapidly, leaving a tube that emits only on the argon lines.

Tubes degrade by gradually falling in output power, reaching the end of their useful life when output falls below application requirements or the tube fails altogether for other reasons. Average life falls within the rated ranges, although there is considerable variation between individual tubes. Laser operation can be restored by replacing the tube. Unless the tube is extremely worn from long use or has suffered severe physical damage, it can be refurbished, as described below, to once again produce rated output.

Maintenance and Adjustments Needed Periodic realignment may be needed because of the sensitivity of ion lasers. Cooling and electrical systems should be checked regularly. As mentioned above, tube replacement or refurbishment is required after prolonged operation. Optics not sealed to the tube must be cleaned periodically, particularly for high-power lasers.

Mechanical Durability Ceramic and glass ion lasers tend to be brittle, making them vulnerable to mechanical damage. New designs have offered some improvements, but the need for ceramics to withstand hostile conditions in the laser tube means that fragility remains a concern.

Failure Modes and Causes Mechanical damage or failure of electrical components can cause failure of an ion laser. Under normal circumstances, however, the most likely problem is degradation of the laser tube caused by several factors, most related to the high discharge currents in the tube bore.

- Depletion and contamination of the laser gas, caused by trapping of gas atoms under material sputtered from the laser tube, and by sputtering of material into the gas. Getters are used to collect some contaminants. Many lasers have automatic gas replenishment.

- Erosion of material from the laser bore, which may be deposited elsewhere in the tube, perhaps blocking the bore.
- Contamination and erosion of the cathode by ion sputtering.
- Contamination of the optics, causing excessive optical losses.
- Degradation of optics exposed to hard ultraviolet radiation produced in the discharge.
- Misalignment of the resonator, caused by temperature or mechanical effects, lowering the gain until the laser can no longer oscillate.

All six types of degradation can occur simultaneously. Mechanisms related to sputtering, such as bore erosion and contamination, tend to increase in severity with current density in the bore.

Possible Repairs The best course for ion-laser users with a dead tube is to replace it, but that does not automatically mean buying a new tube. Several companies now offer repair, refilling, and refurbishing of ion-laser tubes. Refilling with fresh laser gas is far from enough. A thorough refurbishing requires disassembly of the tube, replacement of some components, repair and cleaning of others, then careful reassembly and refilling with fresh gas (Wiedemann, 1984). Generally, the refurbished tubes are burned in for several hours. A well-refurbished tube can give 80 percent or more of the operating life of a new tube for about half the cost, and in many cases can be repeatedly refurbished if mechanically sound.

Refurbishing tends to be a small, specialized business offered by companies independent of tube makers, who generally would rather sell a new tube. So far, most refurbishers have concentrated on a few models they know well. Users should check current industry directories or listings of refurbishing services and inquire which tubes the services handle most often. So far the business has concentrated on custom refurbishing of customer-supplied tubes, but if adequate supplies of used tubes become available it may become possible to buy refurbished tubes outright.

Commercial Devices

Standard Configurations Some elements of ion-laser design are common to many manufacturers, such as the separation of the tube-containing laser head from the power supply. Otherwise, however, configurations and design approaches can vary widely.

One variable is the type of cooling. Water cooling is essential for high-power operation and is common in moderate-power lasers because of its simplicity. It is also the usual choice for the comparatively inefficient krypton laser. Air cooling is the rule for argon lasers under 200 mW and can also be used at moderate powers, up to a few watts with sufficiently fast air flow. The simplicity of air cooling makes it attractive for industrial applications.

Lasers tend to be packaged either for laboratory applications or for incorporation into laser systems for applications such as exposure of printing plates, high-speed printing, or medical treatment. Laboratory models normally are designed for maximum flexibility and leave intracavity space for accessories such as etalons. Industrial versions tend to be packaged functionally, for incorporation inside a larger system package, and tend to make little or no provision for inclusion of optional equipment. Many laboratory models are intended to be tunable so they can emit at any of several possible single wavelengths and are packaged with the option to oscillate in a single longitudinal mode. Industrial versions may emit at one or several wavelengths, but they are rarely tunable and have little call to oscillate in a single longitudinal mode.

Some companies take a modular design approach that lets them offer many different models. The same power supply may be used to drive several different tubes, often including tubes with different active gases. The same laser head housing may hold different tubes or be used with different optics to obtain different performance from similar or identical tubes.

As the industrial market has grown, manufacturers have shown increasing interest in designing lasers that could directly replace products made by other companies. However, optical, mechanical, and electrical characteristics are still far from standardized.

Options Principal options offered on commercial ion lasers include

- Modelocking
- Cavity dumping
- Choice of TEM_{00} or multiple transverse mode operation
- Selection of a single wavelength or line for laser oscillation
- Etalons to limit oscillation to a single longitudinal mode
- Temperature stabilization of components to enhance laser stability
- External power monitors for light stabilization of output
- Choices of power supplies and (in some cases) of cooling method
- Polarization rotators
- Operation on ultraviolet or other wavelengths (a choice that normally must be made when the laser is purchased and cannot be retrofitted onto an existing laser)

Special Notes The ion-laser market has been exceptionally active recently, with new entries in laser manufacturing, the emergence of serious tube-refurbishing efforts, and continuing refinement of the technology for both high- and low-power models. Users should keep alert for changes in the market and for manufacturers interested in targeting their special needs.

Pricing Some laser makers have boosted their prices in the past couple of years, but the increasing competition may bring them back down again. Some representative prices as of mid-1984 for argon lasers are listed in Table 8-1. The variations reflect differences among manufacturers, performance, lifetimes, and other factors.

Krypton lasers cost much more per watt than argon: a 0.75-W multiline krypton laser has about the same price as a 5-W multiline argon laser. Mixed-

TABLE 8-1 Representative Prices for Argon Lasers

Laser	Price
5 mW, multiline blue-green	\$3000
5 mW, 488 nm	\$4700
10 mW, 488 nm	\$5000–\$6000
40 mW, multiline blue-green	\$7000
50 mW, multiline blue-green	\$5500
5 W, multiline blue-green	\$12,300–\$25,000
20 W, multiline blue-green	\$35,000–\$50,000
2–3 W, ultraviolet	\$50,000

gas lasers have prices comparable to a krypton laser of equivalent power. The option to produce ultraviolet output from argon or krypton costs an added \$1000 to \$2500.

Suppliers Over a dozen companies make argon lasers, and the number is growing. They range in size from small startups to overseas giants. Some of the companies which have been in the field longest are best known for their scientific lasers, but other veteran companies have concentrated on the industrial market—a market now being targeted by several new companies as well.

Fewer companies make krypton lasers than argon types, reflecting the much smaller market for krypton, particularly in industry. There is only a single maker of mixed-gas lasers at this writing, although other companies have offered mixed-gas lasers in the past.

Applications

Ion lasers have become standard sources of several milliwatts to 20 W of visible light for a variety of applications. The blue-green output of argon is particularly valuable because many materials respond more strongly to it than to the less-energetic red photons from krypton or helium-neon lasers. Argon competes with helium-cadmium lasers at powers of several milliwatts but is unrivaled at much higher powers. In the red, krypton can offer higher powers than helium-neon. Some of krypton's weaker lines are also valuable because they are otherwise unobtainable except from dye lasers. Both argon and krypton are also valuable sources of continuous-wave near-ultraviolet light. Major applications of ion lasers include

- Production of halftone patterns during color separation for printing of full-color material
- High-speed printing of computer output
- Exposure of printing plates, films, and other materials used in printing and publishing
- Mastering disks for videodisk and audiodisk players
- Medical treatment, particularly in ophthalmology, where argon laser light can penetrate readily to the retina to treat diabetic retinopathy

- Measurement in medical laboratories and in research applications
- Pumping dye lasers to obtain tunable continuous-wave output or picosecond pulses
- Entertainment and display applications requiring visible light
- Special-purpose optical data storage systems requiring high-speed recording

BIBLIOGRAPHY

G. N. Alferov, V. I. Donin, and B. Ya. Yurshin: "CW argon laser with 0.5 kW output power," *JETP Letters 18*:369–370, 1973.

William B. Bridges: "Laser oscillation in singly ionized argon in the visible spectrum," *Applied Physics Letters 4*:128–130, 1964; erratum *Applied Physics Letters 5*:39, 1964.

William B. Bridges: "Ionized gas lasers," in Marvin J. Weber (ed.), *CRC Handbook of Laser Science & Technology*, vol. II, *Gas Lasers*, CRC Press, Boca Raton, Fla., 1982, pp. 171–269.

Wayne S. Mefferd: U.S. Patent 4,376,328, "Method of constructing a gaseous laser," March 15, 1983.

Orazio Svelto: *Principles of Lasers*, 2d ed., Plenum, New York, 1982.

Rudi Wiedemann: "Reprocessing argon-ion lasers," *Lasers & Applications 3*(2):81–84, February 1984.

CHAPTER

9

helium-cadmium lasers

The helium-cadmium laser is the best known member of a family of lasers which emit on lines of ionized metal vapors. One of the first metal vapor lasers to be discovered, He–Cd can produce continuous powers up to about 100 milliwatts (mW) at 442 nanometers (nm) in the blue, or powers to about 20 mW on an ultraviolet line at 325 nm. These characteristics are attractive for many applications and have stimulated further development that has extended operating lifetimes of He–Cd tubes to several thousand hours.

The metal vapor ion lasers were among the first types discovered in a systematic series of experiments intended to search for laser lines from a variety of materials. William Silfvast, then a graduate student at the University of Utah, and his faculty advisor Grant Fowles placed various metals in a discharge tube and observed what happened when electrical pulses were passed through the tube. Their first paper reported laser action on lines of six metals (Silfvast et al., 1966). Later research by Silfvast and others at Bell Laboratories showed that cadmium was a particularly efficient laser material and that addition of helium enhanced efficiency and permitted continuous-wave operation.

Metal vapor laser development has concentrated on helium-cadmium because of its useful wavelengths and high efficiency. Silfvast also developed the related helium-selenium laser, which emits on over 20 lines, mostly in the visible (Silfvast, 1973). He–Se was put on the market briefly in the 1970s, but it has never received as much development as He–Cd and is no longer marketed actively. Reference tabulations (e.g., Bridges, 1982) list many other metal ion lasers which have never received much development.

Internal Workings

Active Medium Singly ionized cadmium vapor has over a dozen laser lines in and near the visible and can operate as a laser under a variety of conditions. Operation is most efficient when several millitorr of cadmium vapor is sealed in a tube with helium gas at pressure of about 6 torr (800 Pa). This is the usual configuration of commercial He–Cd lasers, in which the metal must be heated to about 250°C to produce adequate vapor pressure.

Although laser operation is possible over a range of conditions, long-term continuous-wave operation at reasonable efficiency requires the proper balance of helium and cadmium pressures. The positively charged Cd^+ ions migrate toward the negative electrode (cathode) and must be replenished from a reservoir of the metal within the tube.

Energy Transfer Excitation energy for the helium-cadmium laser comes from a direct-current discharge passing through the laser tube. Typical discharges are around 1500 V, with current densities in the small-diameter bore on the order of 4 A per square centimeter of cross section. Helium atoms in the laser gas absorb energy from the discharge and then transfer that energy to cadmium ions.

The energy levels involved for cadmium and helium are shown in Fig. 9-1, which indicates the close relationship of the blue and ultraviolet transitions of Cd^+. As in other metal vapor ion lasers, there are two dominant mechanisms by which energy is transferred between the two species: Penning ionization and charge-transfer ionization.

For He–Cd, the dominant process is Penning ionization, in which energy from an excited helium atom ionizes a cadmium atom:

$$He^* + Cd \rightarrow He + Cd^+ + e^-$$

Because helium's lowest excited energy level is well above the ground state of Cd^+, Penning ionization produces excited cadmium ions. The metastable helium excited states transfer energy to $2D$ levels of Cd^+ which have lifetimes on the order of 100 nanoseconds (ns). The lower $2P$ levels of Cd^+ also receive some energy but are quickly depopulated because of their 1-ns lifetime. The result is a population inversion between $2D$ and $2P$ states of cadmium, allowing laser action on either the 442-nm or 325-nm lines.

Charge-transfer ionization is of secondary importance in the traditional positive-column He–Cd laser, although it is dominant in some metal vapor ion lasers. In this case, a helium ion captures an electron from a cadmium atom to create an excited cadmium ion

$$He^+ + Cd \rightarrow He + (Cd^+)^*$$

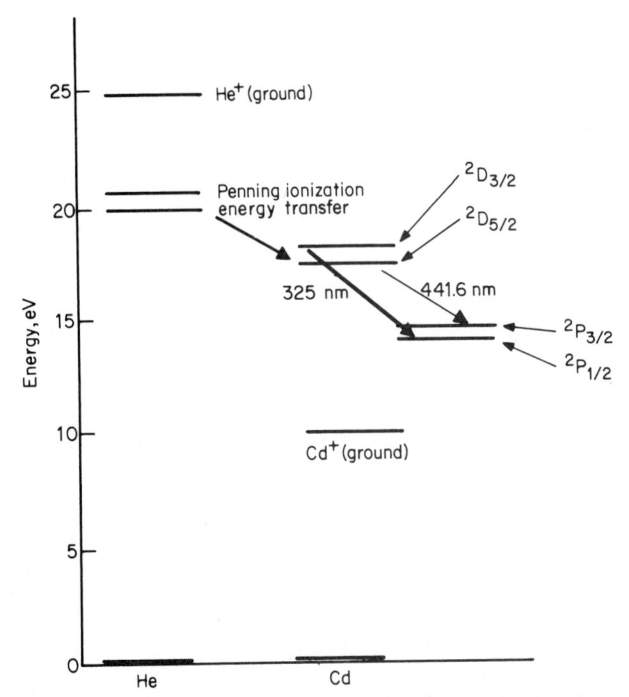

Figure 9-1 Energy levels in helium and cadmium, showing those involved in 441.6- and 325-nm laser transitions.

This is the dominant excitation process in the helium-selenium laser, because a selenium energy level is resonant with the helium ion state.

Internal Structure

The internal structure of the helium-cadmium laser bears a general resemblance to two other discharge-driven gas lasers: the helium-neon laser and the argon ion laser. In all three types, the discharge passes through the length of a long, thin discharge bore which confines the current to increase current density and hence enhance laser efficiency. The operating characteristics and efficiency of helium-cadmium lasers are intermediate between helium-neon and argon lasers, so many of the features of the He–Cd laser fall partway between those of He–Ne and argon lasers.

The need to maintain adequate concentrations of cadmium vapor for laser operation leads to some of the novel aspects in design of He–Cd laser tubes such as the one shown in Fig. 9-2. A reservoir of cadmium metal is required, which must contain about a gram of the metal for each thousand hours of tube operation. This reservoir is placed at the anode end of the discharge bore and is heated to provide the necessary metal vapor pressure. As the discharge ionizes the metal vapor in the tube, the positively charged ions migrate toward the negative electrode, the cathode, a phenomenon called *cataphoresis*. This process helps to distribute cadmium vapor uniformly in positive-column He–Cd lasers, but it also can cause deposition of metal toward the cathode as the vapor cools.

Metal deposition in the wrong places can cause serious problems, including obstruction of the cavity optics, blockage of the discharge bore, and poisoning of the cathode. To limit these problems, He–Cd tubes normally include a condenser outside the laser bore. In the tube design shown in Fig. 9-2, an additional cold trap for cadmium ions is placed in front of the cathode (Dowley, 1982).

As in the helium-neon laser, the optimum helium pressure (for maximum power) depends on bore diameter, with the product of pressure times bore diameter staying roughly constant. In a typical He–Cd laser, output is highest when helium pressure is about 6 torr (800 Pa) and when cadmium pressure is around 0.1 percent of that level, i.e., several millitorr. Metal is depleted by the processes described earlier. Helium can also be depleted, by cathode sputtering, burial under condensing metal, and wall diffusion. Thus a reservoir is needed for helium as well as for cadmium. These reservoirs may be in separate branches to the tube or in structures coaxial with the discharge bore. The tubes also normally contain a getter to remove contaminants that could poison the laser tube, a standard feature in sealed gas lasers.

The optics on He–Cd lasers are generally similar to those on He–Ne and ion lasers. In some designs, output is coupled from the sealed laser tube through Brewster angle windows to external resonator optics. In other designs, resonator mirrors are hard-sealed to the tube, as shown in Fig. 9-3, which illustrates coaxial tube structure. As in other low-gain lasers, the rear mirror in a He–Cd cavity is a total reflector, and the front mirror transmits only a small fraction of the light circulating within the laser cavity. Stable resonators are the rule, with typical cavity lengths ranging from 40 cm to a little over a meter. The cavity optics provide selection between the 442- and 325-nm lines.

Inherent Trade-offs

Although the ultraviolet line of He–Cd may be more potent in producing certain effects than the blue line, output is weaker and tube lifetime is shorter

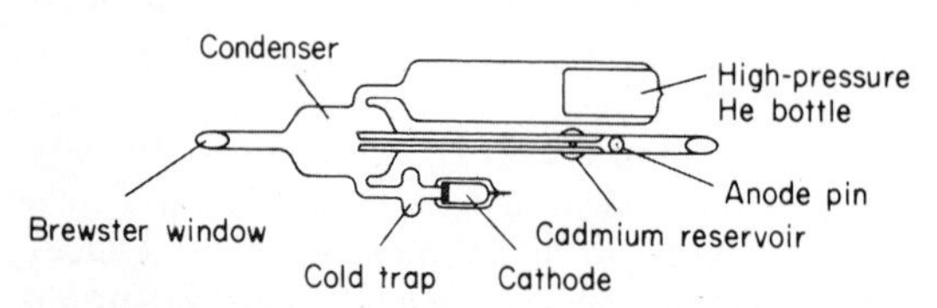

Figure 9-2 Simplified diagram of a He–Cd tube with high-pressure helium bottle to the side, and condenser and cold trap to catch cadmium metal. Output is coupled to external optics through Brewster windows. *(Courtesy of Liconix.)*

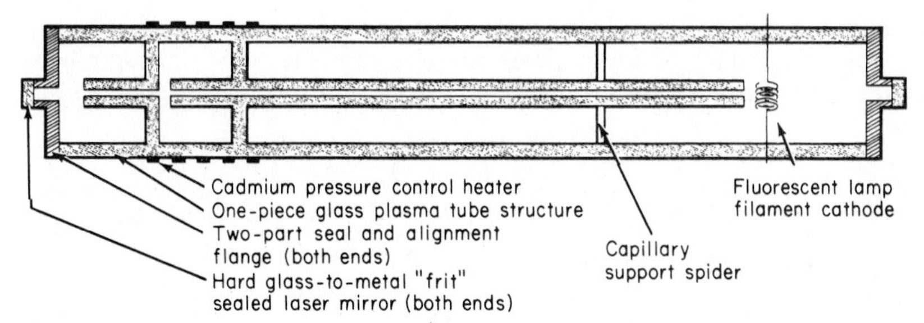

Figure 9-3 Coaxial-tube design, in which optics are bonded directly to the tube. *(Courtesy of Omnichrome.)*

when operation is in the ultraviolet. If the same tube design is used at the two wavelengths, ultraviolet power will typically be 15 to 25 percent of that at 442 nm. Tubes used on the ultraviolet line are rated for about half the lifetime of those used on the blue line.

Variations and Types Covered

Commercial He–Cd lasers operate on one of the two strongest cadmium transitions, at 325 or 442 nm, but the cadmium ion also can emit at several other visible wavelengths. Simultaneous oscillation at 442 nm and at two other visible Cd^+ lines—537.8 and 636.0 nm—is the basis of a so-called white-light laser. The wavelengths are distributed through the visible spectrum in a way that makes it possible to balance color seen by the human eye by adjusting relative intensities. Such white-light He–Cd lasers have been demonstrated in the laboratory (Schuebel, 1970; Fujii et al., 1975; Otaka et al., 1981; Wang and Reid, 1981), but so far limited lifetimes and problems in controlling relative intensities of the different lines have made commercial devices impractical (Sasaki et al., 1983). White-light He–Cd lasers require different tube designs and optics than conventional blue or ultraviolet types and will not be described here because they are not available commercially.

Some development work has been done on other members of the metal vapor ion laser family, notably helium-selenium and helium-zinc. However, because they are not on the market, they will not be described here. Commercial copper vapor and gold vapor lasers emit on lines of the neutral species and operate in ways rather different from He–Cd and related lasers; they are treated separately in Chap. 12.

Beam Characteristics

Wavelength and Output Power He–Cd lasers emit at 441.6 nm in the blue or 325 nm in the ultraviolet. The cavity optics, which may be permanently bonded to the tube or externally mounted, select one wavelength. Continuous output on the blue line ranges from the milliwatt range to about 50 mW in TEM_{00} mode, and to above 100 mW for multimode output. However, those high multimode powers are specified only for products manufactured overseas. These products may be hard to obtain in the United States. Rated continuous output on the less-efficient ultraviolet line is 1 to 10 mW, typically in TEM^*_{01} mode.

Efficiency Overall wall-plug conversion efficiency for blue emission from a He–Cd laser runs from 0.014 to about 0.002 percent, higher than argon ion but below helium-neon. Efficiency of He–Cd on the ultraviolet line is about one-fifth that in the blue. As in many other gas lasers, efficiency tends to increase with output power in commercial He–Cd lasers, until other factors begin to limit output power.

Temporal Characteristics Helium-cadmium lasers normally emit continuously. The need to heat the tube to about 250°C to achieve the required cadmium pressure discourages turning the laser off; He–Cd lasers may have a "standby" mode that leaves the tube nearly ready to operate but does not produce an output beam. External modulators are used if the beam must be turned off and on quickly and often.

Spectral Bandwidth He–Cd lasers have oscillation bandwidths of about 3 gigahertz (GHz), corresponding to 0.002-nm spectral linewidth at 442 nm. This is about six to eight longitudinal cavity modes in normal He–Cd cavities.

Amplitude Noise Typical amplitude noise specified at 442 nm is in the 1 percent range between 10 Hz and 10 megahertz (MHz) with maximum noise specifications 2 to 3 percent on the blue line and as high as 5 percent on the ultraviolet line. Output-power stability, which measures long-term fluctuations, is usually specified between 2 and 5 percent.

Beam Quality, Polarization, and Modes He–Cd lasers have good beam quality, with TEM_{00} output normal at 442 nm except at unusually high powers, and TEM^*_{01} output at 325 nm. Beams normally contain several longitudinal modes, but a Fabry-Perot etalon can restrict oscillation to a single longitudinal mode under carefully controlled conditions. Models with Brewster angle windows on the tubes produce output with linear polarization ratios greater than 500 to 1; other models produce unpolarized beams.

Coherence Length Coherence length for a He–Cd laser emitting with its natural 3-GHz linewidth is about 10 cm. Restriction of oscillation to a single longitudinal mode, with linewidth on the order of megahertz, can stretch coherence length to the order of 100 m.

Beam Diameter and Divergence TEM_{00} 442-nm beam diameters are 0.2 to 1.2 mm, with the smallest diameters measured at a beam waist at the output mirror. Divergences at 442 nm are 0.5 to 3 milliradians (mrad). Divergence at the 325-nm line is similar, 0.4 to 3 mrad, with diameters 0.135 to 1.2 mm, again with the smallest values measured at a beam waist at the output mirror. As with other lasers, beam divergence is inversely proportional to beam diameter, so the smallest diameters come only at a cost of comparatively high divergence, and thus cannot be realized away from the laser except with focusing optics.

Stability of Beam Direction Because He–Cd lasers are often used in reprographic applications where beam directional stability is vital, this quantity is specified on many data sheets. Beam-pointing stability is specified at ± 10 to 50 μrad/°C, with the explicit inclusion of temperature reflecting the crucial influence of temperature on alignment stability of the laser cavity. Stability of beam position and divergence may also be specified, with typical values ± 0.1 mm (at the output mirror) and ± 1 mrad.

Suitability for Use with Laser Accessories He–Cd lasers often are used with external modulators to adjust output intensity for applications such as printing and writing.

Modelocking is preferable to *Q* switching for generating short pulses with high peak power. Energy-storage capacity of He–Cd lasers is too low for practical *Q* switching, but the 3-GHz linewidth is large enough to generate transform-limited modelocked pulses shorter than 1 ns. Modelocked pulses could be used for harmonic generation, although the 442-nm blue line is close to the short-wavelength limit for practical frequency doubling in crystals (the 325-nm wavelength is much too short). However, He–Cd lasers are rarely used in harmonic generation because their low average power makes harmonic-generation efficiency much lower than for higher-power lasers.

Operating Requirements

Input Power During normal operation, a He–Cd laser emitting a few milliwatts in the blue draws about 165 W of electrical power, while one emitting 50 mW in the blue will require about 350 to 560 W. Similar input powers will produce roughly one-fifth as much light on the 325-nm ultraviolet line. Heating of the cadmium metal during tube warm-up to produce adequate metal vapor pressure may require an added 10 to 50 W of electrical power. Most models operate from a 115-V ac line, but some higher-power models may have 220-V operation as an option or requirement.

Cooling Electrical input to a He–Cd laser is typically a few watts per centimeter of tube length, leading to heat-dissipation needs of 100 W and up from the tube. Convection air cooling is adequate for smaller tubes; higher-power models may need forced-air cooling.

Consumables He–Cd lasers are built as sealed units and require no consumables per se.

Required Accessories Normally He–Cd lasers require no special accessories other than those needed for a specific application, such as an external modulator if the beam is to be modulated.

Operating Conditions and Temperature Specified operating temperatures for He–Cd lasers are 10 to 40°C, with storage temperatures specified at 0 to 50°C, or in some cases 0 to 80°C. Specified maximum humidity during operation is 90 to 100 percent.

All He–Cd lasers must be warmed up from a cold start. About 5 minutes is needed to heat the cadmium metal to produce enough vapor for operation at a reasonable fraction of normal laser output, as shown in Fig. 9-4. Up to half an hour may be needed to fully stabilize operation. Such delays are avoided in some models by a standby mode, which keeps the tube heated and cadmium pressure up, without generating a beam, cutting warm-up time to a couple of minutes.

Mechanical Considerations Standard commercial He–Cd lasers include a laser head and separate power supply. The laser heads are 60 to 120 cm long, with round or rectangular cross section that measures 10 to 15 cm in most models. Low-power heads weigh a few kilograms, while those with output in the 50-mW range weigh up to about 16 kg. The highest-power lasers, which are not in common use, may have heads that are heavier and larger.

Most power supplies are boxes 10 to 15 cm high, 30 to 50 cm across, and about as deep as wide. Moderate-power versions weigh 10 to 20 kg.

Safety The main hazards posed by the 442-nm blue He–Cd line are to the retina of the eye. The eye appears to be more vulnerable to retinal damage from long-term exposure to the He–Cd line than to exposure to longer visible wavelengths, with the threshold perhaps an order of magnitude lower than for the 632.8-nm He–Ne line (Sliney and Wolbarsht, 1980, p. 136). Thus users should be particularly cautious, even though He–Cd power levels may appear low.

The 325-nm line lies in the ultraviolet-A region and presents different hazards. Some retinal damage has been produced by modelocked pulses at that wavelength (Sliney and Wolbarsht, 1980, p. 109), but standard measurements indicate that under 1 percent of the light at that wavelength entering the eye reaches the retina. The lens absorbs strongly at that wavelength, and there is some indication that long-term exposure to ultraviolet-A radiation

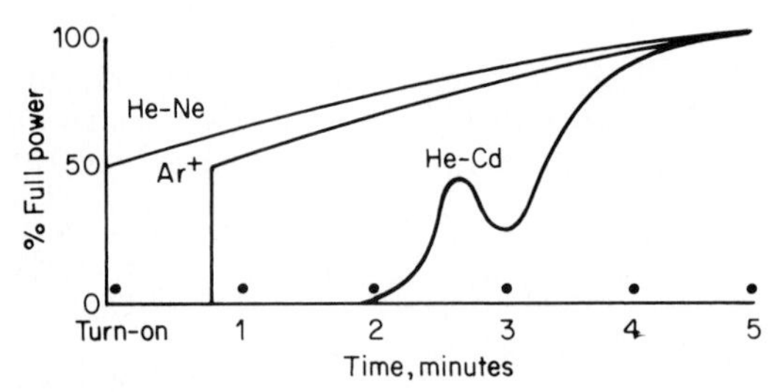

Figure 9-4 Types of power fluctuations seen in He–Cd lasers during initial warm-up, as compared to those of He–Ne and argon ion lasers. *(Courtesy of Omnichrome.)*

could contribute to cataracts. Neither He–Cd line presents serious hazards to the skin or other parts of the body except at power densities higher than normally produced by He–Cd lasers.

The high voltages—one to several kilovolts—applied across the He–Cd discharge tube are a potential shock hazard, but the direct-current levels are low enough that they do not pose severe electrocution hazards (although they can give one a nasty jolt). The high tube temperature needed to vaporize the metal presents a potential burn hazard, but in commercial lasers the hot tube is packaged safely out of reach.

Reliability and Maintenance

Lifetime Extensive work has gone into extending He–Cd laser tube lifetime (see, for example, Dowley, 1982). Commercial blue models are rated for 3000- to 6000-hour operating lifetimes, with ultraviolet models often rated for lifetimes about half as long. He–Cd output gradually degrades in normal operation at a rate of about 3 to 10 percent per 1000 hours. Manufacturers define tube failure as the point when it no longer meets published specifications, or when output power drops below a certain fraction of the specified level. This is not necessarily the point at which the customer can no longer use the tube.

Other components normally outlive the tube, so laser lifetime can be extended by simply replacing the tube.

Maintenance and Adjustments Some models may require periodic adjustment and/or cleaning of optics in the laser cavity external to the tube, while other types require no active maintenance during tube lifetime.

Mechanical Durability Like other glass tubes, He–Cd tubes are fragile, but they can be packaged to minimize that problem. One company specifies that its packaged lasers can survive a shock of 20*g*'s.

Failure Modes and Causes Several factors contribute to degradation of He–Cd tubes:

- Discharge contamination
- Growth of cadmium deposits on the beam path, bore, or optical surfaces
- Depletion of helium gas by cathode sputtering, burial under condensing cadmium metal, and wall diffusion

Tube designers have found various mechanisms to reduce these problems and extend tube lifetime, but they remain limits on operation.

Possible Repairs Because it is the tube that fails, the simplest way to fix a He–Cd laser is to replace the tube. Replacement tubes typically cost one-third to one-half the price of a complete new laser and generally can be installed on user premises.

Commercial Devices

Standard Configurations The standard configuration for He–Cd lasers is a laser head with separate power supply. The laser usually is factory-built to emit in the blue or ultraviolet, but if the optics are external to the tube they can be changed. Major manufacturers offer two parallel product lines which differ primarily in packaging—one intended for stand-alone use in the laboratory, the other for incorporation into larger laser-based systems by equipment manufacturers.

Options He–Cd lasers come with few options, with the main user choice being the selection of wavelength. Multimode 442-nm operation at powers higher than available in TEM_{00} mode is optional on some models. Other possible options include output-power stabilization and standby mode.

Pricing As of 1985, a low-power 3-mW He–Cd laser emitting at 442 nm would sell for about $3750, while a model emitting 50 mW at the same wavelength would carry a price tag in the $12,500 range. Lasers emitting on the 325-nm line tend to be both more expensive and less powerful. Replacement tubes cost about half as much as a new laser for low-power models, with the fraction dropping to about one-third for higher-power models.

Suppliers Somewhat over a thousand He–Cd lasers are sold around the world each year, with sales divided among a handful of manufacturers. The two major suppliers are in the United States; there are a couple of others overseas. Entry to the field is limited by the need for expertise to achieve long tube lifetimes. Despite the small number of manufacturers, there is active competition, both among makers of He–Cd lasers, and between them and makers of argon ion and He–Ne lasers.

Applications

Uses of helium-cadmium lasers fall into three major categories: information-handling systems, inspection and measurement, and research.

The largest numbers of He–Cd lasers are used in information handling because of the laser's combination of short wavelength and moderate output power. Because the response of many light-sensitive materials drops off sharply with increasing wavelength, a few milliwatts at 442 nm may do the job of tens of milliwatts at the 633-nm He–Ne line. In addition, the shorter-wavelength light can be focused onto a smaller spot, allowing higher-density recording and finer resolution. Argon ion lasers offer the same wavelength advantages but are more expensive than He–Cd, and their higher power is not needed for many applications.

The major information-handling application of He–Cd lasers has been in printing, where the laser scans a photoconductive drum, writing images that are transferred to paper by an electrostatic toner process very similar to that used in photocopiers. He–Cd lasers are used mainly in high-speed, high-performance printers which can churn out up to 100 pages a minute and can cost up to a few hundred thousand dollars. (Less-expensive semiconductor or He–Ne lasers are used in lower-performance laser printers.) He–Cd lasers also are used in typesetting systems, where the tightly focused laser spot writes fully formed characters on light-sensitive paper with resolution so high that no spots are visible. He–Cd lasers also are used in color separation for printing, where they create the halftone patterns used to generate the images printed in different colors to produce a full-color image.

Another fast-growing information-processing application is optical recording, either of digital data or of audio or video signals. He–Cd lasers can master original audio- and videodisks, which are reproduced by sophisticated pressing techniques. The small spot size possible with He–Cd lasers makes them a candidate for high-performance optical data storage systems, although the vast bulk of optical disk data storage will rely on less-costly semiconductor lasers.

He–Cd laser wavelengths are particularly valuable for certain measurement and inspection applications. He–Cd lasers are used in some inspection of printed circuits. Inspection of U.S. paper currency is another application, because the special paper on which the currency is printed has a unique fluorescence response at the He–Cd line. The ultraviolet line can be a valuable light source for fluorescence measurements.

Research applications cover a broad range, with many involving fluorescence induced by the 325-nm ultraviolet line. Another application is holography, for which the long coherence length of the single-line He–Cd laser is a better choice than the multiline output of argon lasers.

BIBLIOGRAPHY

William B. Bridges: "Ionized gas lasers," in Marvin J. Weber (ed.), *CRC Handbook of Laser Science & Technology*, vol. 2, *Gas Lasers*, CRC Press, Boca Raton, Fla., 1982, pp. 17–269.

Mark W. Dowley: "Reliability and commercial lasers," *Applied Optics 21*:1791–1795, May 15, 1982.

K. Fujii, T. Takahashi, and Y. Asami: "Hollow-cathode type CW white light laser," *IEEE Journal of Quantum Electronics QE-11*:111–114, March 1975.

M. Otaka et al.: "He–Cd^{+} whitelight laser by a novel tube structure," *IEEE Journal of Quantum Electronics QE-17*(3):414–417, March, 1981.

Wakeo Sasaki, Hideshi Ueda, and Tatehisa Ohta: "Spectral quality of He–Zn II and He–Cd II hollow-cathode metal-vapor lasers in the magnetic fields," *IEEE Journal of Quantum Electronics QE-19*(8):1259–1269, August 1983.

Wolfgang K. Schuebel: "Transverse discharge slotted hollow-cathode laser," *IEEE Journal of Quantum Electronics QE-6*:574–575, September 1970.

William T. Silfvast, "Metal-vapor lasers," *Scientific American 228*(2):89–97, February 1973.

William T. Silfvast, G. R. Fowles, and B. D. Hopkins: "Laser action in singly ionized Ge, Sn, Pb, In, Cd, and Zn," *Applied Physics Letters 8*:318–319, 1966.

David Sliney and Myron Wolbarsht, *Safety with Lasers and Other Optical Sources*, Plenum, New York, 1980.

S. C. Wang and R. D. Reid: "Parametric performance of a hollow-cathode white laser," Paper WK3 at Conference on Lasers and Electro-Optics, Washington, D.C., June 1981.

CHAPTER

10

carbon dioxide lasers

The carbon dioxide laser is one of the most versatile types on the market today. It emits infrared radiation between 9 and 11 micrometers (μm), either at a single line selected by the user or on the strongest lines in untuned cavities. It can produce continuous output powers ranging from well under 1 W for scientific applications to many kilowatts for materials working. It can generate pulses from the nanosecond to millisecond regimes. Custom-made CO_2 lasers have produced continuous beams of hundreds of kilowatts for military laser weapon research (Hecht, 1984) or nanosecond-long pulses of 40 kilojoules (kJ) for research in laser-induced nuclear fusion (Los Alamos National Laboratory, 1982).

This versatility comes from the fact that there are several distinct types of carbon dioxide lasers. While they share the same active medium, they have important differences in internal structure and, more important to the user, in functional characteristics. In theory, the structural variations could range over a nearly continuous spectrum, but manufacturers have settled on a few standard configurations which meet most user needs. Thus users see several distinct types, such as waveguide, low-power sealed-tube, high-power flowing-gas, and pulsed transversely excited CO_2 lasers. This chapter covers all these major types.

Internal Workings

The active medium in a CO_2 laser is a mixture of carbon dioxide, nitrogen, and (generally) helium. Each gas plays a distinct role.

Carbon dioxide is the light emitter. The CO_2 molecules are first excited so they vibrate in an asymmetrical stretching mode. The molecules then lose part of the excitation energy by dropping to one of two other, lower-energy vibrational states as shown in Fig. 10-1. These two decay paths are the two principal laser transitions: a shift to a symmetrical stretching mode accompanied by emission of a 10.6-μm photon, or a shift to a bending mode accompanied by emission of a 9.6-μm photon. Superposition of changes in the molecules' rotational states on the vibrational transitions yields large families of laser lines surrounding the 9.6- and 10.6-μm transitions. Once the molecules have emitted their laser photons, they continue to drop down the energy-level ladder until they reach the ground state.

The nitrogen molecules help to excite CO_2 to the upper laser level. The lowest vibrational state of N_2 is only 18 inverse centimeters lower in energy than the asymmetric stretching mode of CO_2, a difference that is less than one-tenth the mean thermal energy of room-temperature molecules, and hence insignificant from a practical standpoint. This lets the nitrogen molecules absorb energy and transfer it to the carbon dioxide molecules, thereby raising them to the upper laser level.

Carbon dioxide molecules can also reach the upper laser level in other ways. They can directly absorb energy from electrons inserted into the gas in a discharge or electron beam. An alternative way of producing the population inversion needed for laser operation is to rapidly expand hot, high-pressure laser gas into a cool near-vacuum; this is the basic principle behind the gas-dynamic carbon dioxide laser. In practice, the presence of N_2 significantly enhances laser operation, and that gas is almost always present in CO_2 lasers.

Helium plays a dual role. It serves as a buffer gas to aid in heat transfer and helps the CO_2 molecules drop from the lower laser levels to the ground state, thus maintaining the population inversion needed for laser operation.

The optimum composition and pressure for the gas in a CO_2 laser varies widely with the laser design. In a typical flowing-gas CO_2 laser, the total

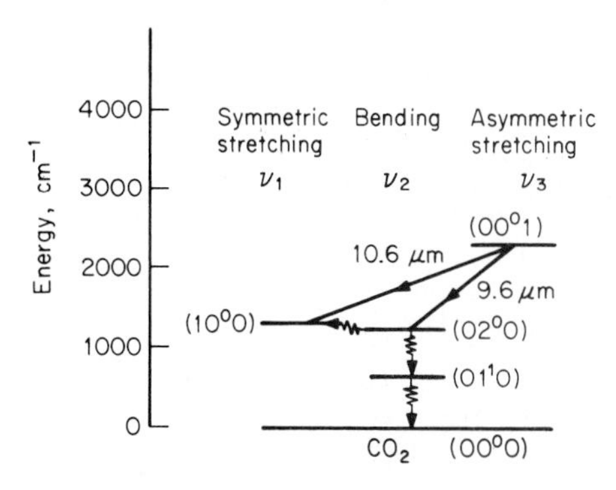

Figure 10-1 Energy-level structure in the CO_2 laser, showing the relevant vibrational modes of the CO_2 molecule. Numbers in parentheses indicate the excitation levels of the symmetric stretching, bending, and asymmetric stretching vibrational modes, respectively, of the molecule.

pressure might be around 15 torr (2000 Pa), with 10 percent of the gas CO_2, 10 percent N_2, and the balance helium. In general, the concentrations of nitrogen and carbon dioxide are comparable, but much lower than that of helium. Low pressures are needed for continuous operation, but pulsed CO_2 lasers can be operated at pressures well above 1 atm.

In some cases, other gases may be added to the laser mixture. For example, hydrogen or water can be added to the gas in a sealed tube to promote regeneration of CO_2 during laser operation, and sometimes carbon monoxide is added for similar reasons. It is even possible to operate pulsed lasers with a 50:50 mixture of air and carbon dioxide, albeit at reduced output power.

Internal Structure

The classification of carbon dioxide lasers into types is based on their internal structure. There are several key parameters involved, including gas pressure, gas flow, type of laser cavity, and excitation method. Many different combinations have been explored in the laboratory, but only a few have found their way into practical use.

- *Sealed-tube lasers:* A sealed gas laser sounds simple to operate: just fill the tube with the proper gas mixture, seal it, and fire away. In practice, life is not that simple because the electrical discharge in the tube breaks down the CO_2. Left by itself, a sealed CO_2 laser with an ordinary gas mixture would stop operating within a few minutes.

One solution is to add hydrogen or water to the gas mixture, so it could react with the carbon monoxide produced by the discharge to regenerate carbon dioxide. Alternatively, a 300°C nickel cathode can act as a catalyst to stimulate the recombination reaction. Such measures make it possible to produce sealed CO_2 lasers which can operate for as long as several thousand hours before their output seriously degrades.

Output powers of sealed CO_2 lasers generally are limited to the hundred-watt range by a couple of problems. One is that output power is inherently limited to only around 50 W per meter of tube length. The other is the difficulty in properly cooling the laser tube without flowing gas. Several companies make sealed CO_2 lasers, some of which incorporate the waveguide design described below.

- *Longitudinal (or axial) flowing gas lasers:* The obvious way to solve the problems of the sealed CO_2 laser is to flow the gas through the laser tube, as shown in Fig. 10-2. The oldest approach is to pass the gas through the length of the laser tube longitudinally or along the tube's axis, hence the name. Generally the electric discharge that excites the gas is also applied along the tube's axis. The gas pressure is low, and gas consumption can be reduced further by recycling options on many lasers.

Axial-flow CO_2 lasers produce continuous-wave output that is roughly linearly proportional to the tube length. Typical output limits are 40 to 80 W/m. The laser beam can be folded or bent with mirrors through multiple tube segments, avoiding the need for unwieldy packages, and the design is simple enough

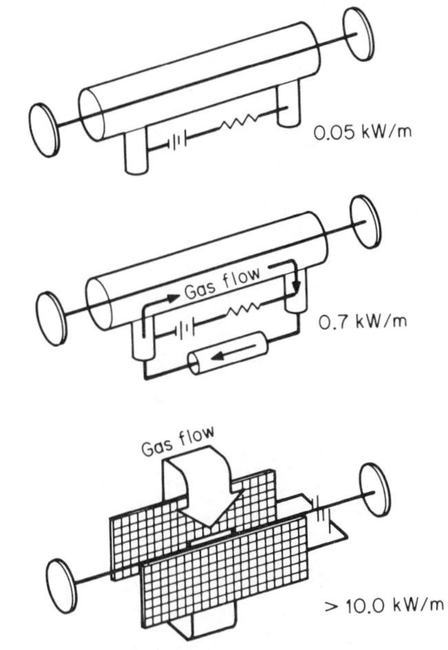

Figure 10-2 Basic structures of *(a)* sealed-discharge CO_2 laser, *(b)* axial-flow laser, and *(c)* transverse-flow laser. Numbers are output power per meter of tube length. *(Courtesy of Combustion Engineering.)*

that it remains common for CO_2 lasers emitting less than a couple of kilowatts. Higher powers are impractical, however; obtaining 8.8-kW output required a tube 750 ft (250 m) long! (Horrigan et al., 1969)

- *Transverse-flow lasers:* Much higher powers, on the order of 10 kW per meter of active medium length, are possible if the gas flows in a direction perpendicular to the laser cavity axis, as shown in Fig. 10-2. The electrical discharge that powers the laser is also applied transversely to the laser axis and is perpendicular to the gas flow. The gas flows much faster than in an axial-flow laser, quickly removing excess heat and dissociation products. The gas is generally recycled by passing it through a system which regenerates CO_2 and adds some fresh gas to the mixture. The cavity length of transverse-flow lasers is comparatively short, and this can lead to beam-quality problems.

The transverse-flow design is standard for most commercial multikilowatt CO_2 lasers. Aerodynamics plays a critical role in performance. A drawing of the largest CO_2 laser offered as a standard product, a 15-kW system built by Combustion Engineering's laser group in Somerville, Massachusetts, identifies one key component as a wind tunnel. Variations on the transverse-flow design have been used in some high-power CO_2 lasers custom-made for laser-weapon research.

- *Gas-dynamic lasers:* Transverse flow is also used in another type of high-power CO_2 laser, the gas-dynamic laser, shown in Fig. 10-3. In the gas-dynamic laser, the excitation energy comes from heat applied to the laser gas, which is initially at a pressure of several atmospheres. (Both the heat and some components of the laser gas may come from combustion of hydrocarbon fuels.) The hot gas is then expanded through a nozzle into a low-pressure chamber. The rapid cooling of the fast-moving gas produces a population inversion—

i.e., more CO_2 molecules in the upper laser level than in the lower one. A laser beam is extracted from the gas by placing a pair of mirrors on opposite sides of the expansion chamber.

At the end of the 1960s, the gas-dynamic laser was an important breakthrough that made it possible for the first time to reach power levels of 100 kW or more. Such powers are required only for military applications, and because of their complexity and high power, gas-dynamic lasers have never entered the commercial world. It now looks as if they are impractical for military field use as well.

- *Waveguide lasers:* If the inner diameter of a CO_2 laser tube is shrunk to a couple of millimeters and the tube is constructed in the form of a dielectric waveguide, the result is a "waveguide" laser such as is shown in Fig. 10-4. The waveguide design limits diffraction losses that would otherwise impair operation of a narrow-tube laser. The tube can be sealed (with a gas reservoir separate from the waveguide itself) or allow for flowing gas. The gas can be excited by an electrical discharge or by an intense radio-frequency field that can pass through the dielectric waveguide material and hence avoid the need for metal inside the waveguide structure.

The waveguide laser is very attractive for powers on the low end of the CO_2 range, continuous powers from under a watt to about 50 W. It provides a good-quality, continuous-wave beam and can readily be tuned to many discrete lines in the CO_2 spectrum. Its most conspicuous advantage is its small size, comparable to that of a helium-neon laser. Waveguide CO_2 lasers are inexpensive, starting at a few thousand dollars, and can cost as little as a few hundred dollars per watt for higher-power models bought in quantity.

- *TEA lasers:* Discharge instabilities make continuous-wave operation of a transversely excited CO_2 laser impractical at gas pressures above about 100 torr (13.3 MPa). However, it is possible to produce pulses lasting tens of nanoseconds to microseconds. Such lasers are called *transversely excited atmospheric* (TEA) lasers because they operate at or near atmospheric pressure, although sometimes the term is applied to pulsed transversely excited CO_2 lasers which operate at higher or lower pressures.

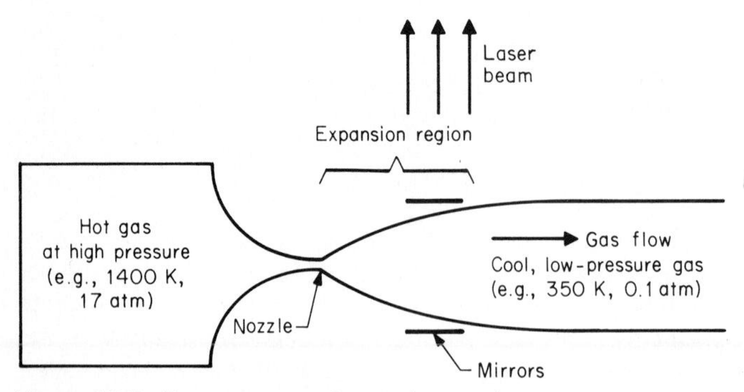

Figure 10-3 Basic structure of a gas-dynamic laser.

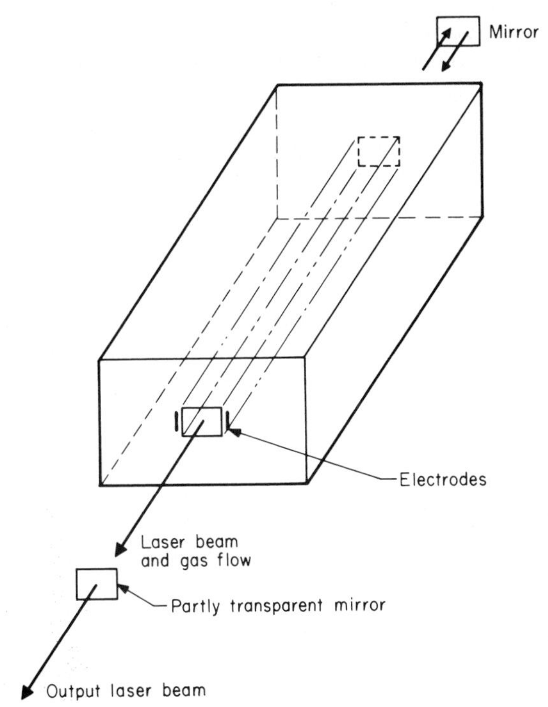

Figure 10-4 Basic structure of a waveguide CO_2 laser.

The TEA lasers' prime attractions are the generation of short, intense pulses and the extraction of high power per unit volume of laser gas. High-pressure operation also broadens the laser's emission lines, permitting the use of modelocking techniques to generate pulses lasting only about 1 ns. At pressures of around 10 atm, the broadening is sufficient to allow near-continuous tuning over most of CO_2's wavelength range.

The basic TEA design is adaptable, and commercial models range in size from tabletop versions to dumpster-sized behemoths. The same basic design can be used with a variety of laser gases, and some companies offer "multigas" lasers which can be adapted for use as carbon dioxide, excimer, chemical, or carbon monoxide lasers simply by switching optics and gases. Some models intended for low-power operation can operate with sealed tubes, a feature that is particularly important when using gas fills of special isotopic composition.

Optics Low- and moderate-power carbon dioxide lasers typically have the usual pair of laser mirrors, one totally reflecting and one partially transparent. There are a couple of basic variations on this design. One is the use of a pair of metal mirrors, which have high power-handling capacity; one mirror reflects all the incident light back into the laser cavity, while the other is cut so that it lets some light escape to form the laser beam. Another variation is the use of Brewster angle windows on the ends of the discharge tube, allowing the use of external optics which can configure the laser as an oscillator or an amplifier (i.e., to amplify the output from another laser).

The transmissive materials used as windows for the 10-μm output of carbon dioxide lasers tend to be very different from ordinary optical glass. Many are not transparent to visible light. Some of the materials most transparent at 10 μm are hygroscopic alkali halides such as sodium chloride, which absorb atmospheric moisture. Other infrared window materials are toxic.

Problems become particularly serious at high powers, where low absorption and high thermal capacity are critical. To avoid having the laser beam pass through transparent materials, engineers have developed "aerodynamic" windows. These are holes in the laser cavity, by which gas flows rapidly enough that air cannot enter to contaminate the laser medium. Such aerodynamic windows are used in the 15-kW laser.

Cavity Structure and Length The types of laser cavity structure have been described above. Some longitudinal-flow lasers can have very long cavity lengths, containing many separate segments connected optically in a folded arrangement that lets the laser package be much shorter than the cavity length. Cavity lengths for most other types are on the order of a meter, depending on power, and in some cases the cavity may be folded. Transverse-flow designs also have cavity lengths on the order of a meter, but as mentioned above yield much higher power per unit length.

Inherent Trade-offs

In addition to the usual trade-offs encountered with most types of lasers, there are a few specific to carbon dioxide lasers.

- High powers are available only at certain wavelengths, where the laser transitions are strongest. Other transitions in the laser emission region have lower power. Generally the highest powers are obtained from cavities which do not attempt to select a specific wavelength; emission from such lasers normally is dominated by the strongest transition wavelengths.
- In TEA lasers, generally the faster the repetition rate, the lower the energy and peak power per pulse, when the laser is operated at its peak ratings. The average power usually decreases above a certain repetition rate.
- In many types of CO_2 lasers, cavity length increases with output power.

Beam Characteristics

Wavelength and Output Power The nominal operating wavelength of the CO_2 laser is usually written as 10 μm, 10.6 μm, or 9 to 11 μm. The actual emission spectrum is complex. The CO_2 laser has two principal vibrational transitions, at 9.6 and 10.6 μm. Many closely spaced rotational transitions are superimposed on the vibrational transitions, producing a total of roughly 100 possible distinct emission lines, as shown in Fig. 10-5.

A diffraction grating or other tuning element can be put into the laser cavity to select a single line in the CO_2 spectrum, or the laser can operate untuned, emitting a higher total power on multiple lines. Because laser amplification builds up the power on the strongest emission line, multiline lasers are often said to operate at 10.6 μm, the strongest transition, although other wavelengths are also present.

Single-line output is not available or required on lasers designed to provide high power for materials working, but it is available on lower-power continuous-wave lasers and on many TEA models. Single-wavelength output on the strongest CO_2 line is generally about half the multiline output. There is some additional loss of power in tuning to other lines, but not as much as might be expected. Depending on the laser model, some 60 to 80 lines can emit at least half as much power as the strongest wavelength in single-line operation. Increasing the gas pressure to about 10 atm broadens the individual lines enough to permit continuous tuning of TEA lasers across much of the CO_2 emission range. Because the laser transitions are vibrational, their wavelength can be changed by substituting different isotopes into the carbon dioxide molecule.

Maximum output (typically measured in multiline operation) depends on the type of CO_2 laser. Waveguide models can produce continuous-wave output up to about 50 W, while commercial sealed CO_2 lasers can reach about twice that level. Axial-flow lasers can reach the kilowatt range, while transverse-flow lasers can produce many kilowatts. (Gas-dynamic lasers also operate in the multikilowatt range but are not available commercially.) Average powers of TEA lasers range up to a few hundred watts.

Temporal Characteristics of Output Outside of TEA lasers, all types of CO_2 lasers normally produce a continuous beam. However, pulsing such lasers can produce a power "spike," desirable for many materials-working applications, lasting 0.1 to 1 ms with peak power 5 to 10 times the continuous-wave level.

Commercial TEA lasers can deliver multiline pulses from the millijoule range to about 75 J at rates ranging from single-shot to 300 per second. Shortest pulses last around 40 ns, but pulse durations of a microsecond or longer are possible with suitable adjustments of the gas mixture and electrical

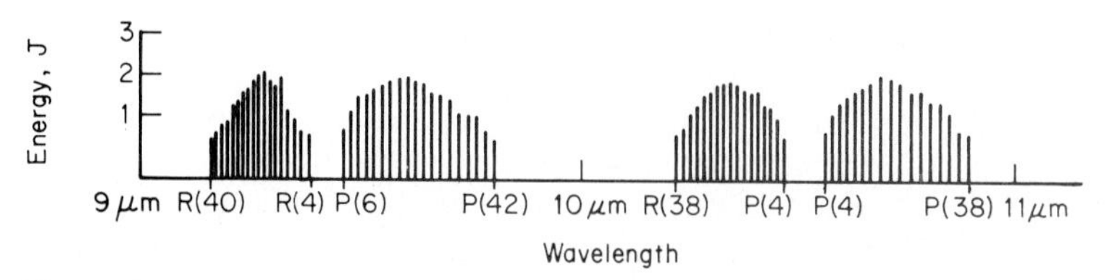

Figure 10-5 Spectrum of wavelengths produced by a transversely excited atmospheric pressure CO_2 laser, showing pulse energy emitted at each line in single-line operation by a commercial laser. *R* and *P* denote rotational sublevels for each of the two main vibrational transitions of the carbon dioxide laser. *(Courtesy of Lumonics Inc.)*

discharge. There are typically two elements to the pulse, a short peak arising from direct electrical excitation of the gas, and a longer-duration falloff (of lower power) as a result of transfer of energy from excited N_2 to CO_2 molecules which then emit light.

Because of the high pressure of the gas in TEA lasers, the gain bandwidth is about 4 gigahertz (GHz), enough to allow the use of modelocking techniques to generate subnanosecond pulses. (The theoretical limit on modelocked pulse duration is that it should be inversely proportional to gain bandwidth; thus 4-GHz bandwidth theoretically permits generation of quarter-nanosecond pulses.) A certain jitter in pulse timing (generally around 20 to 30 ns) and a small misfire rate (generally well under 1 percent) are inherent in the electronics of TEA lasers. Pulses can be triggered remotely or internally in most models.

Efficiency The overall efficiency of carbon dioxide lasers typically runs between 5 to 20 percent, not good when compared to other types of electrical equipment, but higher than most other lasers. Efficiencies are lower if operation is on a single line, if the electrical power supply is inefficient, or if the optics are particularly inefficient in extracting energy from the laser cavity. Users concerned with efficiency should make sure that it is defined explicitly—wall-plug efficiency is always lower than efficiency in converting energy deposited in the active medium into laser emission.

Spectral Bandwidth and Frequency Stability The spectral bandwidth is irrelevant to people using carbon dioxide lasers as heat sources. With multiline output, the bandwidth is large—although limited to the 9- to 11-μm emission of the laser—and unspecified in manufacturers' literature. The natural bandwidth of a single emission line is on the order of 100 megahertz (MHz) at low gas pressures, but it broadens as pressure increases until eventually the lines overlap. At such high pressures, the laser operates only in pulsed mode. Sophisticated optical cavities can reduce the spectral bandwidth well below the natural linewidth of a single emission line.

Lasers tunable to individual transitions in the CO_2 emission band normally can operate stably on that transition at low pressures. Accessories can boost frequency stability to between one part in a million and one part in a billion for continuous-output lasers. Similar stabilities can be achieved with pulsed lasers operating at higher pressures if suitable optics are used.

Amplitude Noise and Pulse-to-Pulse Variations Power levels of both pulsed and continuous-wave carbon dioxide lasers are subject to fluctuations which can be reduced by stabilization techniques. Generally materials-working lasers have fluctuations in continuous output limited to a few percent, with some models including active stabilization to provide even tighter control. Unstabilized waveguide lasers can have amplitude variations of ± 10 percent over a few hours and ± 4 percent on a scale of minutes, but better results can be obtained with stabilization. Pulse-to-pulse variations for repetitively pulsed TEA lasers are often specified at within a few percent, but variations could be larger for lasers producing high-energy pulses at low repetition rates.

Grating-tuned lasers can be made more stable than multiline lasers; for example, specifications of one model claim power can stay constant to ± 0.25 percent over several hours.

Beam Polarization, Quality, and Modes The design of the laser cavity and the resonator optics determine the modes in which a laser can oscillate and these modes, in turn, play a major role in determining beam quality. There are four basic transverse modes of emission common in carbon dioxide lasers:

- *Multimode emission,* in which the laser cavity simultaneously supports many different modes of oscillation. This extracts the most output power from a laser cavity, but the beam tends to be large in diameter and rapidly diverging, with some unevenness in quality.
- *TEM_{00} emission,* in which the laser cavity allows oscillation only in its fundamental (TEM_{00}) mode—producing an intensity pattern with a peak in the center and a decline in intensity to the edges. The lower intensity at the edges limits the amount of energy that can be extracted from the laser cavity, but the resulting beam is of better quality than a multimode beam. Although there can be exceptions, a small-diameter laser operating in TEM_{00} mode typically will emit a quarter to a half as much power as it would in a multimode beam, and the beam diameter and divergence will be reduced proportionately.
- *Unstable-resonator emission,* in which the cavity optics produce a beam with a doughnutlike cross section. Despite the name, lasers with such cavities can operate stably. They are attractive for CO_2 lasers because they offer good beam quality while extracting more energy from the active medium than is possible with TEM_{00} mode emission, particularly for a large-diameter laser. Output power with an unstable resonator typically is one-half to two-thirds that of a multimode beam. The unstable-resonator design also permits use of totally reflecting metal mirrors which can handle high levels of optical power.
- *Waveguide lasers* have a mode structure determined by the nature of the waveguide rather than by the resonator mirrors. The lowest-order EH_{11} waveguide mode is functionally the same as the TEM_{00} mode of non-waveguide lasers once the beam emerges from the laser. The laser cavity's small diameter allows efficient extraction of energy from the laser gas.

If a continuous-wave CO_2 laser is tuned to emit at a single wavelength, normally it will emit in a single longitudinal mode. This occurs even without internal line-narrowing optics because the long laser wavelength leads to spacing of longitudinal modes that is broader than the gain bandwidth on low-pressure CO_2 lines. A conventional stable-resonator CO_2 laser emitting a single line produces a TEM_{00} beam; unstable-resonator and waveguide CO_2 lasers also can be tuned to a single line.

Multiline CO_2 lasers normally emit randomly polarized light unless their tube designs incorporate Brewster angle windows. The optics used for grating tuning impose linear polarization on the beam (although the angle of polarization may change as wavelength is changed), so single-line lasers emit linearly polarized beams.

Coherence Length Coherence length generally does not matter for most CO_2 laser applications. The broad emission bandwidth of multiline lasers leads to a very short coherence length. Single-line CO_2 lasers have coherence lengths on the order of a meter, or longer if line-narrowing accessories are used.

Beam Diameter and Divergence The shape and length of the laser cavity and the nature of the resonator optics determine beam diameter and divergence. Thus these quantities differ among CO_2 laser types. Typical ranges are

- Axial-flowing gas: 5- to 70-mm diameter, 1- to 3-mrad divergence
- Sealed tube: 3- to 4-mm diameter, 1- to 2-mrad divergence
- TEA: 5- to 100-mm diameter, 0.5- to 10-mrad divergence
- Waveguide: 1- to 2-mm diameter, 8- to 10-mrad divergence

In general, except for waveguide lasers, beam diameter and divergence in single-mode operation are 20 to 50 percent of the values for multimode output. Some large-diameter laser tubes produce beams with oval cross sections.

Stability of Beam Direction Pointing stability is unmentioned in most data sheets, but one company specifies stability of ±0.15 mrad for its 150- to 1200-W axial-flow lasers, which are designed for materials-working applications.

Suitability for Use with Laser Accessories Carbon dioxide lasers are versatile enough to be used with a number of accessories. However, their wavelength is enough longer than those of other major lasers that they require special components, and many CO_2 laser applications do not require special accessories. Important accessories for CO_2 lasers include:

- *Q* switches to produce microsecond or nanosecond pulses from continuous-wave beams
- Shutters or pulse choppers to produce longer pulses from continuous-wave beams
- Modelockers to produce nanosecond pulses from lasers which operate at pressures high enough (near atmospheric pressure) to produce the required broadening of laser lines
- Harmonic generators to produce integral multiples of the laser frequency (or equivalently to reduce the wavelength by an integral factor) remain essentially at the laboratory stage
- Optically pumped far-infrared lasers, in which the energy from a carbon dioxide laser is used to produce laser emission at a longer wavelength in the far-infrared or submillimeter region as described in Chap. 15.

Operating Requirements

Input Power The input power for CO_2 lasers, like the output power, ranges over a few orders of magnitude. The smallest CO_2 lasers can plug into an ordinary wall socket; the largest consume prodigious amounts of power. All CO_2 lasers require high voltages—either continuous dc or periodic pulses—which must be generated from normal current supplies and applied to the laser gas. Typical requirements depend on the type.

Waveguide lasers generally draw 2 or 3 A from a 110-V ac source. In discharge-driven waveguide lasers, that energy goes to produce a current of a few milliamperes at about 15 kV for application to the gas. Alternatively, the current may power a radio-frequency transmitter operating at 20 to 30 MHz, which applies its energy to the dielectric waveguide structure.

Sealed-off CO_2 lasers of conventional design require somewhat higher input power to produce comparable output. Some models require dc currents of 6 to 10 mA and potentials of 8 to 20 kV to produce outputs of 1 to 18 W.

Continuous-wave flowing-gas CO_2 lasers require peak electrical input capacity in the range of 10 to 20 times their optical output, and may require extra startup power. This energy goes to power vacuum pumps, cooling equipment, and other accessories as well as the laser tube itself. Some models can operate from single-phase 110-V ac supplies, but large versions require 440-V, three-phase sources. One 1.2-kW laser, for example, requires 22-kV·A capacity from a 440-V supply, which reduces to 15 kV·A after startup.

TEA lasers have fairly complex power requirements because of the nature of their pulsed operation. Typically, some energy in the form of electrons or ultraviolet photons is discharged into the laser gas slightly before the main pulse to make it possible to obtain higher output power. Energy for the main pulse typically comes from a storage capacitor which accumulates a charge from a dc power supply operating at tens of kilovolts and tens of milliamperes. Some of the highest-power TEA lasers use more complex—and more expensive—pulse-generating electronics and may have overall voltages above 100 kV. Wall-current requirements for the charging power supply range from a couple of amperes at 110 V to 50 A of three-phase 220-V power.

Cooling Forced-air cooling is used in some small CO_2 lasers. Higher-power lasers require water cooling, and the largest devices have sophisticated multicycle cooling systems. Typical water-flow requirements are 2 L/min for a 150-W axial-flow laser and up to 40 L/min for a 1.2-kW model with a water-Freon-dielectric oil cooling system.

Consumables The prime consumable material used with CO_2 lasers is the laser gas. Even low-power models with sealed tubes require periodic gas replacement, on the order of every thousand hours of operation. Higher-power lasers require continuous replenishment of the laser gas. Gas consumption is roughly proportional to power level for lasers of the same design. For example, one continuous-output 150-W laser each hour needs 40 L of helium, 4 L of CO_2, and 7 L of N_2, against hourly consumption of 425 L He, 54 L CO_2, and 127.5 L N_2 for a 1.2-kW laser.

Some TEA lasers need dry air for spark-gap switching. Note also that cooling water may be needed, as indicated above.

Required Accessories Carbon dioxide lasers are offered in configurations ranging from complete materials-working systems to "kits" which contain little beyond the laser cavity itself. While complete systems normally are almost ready to run after being hooked up to power, water, and gas supplies, the less-complete systems or kits may also need high-voltage supplies, cavity optics, vacuum pumps, and other equipment.

Operating Conditions and Temperature Carbon dioxide lasers are designed to operate at room temperature and can function well in normally clean environments. Many models are intended for industrial use. Extremes of dirt and/or vibration in factories have caused problems in some cases, and such protective measures as air filters may be needed in particularly severe environments.

Mechanical Considerations Waveguide CO_2 lasers are impressively compact devices, with heads as light as a few kilograms and power supplies not much heavier. The smallest heads are shorter than 40 cm, and have square cross sections only about 5 cm across. The compactness of these devices has led to their use in portable military range finders.

Some TEA and most sealed-off continuous CO_2 lasers are compact enough to fit on a laboratory bench; a few are very compact. Flowing-gas and the largest TEA lasers are massive devices that weigh in at a few hundred kilograms to a few tons and require at least a few square feet of floor space. The biggest commercial CO_2 laser, a 15-kW metal-working system, takes up 1000 ft^2 (nearly 100 m^2) of floor space altogether.

Safety There are two significant hazards with CO_2 lasers: high voltages and the laser beam.

The tens of kilovolts needed to operate the laser are by far the more deadly hazard. In complete laser systems, these voltages generally are only accessible inside the laser cabinet, but in other cases (such as when an external high-voltage supply is used), there may be external connections.

The beams from virtually all CO_2 lasers fall under Class IV of the federal laser product standards. That means that the beams could, at least in theory, cause skin or eye burns (the 10-μm wavelength is not transmitted into the eye). The federal laser product code requires elaborate interlocks that make it hard for operators to expose themselves to the beam and for unauthorized persons to operate the laser. However, the invisible infrared beam does pose a potential hazard. Many materials-working CO_2 lasers are shielded by plexiglass enclosures which are transparent to visible light but absorb any stray laser light.

Special Operating Considerations Optics which transmit the 10-μm CO_2 laser beam are made of materials quite different from optical glass. Many materials transparent at 10 μm do not transmit visible light. Unlike ordinary optical glass, many 10-μm window materials are chemically reactive, hygroscopic (water-absorbing), and/or toxic, although suitable coatings can reduce their sensitivity. Ordinary glass totally absorbs 10-μm energy and can be cut with a CO_2 laser.

Discharging capacitors in TEA lasers may produce audible "pops" in normal operation. Such pulsed lasers also can produce electromagnetic interference and may require shielding to avoid interfering with other equipment. Radio-frequency discharges also require shielding to prevent interference with external equipment. Although commercial lasers generally incorporate shielding, users should check that it is sufficient.

Some lasers, particularly TEA types, are built for operation with several different gas mixtures. Thus a single laser cavity would function as a carbon dioxide laser when filled with a CO_2 gas mixture but could also function as other types—typically hydrogen fluoride, excimer, or carbon monoxide—when used with different gas mixtures and different optics. When gas mixtures are changed, the system should be purged before operation with the new mixture, and optics should be changed as required.

Reliability and Maintenance

Laser Lifetime Different factors limit operating lifetimes of different types of CO_2 lasers, including gas lifetime in sealed lasers and degradation of optics and power supplies in all types. Sealed continuous-output CO_2 lasers are designed to operate from one to several thousand hours on a single gas fill. TEA lasers can deliver millions of shots from a single gas fill. Output power drops gradually if the gas is not changed, but refilling (and in some cases cleaning) the tube should restore the laser to original output levels.

Lifetimes of flowing-gas CO_2 lasers normally are not specified, and in practice depend heavily on environmental factors, such as contamination of optics, which may require periodic replacement in materials-working applications.

Maintenance and Adjustments A few basic types of maintenance are common for CO_2 lasers

- Replacement of optics subject to degradation
- Replacement of gas in sealed-tube lasers after 1000 hours or more of operation
- Replacement of high-voltage components such as spark gaps
- Lubrication and other upkeep of essential accessories such as vacuum pumps

Many CO_2 lasers designed for industrial applications have preventive maintenance schedules and come with elapsed-time meters that indicate when maintenance is due.

Mechanical Durability Military contractors have built compact CO_2 lasers designed for field use as range finders, although they were not deployed when this book was written. Otherwise, CO_2 lasers are generally designed for use in a fixed location, not to withstand extensive vibrations and shock.

Failure Modes and Possible Repairs Degradation of the optics, the power supply, or (in sealed-tube lasers) the laser gas can all cause failure in CO_2 lasers. Repair is generally by replacing the affected components.

Commercial Laser Systems

Standard Configurations Outside of safety standards set by the federal government, there are no formal or de facto standards on the configurations of carbon dioxide lasers sold in the United States. Users can buy anything from a bare-bones laser kit to a complete system including beam-direction optics. There are several key elements of a CO_2 laser system which generally are packaged in one or more units:

- The laser cavity itself, including the housing for the laser gas and the electrodes which excite the gas
- The resonator or beam-forming optics
- The power supply, which in pulsed lasers includes two segments, a high-voltage dc supply and a pulse-generating capacitor or network. In continuous-wave lasers only the dc supply is present.
- A control panel
- Gas mixing and monitoring equipment
- Beam-directing optics, which take the beam from the laser to the place it will be used and generally are not included with the laser itself, but only in laser systems packaged for a particular application, such as materials working or surgery

Small lasers intended for laboratory use as well as general-purpose waveguide lasers sometimes are packaged as a single integrated unit containing laser cavity, resonator, and power supply. Larger laboratory lasers are generally packaged in two units, a laser head and a separate power supply. The laser head contains only what is needed on the laboratory bench; the power supply is usually mounted on wheels and rests on the floor. In many lasers designed for industrial applications, the laser head and power supply are housed in a single unit, but the control panel is mounted remotely. The laser may operate in an enclosure transparent to visible light but opaque to the CO_2 laser's 10-μm output.

One alternative approach under development is a compact laser head based on transverse gas flow, designed to deliver 80 to 800 W. This lets the water-cooled head be just 71 cm long and 46 cm in diameter, and weigh 54 kg, small enough to consider mounting on a movable arm. However, a bulky power supply and control cabinet is required, along with water-cooling facilities.

In many TEA lasers, the laser head includes an energy-storage capacitor which is discharged across the laser cavity to produce a laser pulse. The dc power supply which charges the capacitor is mounted remotely. Both TEA and continuous-wave lasers are sometimes offered in models designed for operation direct from an existing high-voltage dc power supply—a money-saving feature for labs which have an available dc supply.

Some companies offer a family of lasers based on a single modular design, with extra laser-cavity modules being added to increase output power. TEA lasers are often available with Brewster angle windows rather than resonator mirrors to allow them to be operated as amplifiers for the beam from a separate external CO_2 laser.

Options A wide range of options is available for CO_2 lasers. Some are intended for specific applications (e.g., materials working), while others simply enhance the laser's versatility or operation. Some of the most important are listed below.

- Materials-handling systems (for moving parts under the laser beam in industrial applications)
- Grating-tuning optics which select one line in the CO_2 emission spectrum
- Cavity optics to select TEM_{00} mode output, or to operate the laser with an unstable resonator
- Beam-delivery systems for surgery and materials working
- External triggering of pulses
- *Q* switches to produce pulses of 1 μs or less from continuous-wave lasers
- Modelockers to generate nanosecond pulses from atmospheric-pressure CO_2 lasers
- Gas cleaning and recycling to reduce gas consumption in flowing-gas lasers
- Temperature stabilization of output
- Power meters for monitoring output power
- Electronics to generate pulses in the 0.1 to 1 ms range from continuous-wave lasers
- Computerized controllers
- Gas-mixing and flow-control systems
- Premixed laser gases
- Brewster angle windows to permit operation of the laser cavity as an amplifier or with external optics

Some of these accessories are available from laser manufacturers, while others are offered by companies specializing in optics or laser accessories. Optical components for application outside the laser cavity are generally offered by companies which specialize in optics.

Pricing The prices of carbon dioxide lasers vary almost as widely as their performance. The values listed below are representative, as of this writing, of those quoted for particular types, with lower-priced versions generally having lower output and fewer options than the higher-priced versions. With the exception of low-power models, CO_2 lasers are almost invariably purchased in small lots, and quantity discounts generally are not mentioned.

- Waveguide: $2500 to $15,000
- Sealed tube: $5000 to $35,000
- TEA: $3000 to $90,000

- Flowing gas: $20,000 to around $1 million (The highest-power and most costly systems generally are built specifically for materials working and include some expensive accessories not normally found on other CO_2 lasers.)

Suppliers Most types of CO_2 lasers are offered by a few suppliers, although many suppliers manufacture only one type of CO_2 laser. Note, however, that specific performance requirements, particularly those on the borders of what is commercially available, may be met only by products of one supplier.

Applications Carbon dioxide lasers can serve many functions, from a comparatively inexpensive source of laser photons to an emitter of precise infrared wavelengths. This versatility has opened up many applications for CO_2 lasers in both industrial and laboratory environments. Some of the major applications of CO_2 lasers and their principal requirements are summarized briefly below.

- *Materials working.* CO_2 lasers are used for materials working, primarily cutting and welding of metals and nonmetals. Nonmetals such as plastics, cloth, and rubber are much more efficient absorbers than metals at 10 μm, so lower laser powers are required. Despite reflectivities that can exceed 90 percent at 10 μm, CO_2 laser metal working is growing because no other commercially available laser can generate as intense continuous-wave output. Cutting and welding generally require continuous-wave powers of 50 W and up, with much higher powers needed for most metals. CO_2 lasers are sometimes used for drilling, but that task is more often done with pulsed neodymium lasers, which produce high peak powers at about 1 μm.
- *Heat treating.* To alter the characteristics of metal surfaces, CO_2 lasers are used for heat treating, generally improving the durability of the metal. This requires a continuous beam, which for the sake of efficiency should contain a high power that can be spread over a large area.
- *Marking letters or symbols.* Pulsed TEA lasers can etch a stencil-like image onto a workpiece in a single (or sometimes multiple) shot, a noncontact technique for permanent marking that is used in a variety of industries. A pulsed laser can write alphanumeric characters as series of dots, but that is generally done with pulsed neodymium lasers.
- *Spectroscopy and photochemistry.* CO_2 lasers can generate narrow-band output in the 10-μm range for spectroscopic applications, and/or high powers for studies of laser-induced chemistry. TEA or waveguide lasers are the commonest types for these applications.
- *Range finding.* For military applications, use of waveguide CO_2 lasers as range finders is in the development and demonstration stages.
- *Laser radar or "lidar."* CO_2 lasers can be used in systems which generate 10-μm pulses and analyze the returns, much as an ordinary radar system in the microwave range. Alternatively, returns from the atmosphere can be analyzed spectroscopically to determine concentrations of various species in the atmosphere.
- *Surgery* can be performed with 10-μm powers in the 50-W range because water in living tissue absorbs that wavelength strongly. Continuous-wave lasers

are used, with waveguide lasers becoming increasingly popular because of their compact size, good beam quality, and low cost.

- *Generation of far-infrared or submillimeter radiation* by using a CO_2 laser to optically pump or energize another gas so that the second gas can emit a laser beam. The use of different gases makes it possible to generate many wavelengths in the far-infrared and submillimeter regions. Commercial far-infrared lasers are described in Chap. 15.
- *Plasma production* using an intense pulse from a TEA laser (often mode-locked) to deposit energy quickly onto a small spot. One specific application has been laser fusion research, although most work in that field is with shorter-wavelength lasers.
- *Directed-energy weapon* concept tests have been conducted with carbon dioxide lasers (Hecht, 1984). Pulsed CO_2 lasers are still considered viable contenders for some laser weapon applications, but beam-propagation problems are thought to make continuous-output CO_2 lasers unsuitable for field use.

BIBLIOGRAPHY

Coherent Inc. Engineering Staff: *Lasers: Operation, Equipment, Application and Design*, McGraw-Hill, New York, 1980. Despite the general title, the main emphasis is on materials working, particularly with carbon dioxide lasers.

W. W. Duley: *CO_2 Lasers: Effects and Applications*, Academic, New York, 1976.

W. W. Duley: *Laser Processing and Analysis of Materials*, Plenum, New York, 1983.

Jeff Hecht: *Beam Weapons: The Next Arms Race*, Plenum, New York, 1984.

F. Horrigan et al.: *Microwaves 8*:68, 1969, cited in W. W. Duley, *CO_2 Lasers: Effects and Applications*, Academic Press, New York, 1976.

Los Alamos National Laboratory: *Inertial Fusion Program Jan-Dec 1981*, Los Alamos National Laboratory, Los Alamos, N.M., Progress Report LA-9312-PR, 1982.

Aris Papayoanou: "CO_2 Lasers for Tactical Military Systems," *Lasers & Applications 1*(3):49–55, December 1982.

John F. Ready: *Industrial Applications of Lasers*, Academic Press, Orlando, Fla., 1978.

P. W. Smith et al.: "Transversely Excited Waveguide Gas Lasers," *IEEE Journal of Quantum Electronics QE-17*(7):1166–1181, July 1981.

Orazio Svelto: *Principles of Lasers*, 2d ed., Plenum, New York, 1982.

William L. Wolfe and George J. Zissis (eds.): *The Infrared Handbook*, Office of Naval Research, Arlington, Va., 1978.

CHAPTER

11

chemical lasers

Chemical lasers are those in which the excited light-emitting species is produced by a chemical reaction. The idea was first proposed at a 1960 conference by J. C. Polanyi, and later published in a journal (Polanyi, 1961), but the first chemical laser did not operate until a few years later (Kasper and Pimentel, 1965). Most chemical lasers operate on infrared vibrational transitions of diatomic molecules. The hydrogen halides are the most important family of chemical lasers, with the archetype and most practical member of the family being the hydrogen fluoride laser. Some chemical lasers rely on reactions which produce excited atomic species, with the most important of these the iodine laser, in which excited iodine atoms are produced by chemical or photochemical reactions.

The main attraction of chemical lasers is their ability to produce very high powers. This is largely because energy can be stored quite efficiently as chemical reactants: a kilogram of chemicals can release much more energy than a kilogram of storage capacitors. If the energy from a chemical reaction can be rapidly deposited in a laser cavity, and much of it can be converted into photons by stimulated emission, the result can be a laser with very high power. The potential for high powers led to extensive military research aimed at developing laser weapons, initially for tactical battlefield use and more recently for space-based ballistic missile defense (Hecht, 1984).

The most attention has gone to three types of chemical lasers: hydrogen fluoride, deuterium fluoride, and iodine. HF emits at 2.6 to 3.3 micrometers (μm) in the infrared, wavelengths where atmospheric absorption is strong. Substitution of deuterium (hydrogen 2) for normal hydrogen shifts wavelength to 3.5 to 4.2 μm, where atmospheric transmission is good, but efficiency is lower and costs are higher because of the need for the rare isotope. Iodine is an atomic laser that emits at a 1.3-μm electronic transition; that transition can be pumped by energy transfer from excited molecular oxygen produced in a chemical reaction (McDermott et al., 1978).

Building-sized demonstration lasers built for military weapons research, such as the one shown in Fig. 11-1, have achieved impressive power levels. According to unofficial reports, a DF laser known as the Mid-InfraRed Advanced Chemical Laser (MIRACL) has produced continuous powers over 2 megawatts (MW). A large HF laser known as Alpha, has been designed to produce 5 MW of continuous output. Chemical oxygen-iodine lasers are a newer technology and have yet to approach those power levels.

Although the vast bulk of investment in chemical laser technology has been for laser weapon research, lower-power continuous-wave and pulsed HF and DF lasers are available commercially. These lasers can be valuable sources of mid-infrared radiation for a variety of research applications, including materials interactions, chemistry, spectroscopy, and biomedicine. They differ in important ways from higher-power military lasers, and after describing the general principles of chemical lasers, this chapter will concentrate on the lower-power types.

Figure 11-1 A major chemical laser built for military demonstrations at San Juan Capistrano, Calif. The structures at left and center house the laser, which unofficial reports indicate could produce continuous powers of a few hundred kilowatts. *(Courtesy of Department of Defense.)*

Internal Workings

Operation of a chemical laser is shown schematically in Fig. 11-2. Input gases which provide hydrogen (or deuterium) and fluorine are passed through a set of nozzles, which mix them together in a region where they react to form HF or DF in a vibrationally excited state. The gas flows rapidly through the nozzles and continues flowing past the reaction zone, passing through an optical resonator cavity that is oriented perpendicular to the flow direction. It is in the laser resonator that infrared photons are extracted from the excited molecules by stimulated emission. The gas is then exhausted by a vacuum pump.

Active Medium Because of the nature of the chemical reaction, the input gases are not the same as the active medium from which the laser beam is extracted. Two reactants are required—one to donate hydrogen, the other to provide fluorine—and other diluent or scavenger gases may be added. Molecular hydrogen is widely used, although in some cases hydrocarbons may be used as H donors. Molecular fluorine is used in some experimental lasers, but F_2 is extremely hazardous, so other F sources are more common. NF_3 (nitrogen trifluoride) is used in some military demonstration lasers, but in commercial chemical lasers SF_6 (sulfur hexafluoride) is used because it is much safer and easier to handle.

In commercial HF lasers, an electric discharge breaks down SF_6 to produce free fluorine, which reacts with H_2 to produce excited HF. Oxygen is mixed with the SF_6 to improve discharge quality and to scavenge the sulfur produced by the discharge, forming SO_2. Helium is also added to the mixture flowing through the laser as a diluent gas. Pressure of the gas mixture flowing through the laser is a few torr (a few hundred pascals).

Energy Transfer Two types of energy transfer can occur in HF chemical lasers: the production of gases ready to react with each other, and the energy

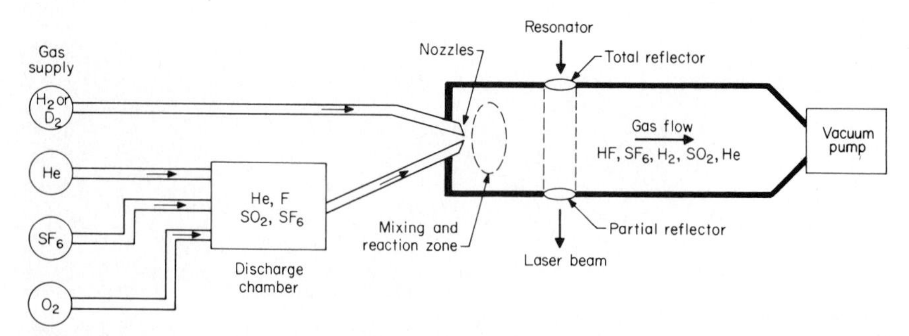

Figure 11-2 Basic elements of a commercial continuous-wave chemical laser include a gas supply, a discharge chamber that produces free fluorine, nozzles which mix the reactants, a mixing region, a laser resonator, and a vacuum pump to collect spent gas. Gas flow is from left to right; the laser beam is perpendicular to the gas flow.

transfer during and after the chemical reaction. In military research lasers, the reactants normally are produced separately, then fed into the laser. However, low-energy commercial lasers generate fluorine internally to minimize gas-handling problems, and similar techniques are used in some experimental lasers.

For continuous-wave lasers operating with SF_6, a direct-current discharge of about 8 kilovolts (kV) is needed to produce fluorine. (Higher discharge voltages or electron beams are needed to break down SF_6 in pulsed lasers because the gas has very high resistance.) The free fluorine atoms then react with hydrogen in a highly exothermic reaction, generating HF in a vibrationally excited state. If the input gases are H_2 and F_2, the laser can sustain a chain reaction

$$H_2 + F \rightarrow HF^* + H$$
$$H + F_2 \rightarrow HF^* + F$$
$$H_2 + F \rightarrow HF^* + H$$

ad infinitum. An electric discharge or electron beam may be used to start the chain reaction, or the laser can operate purely by chemical reaction. Where discharges are used, it is possible to speak of two types of efficiency—electrical and chemical—with electrical efficiencies in some special cases over 100 percent.

The initial energy of HF or DF molecules depends on the reaction that produces them, as indicated in Fig. 11-3. HF and DF molecules have a series of vibrational energy levels, with spacing decreasing at higher energy levels, meaning that the transition wavelength is longer at higher energy levels (e.g., from level 6 to level 5 than from level 1 to level 0). Each vibrational level is split into a number of rotational sublevels. Thus laser action is possible at many discrete wavelengths in a broad band, from 2.41 to 3.38 μm for HF and from 3.49 to 4.19 μm for DF (Chang, 1982).

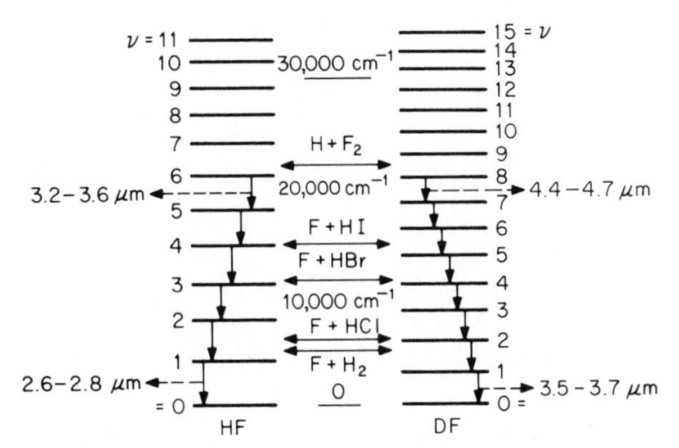

Figure 11-3 Vibrational energy levels of HF and DF, shown with the energies remaining with the HF or DF molecules after certain reactions. *(From Chester, 1976.)*

In practice, various selection processes tend to concentrate emission to the short-wavelength side of the center of those emission bands, even in lasers using the more energetic reactants. The H_2 + F reaction used in commercial HF or DF lasers is the least energetic and thus concentrates emission on lower vibrational transitions, but there are enough rotational levels to allow emission on many lines, as shown in Fig. 11-4.

The populations of various energy levels in pulsed HF and DF lasers are time-dependent. These changing populations can lead to changes in the relative strengths of different lines during the pulse.

Internal Structure

The internal structure of continuous-wave chemical lasers is unusual because even mixing of the chemical reactants as they flow into the reaction zone is crucial to laser performance. Much care must be given to design of the nozzles through which the reactants flow into the laser cavity, particularly in high-energy military lasers. As indicated in Fig. 11-2, after any required preparation (such as electrical breakdown of SF_6 to generate F), the reactants flow through nozzles into the reaction zone, where they react to product HF and continue flowing past the resonator optics and out to a vacuum pump. Gas flow in high-energy lasers may be supersonic, but in commercial models the flow is subsonic. Design of low- and moderate-power HF and DF lasers is generally similar, except that different optics are needed for the different wavelengths.

A rather different configuration can be used for low-power pulsed operation of HF or DF. In this case, a mixture of H_2 and a fluorine source (typically SF_6) is placed in a laser cavity, (optionally) preionized with ultraviolet light, then initiated with a strong electric discharge transverse to the axis of the laser cavity. For the typical case of SF_6, discharge voltage must be about 30 to 40 kV. The approach is similar to transverse excitation of TEA carbon dioxide lasers, and some designs are based on TEA CO_2 lasers. The spent gas is exhausted from the laser cavity, but because only a small fraction is consumed, the HF can be removed and the rest recycled through the laser. Careful removal of the reacted gases and addition of small amounts of SF_6 and H_2 can allow operation of the laser for several hours with stable output from one gas fill.

Optics The resonator on a chemical laser is defined by a pair of mirrors on opposite sides of a volume through which the reacted gases flow. The laser cavity axis is transverse to the flow of the gas, and somewhat "downstream" of the reaction zone. Although diagrams such as Fig. 11-2 typically show a significant distance between reaction zone and optics, this is only an artistic convenience and is not a significant factor in determining dimensions of commercial chemical lasers.

Typically, continuous-wave commercial chemical lasers couple output from the laser cavity using Brewster angle windows of zinc selenide or calcium fluoride. For multiline emission, the resonant cavity is defined by a totally reflecting rear mirror and a partly transparent front mirror. Gain is high

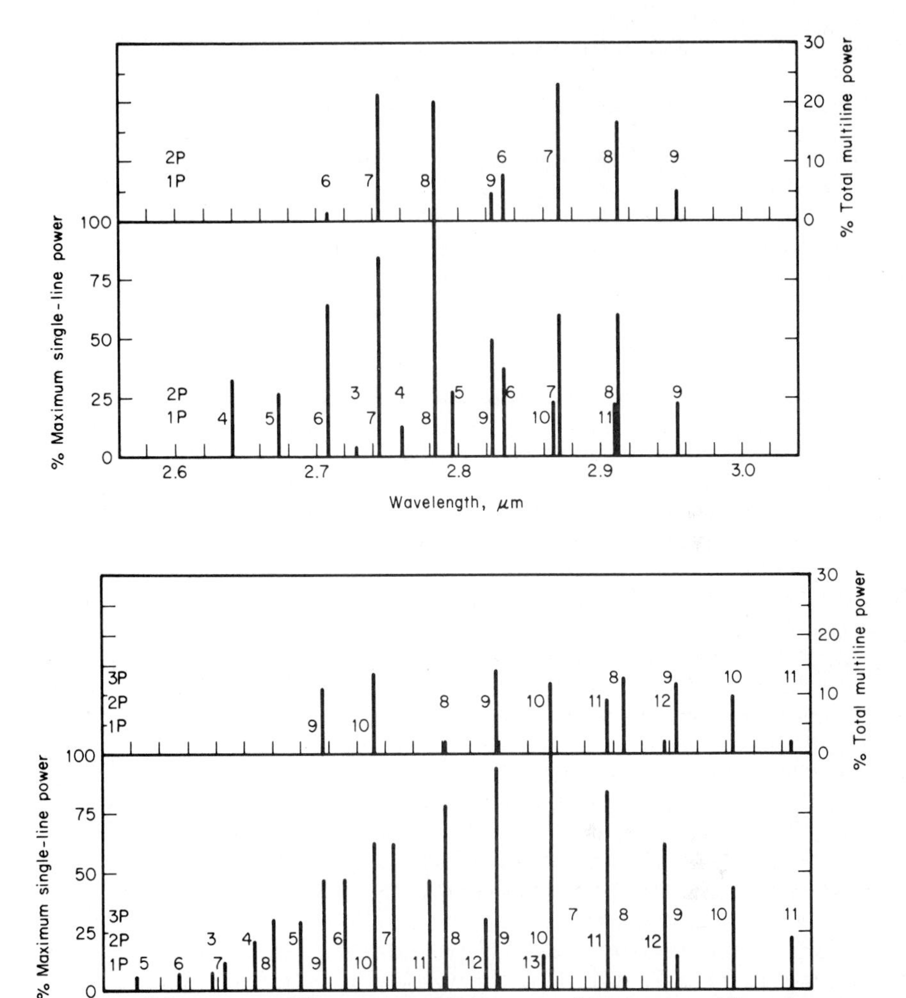

Figure 11-4 *(a)* HF and *(b)* DF emission lines from a continuous-wave chemical laser. *(Courtesy of Helios Inc.)*

enough that a reasonable fraction of the beam can be coupled out of the cavity, although manufacturers consider coupling ratios proprietary. If the laser is operated on a single line, a diffraction grating is inserted into the cavity and angle-tuned to give the desired wavelength; zeroth-order reflection from the grating can provide output coupling.

Different types of optics are used with high-energy chemical lasers because transparent optics are vulnerable to optical damage at high power fluxes. For example, the output mirror typically would have a hole to pass part of the beam, rather than transmitting part of the light falling on its surface.

Pulsed commercial lasers use optics similar in structure to those used with low-power continuous-wave lasers. They are mounted on opposite ends of the laser head, which has transverse electrodes and gas flow.

Cavity Length and Configuration The commonest commercial design for continuous-wave chemical lasers uses a modular cavity structure. A single gas generator feeds one, three, or five flow modules. The gas flow is in parallel, but the modules are in series optically, as can be seen in Fig. 11-5, where the resonator axis is along the length of the housing. Overall cavity lengths are 0.6 to 1 m. The lasers have stable resonators which produce TEM_{00} beams; at low powers unstable resonators seriously degrade output of HF and DF lasers and only improve beam quality slightly. Unstable resonators are more common in higher power continuous-wave lasers, where high power fluxes limit cavity design options.

Unstable resonators are common on commercial pulsed lasers, because they can provide good beam quality, and output-power degradation is not as serious as in continuous-wave models. Cavities of such lasers are on the order of a half a meter to a meter long.

Both pulsed and continuous-wave commercial chemical lasers can accommodate wavelength-tuning gratings in their resonant cavities.

Variations and Types Covered Representative major chemical lasers are listed in Table 11-1, along with the reactions involved. HF and DF are by far the most important practically and are listed separately because of their importance. However, because they are the same except for the hydrogen isotope used, they are functionally interchangeable in many laser devices. Other isotopic substitutions also can change wavelengths of vibrational chemical lasers.

Many chemical lasers, such as HF and DF, are "direct" types, in which the laser emission is by the species produced by the chemical reaction. There

Figure 11-5 Continuous-wave chemical laser is assembled from five modules. Gas flows through them in parallel, but they are arranged in series optically, to form a single laser resonator, nearly horizontal on the page. *(Courtesy of Helios Inc.)*

TABLE 11-1 Major Chemical Lasers

Laser emitter	Typical reactions	Wavelengths, μm
I	$O_2^* + I \rightarrow O_2 + I^*$ (transfer)	1.3
HF	$F + H_2 \rightarrow HF^* + H$ $H + F_2 \rightarrow HF^* + F$	2.6–3.5
HCl	$H + Cl_2 \rightarrow HCl^* + Cl$	3.5–4.1
DF	Similar to HF, with D substituted	3.5–4.1
HBr	$H + Br_2 \rightarrow HBr^* + Br$	4.0–4.7
CO	$CS + O \rightarrow CO^* + S$	4.9–5.8
CO_2	$DF^* + CO_2 \rightarrow CO_2^* + DF$ (transfer)	10–11

also are "transfer" chemical lasers, in which the excited species produced by a chemical reaction transfers its energy to another species, which then emits laser light. The chemical oxygen-iodine laser falls into this category, with excited O_2 transferring its energy to electronic excitation of atomic iodine, which emits at 1.3 μm. There also are transfer lasers in which excited DF transfers energy to CO_2, which emits on its characteristic wavelengths in the 10-μm region.

This chapter concentrates on the HF and DF lasers as the only true chemical lasers available commercially. The iodine laser has come on the market recently, but it is a version in which excitation is by a flashlamp, not chemically. The carbon monoxide laser also can be operated as a chemical laser, but the versions on the market are excited by electric discharges. Both CO and I lasers are covered briefly in Chap. 16.

Beam Characteristics

Wavelength and Output Power Commercial continuous-wave HF lasers are rated for maximum multiline output of several watts to about 150 W at a number of lines between 2.6 and 3.0 μm. When operated with DF, those lasers can produce multiline output powers 60 to 80 percent of those from HF at emission wavelengths of 3.6 to 4.1 μm. Spectral lines and relative intensities are shown in Fig. 11-4.

Grating tuning can restrict output of continuous-wave commercial HF and DF lasers to single lines in their emission bands. Maximum powers for HF are 1 to 50 W; for DF, maximum powers are 0.3 to 25 W. Both multiline and single-line output powers can be adjusted over considerable ranges, with minimum powers 0.1 to 1 percent of maximum power.

Pulsed chemical lasers available commercially emit at the same wavelengths but produce much lower average power—on the order of a watt for HF, and somewhat less for DF, for multiline operation. Multiline peak powers are one to a few megawatts. One model delivers 1-joule (J) multiline HF pulses at a 0.5-Hz repetition rate; another delivers 50-millijoule (mJ) pulses at repetition rates to 25 Hz. Pulse durations are around 50 nanoseconds (ns) for the high-repetition-rate laser and under 200 ns for the slower model. Output power on a single line is about a tenth that of multiline power.

Custom-built laboratory chemical lasers and military demonstration systems have produced thousands of times more power, but that technology is beyond the scope of this book.

Efficiency The infrared output power of commercial continuous chemical lasers is 1 to 2 percent of the electrical power input to the discharge that breaks up SF_6 to generate fluorine. When power requirements of other components such as vacuum pumps are considered, overall electrical-to-optical efficiency becomes 0.25 to 0.5 percent. Somewhat higher efficiencies are possible for pulsed operation with fast discharges.

Much higher electrical efficiencies have been recorded in the laboratory. Indeed, it is possible to have electrical efficiency above 100 percent when the electrical input is used to trigger a highly exothermic reaction between reactants such as F_2 and H_2. However, in such cases it is the chemical reaction, not the discharge, that supplies most of the energy that emerges in the laser beam. Thus it is the chemical efficiency which actually measures overall efficiency, and that figure can run from about 1 to about 20 percent.

Spectral Bandwidth Multiline emission from chemical lasers is on a number of discrete lines in a band about 0.5 μm wide, as shown in Fig. 11-4. Both pulsed and continuous-wave commercial models can be grating-tuned to emit on a single line in multiple modes. Continuous-wave models can be restricted to single-mode operation on a single line, at powers 10 to 30 percent of multimode output on a single line—yielding tens or hundreds of milliwatts. Under laboratory conditions, instantaneous spectral widths may be as small as 1 kilohertz (kHz), with spectral width 1 to 15 megahertz (MHz) when averaged over 0.1-s intervals (Munch et al., 1978).

Amplitude Noise Typical specified values of amplitude stability for multiline continuous chemical lasers are ±5 to 10 percent for HF and ±2 to 3 percent for DF. For single-line operation, amplitude stability below 1 percent has been measured for DF, and slightly above 1 percent for HF, but those are not *specified* values. Specified pulse-to-pulse reproducibility for pulsed lasers is ±2 to 10 percent.

Beam Quality, Polarization, and Modes Beam quality of chemical lasers tends to be limited and can become a serious problem at the high powers in some military demonstration lasers. Pulsed commercial models oscillate in multiple transverse modes. Continuous-wave devices normally oscillate in one to three transverse modes, with about 90 percent of the total power in TEM_{00} mode. Single-line continuous lasers can be made to oscillate in TEM_{00} mode by limiting cavity aperture and beam diameter. Unstable resonators can limit pulsed chemical lasers to a single mode and are often required to achieve reasonable beam quality. The use of Brewster angle windows generates a linearly polarized beam.

Beam Diameter and Divergence Continuous-wave chemical lasers have beams ranging from 2.0 to 3.0 mm in diameter; full-angle beam divergences are 1.7 to 2.0 milliradians (mrad). For one pulsed laser, the output beam is 5 by 10 mm in cross section with divergence of 5 by 12 mrad, although use of an unstable resonator can limit divergence to less than 1 mrad.

Suitability for Use with Laser Accessories Chemical lasers do not store much energy within the laser resonator, but *Q* switching is possible at very high repetition rates (1 MHz) producing peak powers about 10 times the continuous level. The high peak powers possible with pulsed versions should allow the use of nonlinear optical techniques such as harmonic generation, but there has been little work done in that area. There is little interest evident in harmonic generation with either HF or DF, in large part because there are much easier ways to produce the harmonic wavelengths.

Operating Requirements

Input Power The major power consumers in continuous-wave chemical lasers are the power supply for the high-voltage discharge, which draws 2.6 to 24 kW, and the vacuum pump, which draws 5.6 to 15 kW. The smallest models could operate from a 110-V ac line, but 220-V operation is commoner. The largest can require three-phase 440-V lines.

Pulsed models require lower power levels, with the discharge supply drawing as little as a few hundred watts (3 A at 110 V). Additional power is needed for the vacuum pump.

Cooling Pulsed commercial HF and DF lasers can be air-cooled. However, water cooling is required for both laser head and vacuum pump on continuous models. Flow requirements are 1 to 4 gal/min (3.8 to 15 L/min) for the laser head and 1.4 to 3 gal/min (5.3 to 11 L/min) for the pump.

Consumables Continuous-wave chemical lasers consume SF_6, O_2, He, and H_2 or D_2. When the laser is operating at maximum power, consumption rates per minute are

Gas	Standard liters	Standard cubic feet
SF_6	20–70	0.7–2.5
O_2	2–28	0.07–1
He	17–40	0.6–1.4
H_2	17–110	0.6–4
D_2	17–85	0.6–3

Consumption rates drop when operating power of the laser is reduced. If pulsed versions are operated at high repetition rates, SF_6 and H_2 or D_2 are consumed at rates about half to two-thirds those of continuous-wave lasers, if the laser is operated in an open cycle.

Gas costs can quickly become significant. SF_6 costs about 6 cents a standard liter, and its consumption can be the single largest operating cost for an HF laser. Deuterium costs about $1 per standard liter, making DF lasers much more costly to operate. Operating costs of pulsed lasers can be reduced by using gas recirculators that remove the small fraction of HF or DF produced during laser operation and recycle the remainder of the gas, cutting gas consumption to about 1 percent of the open-cycle level.

Required Accessories A vacuum pump is needed to remove the laser gas from the laser. In pulsed models, the gas can be put through a recirculator, as mentioned above, to reduce gas consumption and operating costs, but in continuous lasers the gas must be exhausted. Normally the spent gas is vented to the atmosphere, either directly or through a scrubber that removes HF and SO_2 produced in operation. The need for a scrubber depends on operating power and local pollution requirements.

Mechanical Considerations The smallest chemical lasers are pulsed types, which without a vacuum pump come in a single case as small as 0.7 by 0.5 by 0.3 m weighing under 40 kg. Continuous-wave commercial lasers are much larger, with smaller ones including a control console and vacuum pump in addition to a laser head, and with a separate high-voltage power supply and ballast system needed for the larger models. Combined weights of major components range from about 710 kg to 2550 kg.

Safety Chemical laser wavelengths cannot penetrate the eye, but power densities above 10 W/cm^2 are sufficient to burn the cornea before the eye's blink reflex can protect it. Because chemical laser wavelengths are strongly absorbed by human skin, high-power beams can cause skin burns.

The input gases used with commercial chemical lasers, SF_6 and H_2, are reasonably safe, although H_2-air mixtures become explosive if H_2 concentration exceeds 4 percent. Other fluorine donors are generally bad news; F_2 is both dangerous to handle and toxic in concentrations above 0.1 part per million (ppm). NF_3 is somewhat stabler and less toxic (hazardous levels are 10 ppm), but it is still much more dangerous than SF_6. The waste gases also are health hazards. ANSI standards limit HF exposure to no more than 3 ppm, and SO_2 is considered harmful in similar concentrations (Weast, 1981). HF and most other fluorine-containing compounds used in chemical lasers (except SF_6) also are highly corrosive.

The high voltages used in the discharges that break up SF_6 in pulsed and continuous-wave commercial chemical lasers also are potential hazards.

Reliability and Maintenance

HF and DF chemical lasers are laboratory instruments built in small quantities which handle extremely corrosive gases, and thus are far from industrial reliability standards. Nonetheless, present commercial designs do achieve impressive performance compared with some demonstration military lasers, which can operate for only minutes at a time.

Manufacturer recommendations for care of continuous-wave commercial lasers are

- Dust off optics in place—each run of the laser
- Change vacuum-pump oil—50 to 100 hours of run time
- Clean H_2 or D_2 injector—50 to 100 hours of run time
- Rebuild alumina plasma tube, polishing electrodes and replacing seals—500 to 1000 hours of run time
- Replace laser bodies—1000 to 2000 hours of run time

Commercial Devices

Standard Configurations Pulsed chemical lasers are similar in size and general appearance to TEA CO_2 or rare-gas-halide excimer lasers. All major components except the vacuum pump may be housed in a single case, or the power supply may be mounted remotely, in addition to the vacuum pump.

Continuous-wave chemical lasers come as one, three, or five subunits, with the laser head normally bench-mountable. The vacuum pump is separate, as is the control console; in smaller models the control console includes power supply and ballast, but they are separate rack-mounted units in larger lasers.

Because production is small, manufacturers can readily accommodate special requests.

Options Major options on commercial chemical lasers include gas scrubbers, gas recirculators, special power supplies and vacuum pumps, vacuum mirror mounts (instead of Brewster windows), single-line optics, and unstable resonators.

Pricing Prices of complete pulsed chemical lasers start in the $20,000 range (excluding vacuum pump). Complete continuous-wave lasers run $50,000 to about $140,000, including vacuum pump.

Suppliers The commercial chemical laser business is small, with only a single company specializing in that market. One or two others produce chemical lasers, but get most of their business from other markets. High-energy military demonstration lasers are built by large aerospace contractors.

Applications The only current application for chemical lasers is in research. Most of that research is military-related, involving development of large chemical lasers, testing the effects of chemical laser beams on various materials and potential targets, and range finding over water. These military programs use small commercial chemical lasers, as well as larger custom-built lasers.

Chemical lasers also are used in other research requiring mid-infrared radiation, including chemistry, materials studies, spectroscopy, and biomedicine. Recently, it has been suggested that HF lasers may be better than CO_2

for certain types of surgery (Wolbarsht, 1984). Problems in gas handling and chemical laser technology continue to make HF lasers much more cumbersome to use than CO_2. However, chemical laser specialists believe that emergence of a viable commercial market would justify an investment in careful engineering that would make chemical lasers considerably more practical, perhaps comparable to open-cycle carbon dioxide lasers of medium power.

BIBLIOGRAPHY

T. Y. Chang: "Vibrational transition lasers," in Marvin J. Weber (ed.), *CRC Handbook of Laser Science & Technology*, vol. 2 *Gas Lasers*, CRC Press, Boca Raton, Fla., 1982, pp. 313–409.

Arthur N. Chester: "Chemical lasers," in E. R. Pike (ed.), *High-Power Gas Lasers 1975*, Institute of Physics, Bristol and London, 1976, pp. 313–409.

R. W. F. Gross and J. F. Botts (eds.): *Handbook of Chemical Lasers*, John Wiley & Sons, New York, 1976.

G. N. Hays and G. A. Fisk: "Chemically pumped iodine as a fusion driver," *IEEE Journal of Quantum Electronics QE-17:*1823, September 1981.

Jeff Hecht: *Beam Weapons: The Next Arms Race*. Plenum, New York, 1984.

J. V. V. Kasper and G. C. Pimentel: "HCl chemical laser," *Physical Review Letters 14:*352, 1965.

W. E. McDermott et al., "An electronic transition chemical laser," *Applied Physics Letters 32:*469, 1978.

J. Munch et al.: "Frequency stability and stabilization of a chemical laser," *IEEE Journal of Quantum Electronics, QE-14:*1, January 1978, pp. 17–22.

J. C. Polanyi: *Journal of Chemical Physics 34:*347, 1961.

Robert Weast (ed.): *CRC Handbook of Chemistry & Physics*, 62d ed., CRC Press, Boca Raton, Fla., 1981.

Myron L. Wolbarsht: "Laser surgery: CO_2 or HF, " *IEEE Journal of Quantum Electronics QE-20:*1427, December 1984.

CHAPTER

12

copper and gold vapor lasers

Copper and gold vapor lasers are the most important members of a family of neutral metal vapor lasers which emit in or near the visible region. They are unusual in their high power and high efficiency in this region, and in that their normal operation is at repetition rates of several kilohertz (the internal physics prevent continuous-wave emission). Commercial copper vapor lasers can emit tens of watts in the green and yellow; gold vapor lasers can generate several watts in the red. These high average powers make copper and gold vapor lasers attractive for applications including pumping dye lasers, detection of fingerprints and trace evidence, and certain biomedical procedures.

The copper vapor laser was first demonstrated in 1966 by W. T. Walter, N. Solimene, M. Piltch, and Gordon Gould at TRG Inc. (Walter et al., 1966). The laser's high efficiency was recognized even then, but the need to heat the tube to 1500 to 1800°C to get high enough copper vapor pressures presented severe technical difficulties. In the early 1970s, Soviet workers showed that waste heat from the laser discharge could heat the tube, raising overall efficiency (Isaev et al., 1972). Others studied the use of copper halides, which do not require such high temperatures (Chen et al., 1973).

Commercial copper vapor lasers based on discharge heating have appeared recently, partly as an outgrowth of technology developed for laser isotope separation at the Lawrence Livermore National Laboratory. Copper vapor lasers were chosen as the best candidate for pumping high-power dye lasers tuned to selectively excite uranium 235 in uranium vapor (Lawrence Livermore Laboratory, 1981, 1982). This led to a large investment in copper vapor laser technology, some of which is applicable at the lower powers of commercial lasers.

Other neutral metal vapor lasers use technology similar to that of copper vapor. The most important of these from an applications standpoint is the gold vapor laser, because its 628-nanometer (nm) red output appears useful for photoradiation or photodynamic therapy of cancer (McCaughan, 1983).

Internal Workings

Neutral metal vapor lasers operate in ways similar to each other, but quite different from ionized metal vapor lasers such as helium-cadmium, described in Chap. 9. In neutral metal vapor lasers, a fast electric discharge directly excites metal atoms, producing a population inversion and laser emission. The laser pulse terminates when the lower laser level fills, but laser action can occur again after slower processes depopulate that level. High repetition rates permit high average power.

Active Medium Vaporized copper and neon form the active medium in commercial copper vapor lasers. Pieces of copper metal are put into the discharge tube, where they are heated to about 1500°C to obtain the required 0.1 torr (13 Pa) of metal vapor. Adding an inert gas improves discharge quality enough that sufficient power can be coupled into the active medium; it also may help depopulate the lower state, although electron quenching appears much more effective. Neon at about 25 torr (3300 Pa) is the normal choice, but helium and argon also can be used.

Waste heat from the discharge keeps the inner part of the laser tube hot, but the copper does condense on outer, cooler portions of the tube, so extra metal must be added periodically. Similar processes occur with other neutral metal vapor lasers, although some of the details such as operating temperature differ. Lead vapor lasers, for example, operate at about 900 to 1100°C, and gold vapor must be heated to about 1550 to 1850°C.

The use of copper halides and other compounds to provide copper vapor has been demonstrated in the laboratory. The advantage of lower operating temperature is more than offset by practical problems in operating the laser, and that technology is not in commercial use.

Energy Transfer Energy is delivered to a copper vapor laser by a pulsed high-voltage discharge passing longitudinally along the length of the plasma tube. Electron collisions raise copper atoms to one of the two upper levels shown in Fig. 12-1. At vapor densities of 10^{11} atoms per cubic centimeter, copper atoms remain in the upper levels only about 10 nanoseconds (ns) before dropping back to the ground state, making laser action impossible. Raising copper density to about 5×10^{13} cm^{-3} increases effective lifetime of the upper laser level (relative to the ground state) to 10 milliseconds (ms) because of radiation trapping, although the actual time an individual atom remains excited is unchanged. That is sufficient to permit laser emission.

Copper vapor has two main laser transitions between separate pairs of states, one at 510.6 nm in the green, and one at 578.2 nm in the yellow. Both terminate in lower levels that are not thermally populated, but are metastable. Relaxation times from these lower levels are tens to hundreds of microseconds, depending on operating conditions, long enough that the populations build up and terminate laser emission in under 100 ns. Laser action cannot resume until the lower laser level is depopulated.

The self-terminating nature of the laser makes fast discharge rise time crucial. Operation at high average powers requires high repetition-rate discharges, typically several kilohertz (kHz) in commercial copper vapor lasers, although experimental lasers have passed the 100-kHz mark (Grove, 1982). Energy-transfer processes in other neutral metal vapor lasers are qualitatively similar to those in copper vapor, although different wavelengths and different numbers of emission lines are involved. All share the self-terminating nature that dictates repetitively pulsed operation.

Internal Structure

The high gain of copper vapor, 10 to 30 percent per centimeter, makes optical requirements for laser oscillation easy to meet. The major design

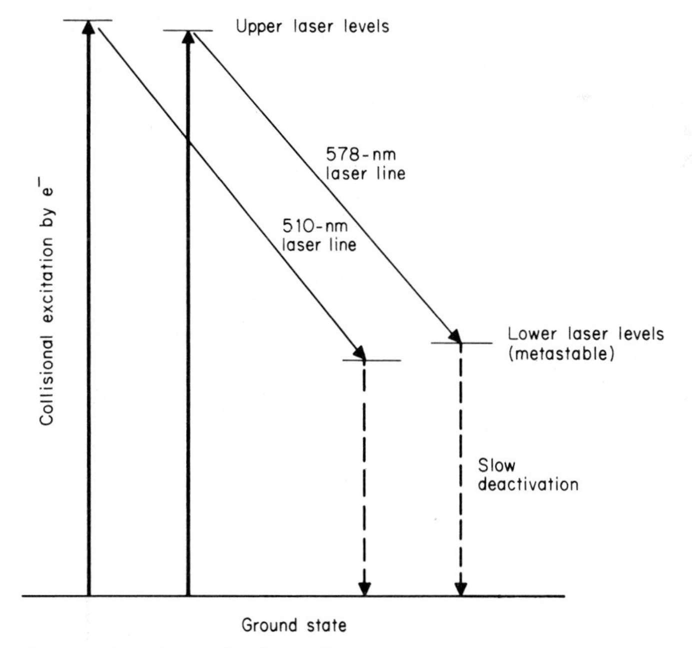

Figure 12-1 Energy levels and laser transitions in copper vapor.

challenges come in sustaining the required high temperatures and vapor pressure, and in delivering fast current pulses at a high repetition rate.

The discharge is contained in ceramic alumina tubes 1 to 8 cm in diameter. External insulation helps the tubes maintain enough waste heat to vaporize metal from pieces of copper put inside, typically near the ends of the bore. Refractory metal electrodes are needed because of the high operating temperatures. Copper migrates out of the hot portion of the tube to condense in cooler portions at a rate on the order of 0.01 g per hour of operation. Schemes for recirculating the escaping copper back into the discharge region have been demonstrated in the laboratory but are not used in commercial lasers. Gold diffuses somewhat faster.

Thyratrons are needed to switch the 10- to 20-kilovolt (kV) discharge pulses at repetition rates in the multi-kilohertz range. This, together with peak currents on the order of a kiloampere, can require special charging circuits.

Optics The copper vapor laser can lase without any mirrors because of its high gain. In practice, the laser tube is sealed with flat glass windows, and the mirrors mounted externally. The rear mirror is a total reflector, with 90 percent transmission a good selection for the output-coupling mirror. An uncoated window makes an entirely adequate output coupler.

Cavity Length and Configuration Metal vapor lasers typically have cavities over a meter long. Both stable and unstable resonator configurations are used.

Output power can be raised above the levels available from a single oscillator by using a master oscillator power amplifier configuration. Output of more than 100 W has been produced by a linear oscillator-amplifier chain of six copper vapor lasers each able to produce about 15 W as an oscillator, but more typically a laser head operated as an amplifier can produce about 50 percent more average power than when operated as an oscillator (Grove, 1982).

Variations and Types Covered

This chapter concentrates on the copper vapor laser, which because of its wavelength and high power has attracted the most interest of any neutral metal vapor laser. Gold vapor lasers are also mentioned, because they are beginning to come on the market for medical applications. Structurally they are similar to each other and to other neutral metal vapor lasers listed in Table 12-1. That table lists the major lines of selected elements; there also are many weaker lines and less efficient lasers listed in more extensive data tabulations (see, for example, Davis, 1982).

Beam Characteristics

Wavelength and Output Power Copper vapor lasers emit at two wavelengths, 510.6 and 578.2 nm. Both lines are produced simultaneously, with relative intensity a function of operating temperature as shown in Fig. 12-2.

TABLE 12-1 Wavelengths of Major Neutral Metal Vapor Laser Lines, with Relative Powers That Might Be Expected from Devices of Comparable Scale*

Element	Wavelength, nm	Relative power	Remarks
Copper	511, 578	1	
Gold	628	0.1–0.3	
	312	Low	Secondary line
Barium	1130	Low	Ba liquid a problem
	1500	0.3–0.5	Ba liquid a problem
Lead	722.9	0.2–0.3	1000–1100°C temperature
Manganese	534	0.2–0.3	Mn vapor a problem
	1290	—	Mn vapor a problem
Calcium	852.4	—	
	866.2	—	

* For copper vapor and the 628-nm gold line, values are for commercial devices; other results are from laboratory experiments. Laser action has been demonstrated experimentally on many other lines.

When the operating temperature is picked to give peak output power, about two-thirds of the power is at 510.6 nm. If only a single line is desired, the two wavelengths can be separated in the output beam. Average power ranges from a few watts to several tens of watts for commercial copper vapor lasers.

Commercial gold vapor lasers produce up to several watts of average power at 628 nm. Wavelengths of other metal vapor lasers and indications of relative power levels are listed in Table 12-1; some lines are shown in Fig. 12-3.

Efficiency Overall "wall-plug" efficiency is 0.2 to 0.8 percent for commercial copper vapor lasers, higher than argon ion lasers. Efficiency figures over 1 percent may be quoted if the measurement compares laser output with energy stored in the discharge circuitry. Gold vapor lasers have 0.1 to 0.2 percent overall efficiency.

Temporal Characteristics Average output power of commercial copper vapor lasers peaks at repetition rates of 4 to 6 kHz, but operation is possible over a somewhat broader range of pulse rates with reduced output power. Optimum repetition rate depends on tube construction, and higher rates are possible with certain tube designs. Individual pulses last tens of nanoseconds and have the types of pulse shapes shown in Fig. 12-4. Rated pulse energies are 1 to 14 millijoules (mJ), with peak powers from tens to hundreds of kilowatts.

Gold vapor lasers have similar pulse length and can operate at similar repetition rates, but output power appears to be highest at somewhat higher repetition rates than copper.

The repetition rates are so high that beams look continuous to the human eye.

Spectral Bandwidth A single line from a copper vapor laser typically has spectral bandwidth of 6 to 8 GHz.

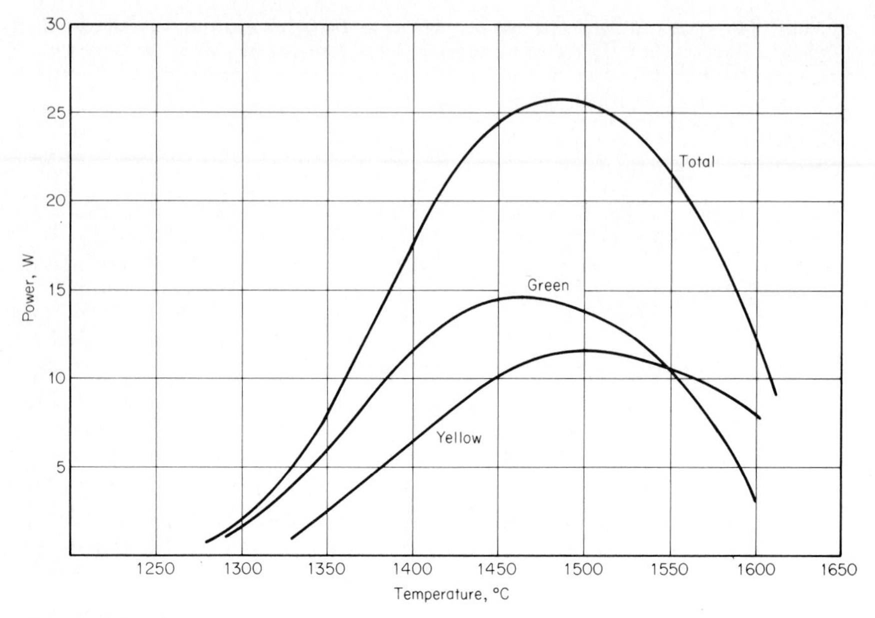

Figure 12-2 Relative strengths of copper vapor lines as a function of temperature. *(Courtesy of Oxford Lasers Ltd.)*

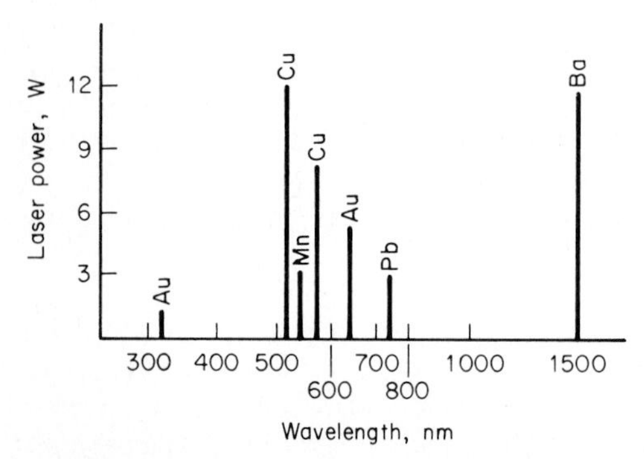

Figure 12-3 Wavelengths and relative intensities of some major neutral metal vapor laser lines. *(Courtesy of Quentron Optics Pty. Ltd.)*

Amplitude Noise and Stability Instantaneous power levels vary widely because of the repetitive pulsing. Pulse-to-pulse amplitude variations normally are unspecified, but one company reports amplitude stability of ±2 percent over 6000 shots. Pulse shape stays reasonably constant if operating conditions stay unchanged, but the shape does vary with changes in repetition rate, gas pressure, resonator temperature, or other factors. Pulse-timing jitter is specified at 3 to 5 ns.

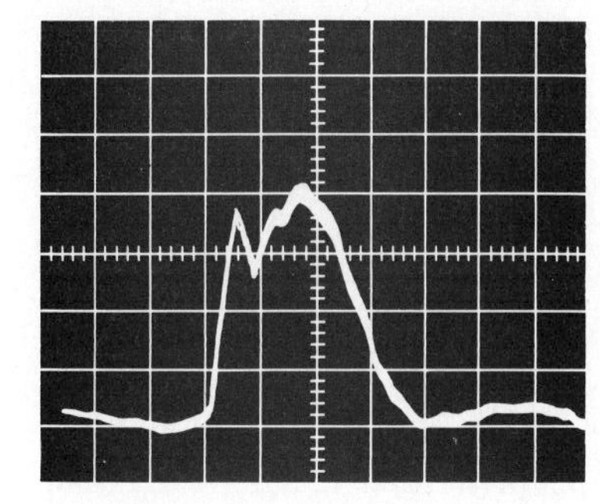

Figure 12-4 Pulse shape from copper vapor laser (20 ns per division). *(Courtesy of Plasma Kinetics, a division of Cooper LaserSonics.)*

Beam Quality, Polarization, and Modes Stable resonators produce a high-order multimode beam with uniform intensity across the beam. Unstable resonators produce a doughnut-shaped mode with much lower divergence (four or five times the diffraction limit), which can be focused to a gaussian spot. However, the low-divergence mode may not contain all the laser energy; some may remain in a much broader divergence beam. Output normally is unpolarized.

Beam Diameter and Divergence Copper vapor lasers have beams 2 to 8 cm in diameter. With stable resonators, beam divergences are 3 to 5 milliradians (mrad). An unstable resonator can produce a mode with 0.3 to 0.5 mrad divergence, but it may not contain all the laser power.

Suitability for Use with Laser Accessories The 578-nm line of copper vapor can be frequency-doubled well in nonlinear crystals, but no good harmonic-generation crystals are available for the 510.6-nm line. Little attention has been paid to the possibility of modelocking, presumably because of the ease of modelocking argon ion and dye lasers. The short energy-storage times in the laser medium makes Q switching impractical.

Operating Requirements

Input Power Operation of a copper vapor laser requires a few kilowatts to about 16 kW of electrical input. The power may be drawn from a 220-V, single-phase ac source for smaller models; larger ones may require three-phase sources of 220 or 440 V.

Cooling Many commercial copper vapor lasers are water-cooled, with flow rates of 1 to 8 L of tap water per minute. Forced-air cooling is used for some small models, such as a 10-W laser which dissipates about 2 kW of heat.

Consumables Copper slowly migrates out of the discharge region, and new metal must be loaded into the laser tube after a few hundred hours of operation. The same phenomenon occurs with gold, which also must be added to the laser tube periodically. The cost of the added gold corresponds to about $20 per 8-hour day of operation, but the gold is not lost—just deposited elsewhere in the tube where it can be recovered.

Buffer gas normally flows slowly through the laser to purge contaminants at a rate of around 1 L/h.

Operating Conditions, Temperature, and Warm-up Copper vapor lasers are designed for operation in normal laboratory conditions and temperatures. Half an hour to an hour is needed to heat the laser tube to the 1500°C temperature needed to produce enough copper vapor for laser operation. This is an unusually long time by laser standards, comparable only to that for lasers with special frequency-stabilization equipment, and much longer than that for ionized metal vapor lasers such as He–Cd, which operate at lower temperatures. Warm-up times needed to reach the 1550° to 1850°C operating temperature of gold vapor lasers are comparable to those for copper vapor.

Mechanical Considerations Copper vapor lasers are large, with laser heads typically 1.5 to 2.5 m long, generally built to be put on an optical bench. The power supply and controls are normally in a separate housing placed on the floor, but in some cases the power supply, controls, and laser head may be combined in a single floor-mounted unit.

Safety The green and yellow lines of copper vapor and the red line of gold vapor are potential hazards to the retina, because the rest of the eye transmits visible light well. Average power levels are high enough to pose serious dangers to the eye. Copper vapor power levels are high enough that they could cause skin burns if the beam were tightly focused, but such lesions are far more likely to heal than retinal damage. Because the two copper vapor wavelengths are widely separated, users should make sure that their safety goggles provide adequate protection at *both* wavelengths, not just one.

The high-voltage, high-current discharges which drive metal vapor lasers can electrocute anyone who comes in contact with them. The repetitively pulsed operation of the laser makes it likely that some components may retain a charge when the laser is turned off, so users and service personnel should be sure all voltages have been discharged before working inside the laser.

The high operating temperature of the laser tube poses thermal hazards—do not touch it unless it has had plenty of time to cool off.

Reliability and Maintenance

Operating conditions of a copper vapor laser are difficult, so much of the research and development devoted to copper vapor lasers has been aimed at improving reliability. Great strides have been made in building practical lasers, but the "hands-off" operating time of copper vapor lasers remains smaller than that of most other types.

Lifetime Some experimental copper vapor lasers have operated for up to a few thousand hours without replenishment of the metal (Grove, 1982), but to sustain maximum output power commercial models require addition of metal every few hundred hours. With regular addition of metal, commercial copper vapor laser tubes have rated lifetimes of 1000 to 3000 hours of operation. Typical manufacturer warranties are 1 year on the laser tube and 500 or 1000 operating hours to 1 year on the thyratron.

Maintenance and Adjustments The most frequent maintenance needed in metal vapor lasers is addition of more metal. Switching thyratrons and plasma tubes must be replaced periodically. Because of the high gain of the laser medium, precise alignment of the optical cavity is not crucial.

Commercial Devices

Standard Configurations Most neutral metal vapor lasers are packaged for general laboratory use, with separate laser head and power supply, but one manufacturer has begun packaging both gold and copper vapor lasers for medical applications, including such accessories as a fiber-optic beam delivery system. One model has a head with two tubes.

Copper and gold vapor lasers are similar in design, and copper vapor can even operate in some gold vapor tubes, if operating temperature is reduced. Operation of gold vapor in copper tubes is more difficult, because it requires temperatures above the level normally used with the copper laser. When other neutral metal vapor lasers are available, they normally are modified versions of copper or gold lasers.

Options Common options include choice of stable or unstable resonator, line selection for copper vapor, and polarizing optics. Because production quantities are small, manufacturers are willing to build customized lasers for special requirements. One version for medical applications is already available, which allows use of copper and/or gold laser tubes.

Pricing As of this writing, commercial copper and gold vapor lasers carry prices of about $30,000 to $100,000, depending on packaging and power levels. Replacement plasma tubes run about $600 to $10,000, with prices based on the exchange of the old tube for a new one.

Suppliers Only a handful of companies make copper vapor lasers, reflecting both the small market and the need for specialized technology. Product lines differ significantly, and manufacturers are located both in the United States and overseas. The use of copper vapor lasers in uranium enrichment has led to stringent export-control restrictions, which are most severe for high-power models, and somewhat limit sales of lower-power lasers.

Applications

The biggest single application for copper vapor lasers has been in pumping dye lasers. The high average power of copper vapor makes it possible to obtain higher wavelength-tunable dye output than when pumping with an argon laser, although the dye output—like the copper vapor—is repetitively pulsed. The use of the visible wavelength for pumping helps extend dye lifetime, although it limits the choice of dyes to those with emission wavelengths longer than the pump line. This set of advantages is particularly important for applications in photochemistry. One special photochemical application, enrichment of uranium-235 in uranium vapor, has been a driving force behind copper vapor development and could be the first laser photochemistry process to be operated at an industrial scale.

Another application that is moving from the research stage to practical applications is the detection of trace evidence for law enforcement. The best-known example is detection of latent fingerprints by illuminating with the green copper vapor line and looking for fluorescence produced at longer wavelengths by substances in the fingerprints. Early work was done with argon lasers, but copper vapor is attracting attention because of its higher average power. The technique has also been extended to detection of other trace evidence, such as bloodstains, and examination of documents to check for forgeries. Most work has been done with the green copper vapor wavelength, but prospects for using the yellow line are being studied also.

A number of other potential applications of copper vapor lasers also are being studied. These include

- Pulsed photography and holography, taking advantage of the short pulse-length of the repetitively pulsed lasers. Coherence lengths appear adequate for pulsed holography when emission is restricted to a single line.
- Measurement of combustion processes, using the short pulse length to separate fluorescence of signals from noise.
- Transmission of light underwater at the 511-nm line, which is close to the wavelength where water is most transparent.
- Color displays.
- Semiconductor processing and research.
- Biomedicine. The 578-nm copper line appears useful in treating "port-wine" birthmarks. Such skin discolorations have been treated with argon lasers, but results have been inconsistent and results hard to predict.

Interest in the gold vapor laser has centered on biomedical applications of its red line. The 628-nm wavelength matches an absorption peak of hemoporphyrin derivative, a substance selectively absorbed by cancer cells which breaks down when it absorbs light, producing by-products which kill the cancer cells. Photoradiation therapy based on this effect is being studied in many laboratories around the world (McCaughan, 1983).

BIBLIOGRAPHY

B. G. Bricks, T. W. Karras, and R. S. Anderson: "An investigation of a discharge-heated barium laser," *Journal of Applied Physics 49*(1):38, January 1978.

B. G. Bricks and T. W. Karras: "Power scaling experiments with a discharge-heated lead vapor laser," paper presented at Lasers '79, Orlando, Fla., Dec. 17–21, 1979.

C. J. Chen, N. M. Nerheim, and G. R. Russell: "Double-discharge copper vapor laser with copper chloride as a lasant," *Applied Physics Letters 23*:514–515, 1973.

Christopher C. Davis: "Neutral gas lasers," in Marvin Weber (ed.) *CRC Handbook of Laser Science & Technology*, vol. II *Gas Lasers*. CRC Press, Boca Raton, Fla., 1982, pp. 3–168. Reference review, see especially pp. 134–139.

Robert E. Grove: "Copper vapor lasers come of age," *Laser Focus 18*(7):45–50, July 1982.

A. A. Isaev, M. A. Kazaryan, and G. G. Petrash: "Effective pulsed copper vapor laser with high average generation power," *JETP Letters 16*:27–29, 1972.

Lawrence Livermore National Laboratory: *Atomic Vapor Laser Isotope Separation*. U.S. Government Printing Office, Washington, D.C., 1982.

Lawrence Livermore National Laboratory: *1980—Selected Highlights Laser Isotope Separation Program*. LLNL Report, Livermore, Calif., June 1981.

James S. McCaughan Jr.: "Progress in photoradiation therapy of cancer following administration of HpD," *Laser Focus 19*(5):48–56, May 1983.

W. T. Walter et al.: "Efficient pulsed gas discharge lasers," *IEEE Journal of Quantum Electronics QE-2*:474–479, 1966.

CHAPTER

13

excimer lasers

The term *excimer laser* does not describe a single device, but rather a family of lasers with similar output characteristics. All emit powerful pulses lasting nanoseconds or tens of nanoseconds at wavelengths in or near the ultraviolet. Most commercial excimer lasers can be operated with different gas mixtures to produce different output wavelengths. The technology is relatively new, and commercial devices are still evolving.

The term *excimer* originated as a contraction of "excited dimer," a description of a molecule consisting of two identical atoms which exists only in an excited state; examples include He_2 and Xe_2. It now is used in a broader sense for any diatomic molecule (and sometimes for triatomic types) in which the component atoms are bound in the excited state, but not in the ground state. That property makes them good laser materials with similar output characteristics. The most important excimer molecules are rare gas halides, compounds such as argon fluoride, krypton fluoride, xenon fluoride, and xenon chloride, which do not occur in nature, but which can be produced by passing an electric discharge through a suitable gas mixture. There are also other types of lasing excimers. Many of these molecules are so similar that they can be made to lase in the same device, which has come to be called an excimer laser. This chapter will use the terminology in the broadest commercial sense, even though some of the species which can be made to lase in excimer laser devices are not really excimers.

Although the excimer laser is one of the newest types on the commercial market, it has gained widespread acceptance in the research laboratory and is attracting increasing interest from potential industrial users. Its quick transition from laboratory demonstration to commercial product is due partially to a coincidence: the same excitation scheme can be used for excimer lasers as for transversely excited atmospheric pressure (TEA) carbon dioxide lasers. A group at the Naval Research Laboratory in 1976 found that they could build a "quick-and-dirty" excimer laser by using suitable optics and gases in a commercial TEA CO_2 laser. Lifetime was initially a problem because of the highly corrosive halogens in the excimer gas mixture, but laser makers soon designed versions specifically for excimer operation. Most excimer lasers sold are used for research and development (Hitz 1985), but laser makers are working on versions for industrial use.

Introduction and Description

Active Medium Excimer lasers contain a mixture of gases at total pressure usually below 5 atm. The bulk of the mixture, 88 to 99 percent, is a buffer gas which mediates energy transfer, normally helium or neon, although argon is used in some cases. The buffer gas does not become part of the light-emitting species. The rare gas that does combine to form excimer molecules in rare-gas-halide lasers is present in much smaller concentrations, typically 0.5 to 12 percent of the total pressure. The halogen donor is normally present in concentrations of 0.5 percent or less; it may be either a diatomic halogen such as molecular fluorine, or a halogen-containing molecule such as hydrogen chloride or nitrogen trifluoride.

The optimum gas mixture is a complex function of operating conditions and gas kinetics and is different for different models of lasers. There are also large differences between different excimer molecules. Argon concentration can run as high as 12 percent in argon fluoride lasers, while a typical xenon fluoride mixture contains about 3% Xe and 0.18% F_2, with the balance buffer gas. Unless users have a special interest in studying gas mixtures, they would be best off relying on the recommendations of laser manufacturers.

Energy Transfer The detailed kinetics of energy transfer in excimer lasers are complex and took years to unravel. The complex interactions among atomic and molecular species in the laser gas, the excitation energy, the walls of the laser tube, and the electrodes can all affect the operating parameters of the laser. The details are far beyond the scope of this book, but it is possible to give a brief overview.

In almost all commercial excimer lasers, energy is deposited in the laser gas by an electrical discharge. The only commercial alternative is pumping with an electron beam. Up to 5 percent of the discharge energy can be converted into laser energy. Electron beams can deposit more energy in the gas and may offer slightly higher conversion efficiency. However, electron-beam generators are large, complex, expensive, inefficient, and limited in repetition rate. With a commercial krypton-fluoride laser, overall "wall-plug" efficiencies to about 2.5 percent are possible, much higher than would be possible with electron-beam pumping. A third alternative, microwave excitation, has been demonstrated in the laboratory (Young et al., 1982); its main promise is in extending pulse length to a couple of hundred nanoseconds, but so far only around 0.1 percent of the microwave energy has been converted to laser output.

To improve energy-transfer dynamics, and avoid discharge arcing, the laser gas is normally "preionized" before the excitation pulse. In commercial excimer lasers, preionization is normally by a pulse of ultraviolet light, but a pulse of x rays or a shot from an electron beam can also be used. The electrons in the excitation pulse then produce the excited rare gas atoms and halogens that react to form excited diatomic molecules or excimers, such as xenon fluoride. (The excited state is often indicated by following the chemical formula with an asterisk, for example, XeF^*.)

Excimer molecules are peculiar things. When electronically excited, the two component atoms attract each other to form a stable molecule. However, in the ground state the two atoms are mutually repulsive or in some cases weakly bound. Thus, as shown in Fig. 13-1, when an excimer drops from the excited state to the ground state, the force between the two atoms changes from attraction to repulsion, and the molecule breaks up.

This unusual energy-level structure makes excimers very good laser materials. Because the ground state essentially does not exist, there is a population inversion as long as there are molecules in the excited state. Thus if excimers exist, so do the right conditions for lasing. The kinetics also lead to very high gain on excimer laser transitions.

Excited-state lifetimes on the order of 10 nanoseconds (ns) set the time scale for excimer laser pulses. The switching times of the components which drive the discharge are also on the same scale. The upper limit on pulse length is functionally limited to the range of several nanoseconds to tens of

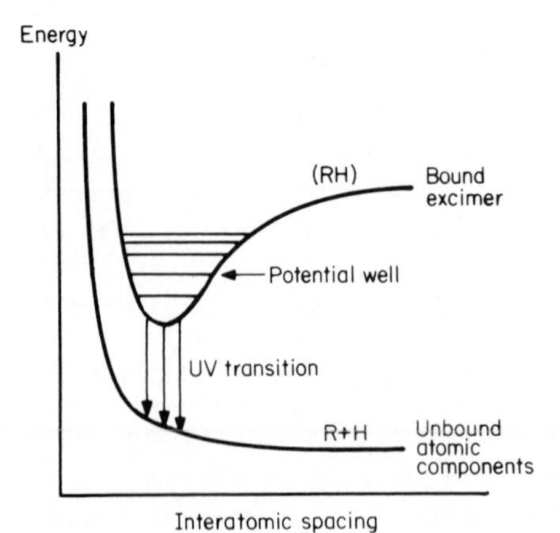

Figure 13-1 Internal energy of excimer molecules as a function of interatomic spacing for excited and ground states. The potential well makes the excited state stable, but the two atoms repel each other at all interatomic spacings in the ground state, so it is unstable. Transitions can occur over a range of energies.

nanoseconds by discharge instabilities. The internal kinetics allow pulse lengths to reach hundreds of nanoseconds with electron-beam or microwave excitation, but that is about the upper limit without elaborate pulse-stretching schemes being developed in the laboratory.

Internal Structure An excimer laser contains a sealed tube filled with laser gas through which an excitation pulse is passed. In discharge-driven lasers, the electrical pulse is perpendicular to the laser beam axis. In an electron-beam-pumped laser, the electrons are more likely to be directed along the beam axis. Normally some of the laser gas lies in a reservoir outside of the excitation region, and the gas may be circulated through the laser for operation at high powers or high repetition rates.

The laser cavity is designed to be sealed and repeatedly refilled, often with different excimer laser gas mixtures. This is necessary because the laser gas degrades during use, lasting thousands to millions of shots in commercial lasers, depending on the type of gas and operating conditions. (Longer gas-fill lifetimes are possible with gas reprocessing.) After the gas is spent, the cavity must be emptied and refilled, with purging and passification needed when changing between gas mixtures. The laser cavity, optics, and electrodes must be designed to resist corrosion by the halogens present in the laser gas. Passive components typically are coated with Teflon, while the electrodes are made of halogen-resistant materials such as nickel.

Optics Excimer lasers have such high internal gain that they are virtually superradiant, and this is a dominant factor in determining the design of their optical cavities. As in most other gas lasers, the rear cavity mirror is highly reflective. However, because of the high gain there is no need to coat the output mirror reflectively. The normal reflectivity of roughly 4 percent from an uncoated optical surface provides enough feedback for laser operation.

Quartz optics are listed as standard for most commercial excimer lasers, but they are used mostly with XeCl lasers because they are etched by fluorine. Also, the transmission of quartz drops at shorter ultraviolet wavelengths, making it necessary to use windows of magnesium fluoride or calcium fluoride for the 157-nanometer (nm) molecular fluorine line. MgF_2 or CaF_2 windows may also be desirable for high-power or high-repetition-rate operation on the 193-nm ArF line. The windows are affixed directly to the laser cavity, and hence they are exposed directly to the laser gas. This can lead to effects such as chemical etching of optics with ArF, or deposition of dust with XeCl. Reflective coatings are deposited on the outside of the rear cavity window to prevent damage from the laser gas.

The atmosphere strongly absorbs short-wavelength ultraviolet radiation. Wavelengths shorter than about 200 nm are known as the "vacuum ultraviolet" because atmospheric absorption is so strong the beams can only propagate in a vacuum. (However, the 193-nm ArF line can propagate in pure helium.) Wavelengths between 200 and 300 nm can be propagated over laboratory scales, but virtually no solar radiation at wavelengths shorter than 290 nm can reach the earth's surface.

Cavity Length and Configuration Excimer laser cavities are generally somewhat under a meter long. The standard configuration is a stable resonator, which generates a fairly large, divergent beam because of the high gain of excimers. Unstable-resonator optics are often available as options to give beams that are smaller and more uniform, although those qualities in some cases may come at the cost of reduced output power. Another possible configuration is injection locking, with pulses from a master oscillator triggering emission from an unstable-resonator "slave" oscillator, as shown in Fig. 13-2.

Inherent Trade-offs Like other repetitively pulsed lasers, excimer lasers produce less energy per pulse once repetition rate passes a certain level. At low repetition rates, average power increases linearly with the number of pulses per second. However, a point of diminishing returns is passed beyond which the energy per pulse decreases with repetition rate unless discharge power is turned up. At somewhat higher levels, the average power actually declines as repetition rate increases. The falloff point depends on both laser gas and configuration and reflects the need for both the gas and the driving electronics to "recover" from the previous pulse.

No single type of excimer laser offers a clear superiority in all characteristics. The krypton fluoride laser offers higher pulse energy and average power than other types, but its operating lifetime is shorter than xenon chloride, and its 249-nm wavelength is not transmitted well by air over long distances. XeCl, in turn, has longer lifetime and the air transmits its 308-nm output reasonably well, but is only about half as powerful as KrF.

Variations and Types Covered Excimer lasers are a generic type, and commercial excimer lasers can operate with several gas mixtures. The best

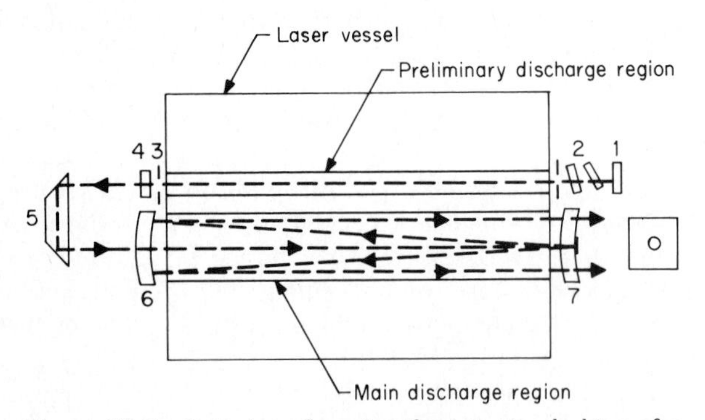

Figure 13-2 Optical configuration for injection locking of a "slave" unstable-resonator oscillator *(bottom)* to a master oscillator *(top)*, showing how power is extracted from most of the unstable resonator cavity. (1) Master oscillator rear reflector; (2) etalons; (3) apertures; (4) master oscillator output coupler; (5) prism; (6) rear unstable resonator optic; (7) front unstable resonator optic. *(Courtesy of Lumonics Inc.)*

known are the rare gas halides: ArF, KrCl, KrF, XeCl, and XeF. Other rare-gas-halide lasers have been demonstrated in the laboratory, but they have not proved commercially viable. Other gases can also be used in some commercial excimer lasers, including F_2 and N_2 in the ultraviolet, N_2^+ and atomic fluorine in the visible, and CO_2 in the infrared, although special optics are required for some of them. This chapter concentrates on rare gas halides and other gases which are used in commercial excimer lasers but not otherwise common; thus N_2 and CO_2 are excluded.

The rare gas halides are not the only excimer lasers to have been demonstrated in the laboratory. There is also a family of mercury halide lasers with similar properties that have ground states in which the atoms are bound together only very weakly. The best known of these is the mercury bromide laser, in which HgBr formed by dissociation of $HgBr_2$ emits at 502 to 504 nm in the blue-green. No commercial versions are on the market, but it has received careful scrutiny because its blue-green wavelength could be useful for communicating with submerged submarines (Burnham and Schmitschek, 1981).

There are several variables among commercial excimer lasers, including wavelength, pulse energy, maximum repetition rate, average power, gas lifetime, and pulse duration. Table 13-1 gives a sampling of these characteristics for selected commercial lasers. Commercial models themselves also differ. Some models offer high-energy pulses at low repetition rate, others provide high average power by delivering moderate-energy pulses at a high repetition rate, and a few try to combine both high-energy pulses and high repetition rate.

Beam Characteristics

Wavelength and Output Power The wavelengths of major excimer laser gases together with pulse energy and average output power from a sampling of 1985-vintage commercial models are shown in Table 13-1. Other ultraviolet and visible wavelengths can be produced by Raman shifting the standard excimer lines.

As a repetitively pulsed laser, the excimer laser has an average output power (in watts) that is the product of the pulse energy (in joules) times the number of pulses per second (hertz). On the strongest lines—ArF, KrF, XeCl, and XeF—typical average powers range from under a watt to 100 W or more. However, lasers with similar average power may have quite different output characteristics. A 10-W average power may be produced by generating 100 pulses of 100 millijoules (mJ) each, or ten 1-J pulses. Although a handful of expensive models deliver high-energy pulses at a high enough repetition rate to produce high average power, more typically the higher-energy pulses are produced at low repetition rates—on the order of one shot per second. Thus a typical laser delivering 1-J pulses would produce only one such pulse per second for an average power of 1 W, while a laser producing average power of 20 W is more likely to generate 100 pulses of 0.2 J each per second.

TABLE 13-1 Pulse Energy, Average Power, and Repetition Rate for Representative Commercial Excimer Lasers*

Laser or gas	F_2	ArF	KrCl	KrF	XeCl	XeF	N_2^+	F
Wavelength, nm	157	193	222	249	308	350	428	624–780
Lambda Physik EMG/200								
Pulse energy, mJ	15	500	—	1000	500	400	—	—
Average power, W	—	4	—	8	4	3	—	—
Repetition rate, Hz	10	10	—	10	10	10	—	—
Lumonics Hyper EX 460								
Pulse energy, mJ	—	180	45	400	200	200	—	—
Average power, W	—	20	6	60	30	27	—	—
Repetition rate, Hz	—	140	130	150	165	150	—	—
Oxford Lasers KX-5								
Pulse energy, mJ	—	75	—	80	40	30	—	—
Average power, W	—	0.3	—	0.4	0.12	—	—	—
Repetition rate, Hz	—	5	—	5	5	5	—	—
Questek 2440								
Pulse energy, mJ	—	300	—	500	300	200	—	—
Average power, W	—	18	—	32	20	12	—	—
Repetition rate, Hz	—	80	—	80	80	80	—	—

* All figures given are maximum values printed on specification sheets. In many cases only one or two of the values can be obtained at any one time; operation at peak repetition rate, for example, may reduce pulse energy and limit average power. The products selected are intended to represent commercial products available in 1985; each company has a broader range of products than shown.

Many laser makers list maximum values for average power, pulse energy, and repetition rate in their specification sheets. However, a quick comparison usually will show that the maximum average power is *not* the product of maximum repetition rate times peak pulse energy. Generally, the maximum pulse energy is produced only at low repetition rates, and beyond a certain repetition rate, the pulse energy starts dropping. The maximum average power normally is reached at a repetition rate slightly above that which can produce the highest-energy pulses. Further increases in repetition rate will cause average power to drop because pulse energy will drop faster than repetition rate increases.

As Table 13-1 indicates, there are significant differences among excimers in the output power available. For example, the relatively inefficient KrCl laser tends to produce pulse energies and average powers only about a tenth those from KrF. Note, however, that the ratios are not constant because the "best" discharge conditions differ among gases.

Although many applications require a specific wavelength, or limit choices to a single excimer, others may need only high-power ultraviolet pulses. In those cases, the choice is often between the 249-nm KrF laser and the 308-nm XeCl laser. KrF generally is a more powerful laser, but XeCl gas fills last longer because they do not contain highly corrosive fluorine, and the longer wavelength is easier to handle optically. Those trade-offs should be factored into any comparison of excimer laser types.

Efficiency Typical wall-plug efficiency for a discharge-driven KrF commercial laser is 1.5 to 2 percent. What is called "electrical" efficiency—the fraction of the energy stored in the discharge circuit that emerges in the laser beam—is typically 2.5 to 3 percent and can reach 4 percent for KrF. Even higher efficiencies come from measurements based on the amount of energy deposited in the laser gas, which is inevitably lower than the energy that drives the discharge. Efficiency based on deposited energy can be 4 to 5 percent for KrF driven by a discharge, and a little higher if pumping with an electron beam. However, any benefits of electron-beam pumping efficiency are more than offset by the high losses involved in generating the electron beam.

Be wary of efficiency measurements in the scientific literature, which typically measure the extraction of energy deposited in the gas, rather than overall efficiency.

Other excimer gases are less efficient than KrF, with the relative efficiency roughly proportional to the relative output power under the same operating conditions.

Temporal Characteristics Pulses from excimer lasers typically last from a few nanoseconds to a few tens of nanoseconds. Pulse lengths vary significantly among different gases used in the same laser, as indicated in Table 13-2. Individual pulses often have structure within them, as shown in Fig. 13-3.

Discharge instabilities in commercial lasers generally limit pulse length to tens of nanoseconds. Pulses as long as 250 ns have been obtained in the laboratory by ballasting the discharge (Hogan et al., 1983). Electron-beam

TABLE 13-2 Differences in Pulse Length for Different Gases Operated in Two Typical Multigas Excimer Lasers

	Pulse length, ns	
Gas	Laser 1	Laser 2
ArF	8–10	17
KrCl	5–8	10
KrF	12–16	23
XeCl	8–12	17
XeF	14–18	—

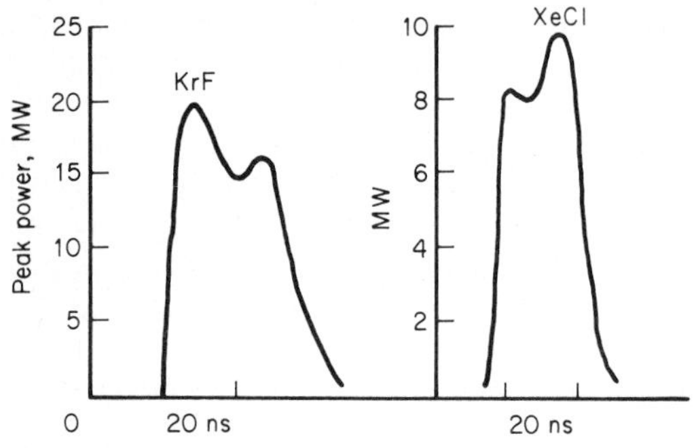

Figure 13-3 Pulse shapes for KrF and XeCl in one commercial laser, indicating the different types of substructures in pulses. *(Courtesy of Tachisto.)*

pumping makes it possible to stretch pulses to the microsecond range, but only at the cost of low repetition rate and efficiency.

The 10-ns upper-state storage time of most excimer molecules normally sets a lower limit on pulse duration. Pulses as short as 0.35 ns have been produced in the laboratory by using saturable absorbers with a small XeCl laser (Varghese and Moody, 1983). Pulses of 1 ns or less have been obtained in the laboratory from a small XeCl laser without saturable absorber. Laser fusion researchers are also looking at ways to compress pulse duration (George, 1982).

The short pulse length of excimer lasers can cause problems if triggering is with a spark gap, which has a jitter of ±20 ns. Because the jitter is comparable to the pulse duration, attempts to synchronize the laser pulses may miss. To avoid that problem, switching is usually with thyratrons, which have jitter of 1 to 2 ns.

Spectral Bandwidth The laser transitions in excimer lasers are broad, due to the lack of well defined energy levels in the ground state, as shown in Fig. 13-1. Thus laser emission takes place over a range of wavelengths. Typical linewidth of an excimer laser is about 0.3 nm, with output wavelength tunable across about 3 nm with some loss of output energy. Standard commercial lasers produce broadband emission on the laser transition, but in some cases optional line-narrowing optics can reduce linewidth to about 0.002 nm.

Amplitude Noise and Pulse Stability Pulse-to-pulse variations can be significant in excimer lasers, with specified values averaging ±5 percent. The figure varies between gases as well as between lasers and is highest for the weakest laser lines. For example, one company specifies amplitude stability of ±5 or 6 percent for gases such as ArF, KrF, XeCl, and XeF in one family of lasers, but ±15 percent stability for the weaker KrCl laser mixture. Some companies offer pulse-stabilization options which can reduce variations to around 1 percent on strong lines.

In the long term, output energy inevitably declines as the gas ages if the mixture is not cleaned and makeup gas added. Generally the decline is a smooth and gradual one. One design option to avoid such declines is to provide initial excess capacity in the laser, and automatically turn up the discharge power to maintain constant output until the excess capacity is exhausted.

Beam Quality, Polarization, and Modes Because of their high gain, stable-resonator excimer lasers have multimode output and poor beam quality. Higher-quality beams are possible if an unstable resonator is used. Beams normally have flat or gaussian profiles (often flat in one direction and gaussian in the other), but some have doughnut-profile beams. Normally output is unpolarized, but polarizing optics can be added.

Coherence Length The broad emission bandwidth of excimer lasers leads to limited coherence length. For a typical laser with linewidth of 0.3 nm at 300-nm wavelength—about 10^{12} Hz—calculated coherence length is about 0.3 mm.

Beam Diameter and Divergence Excimer lasers with standard stable-resonator optics typically have oblong beams roughly 1 by 2 cm across at the output window. Although there are minor variations among different models and gases, beam dimensions almost always fall within a factor of 2 of those values for discharge-driven lasers. Beam divergence is also uneven, typically 2 by 3 or 2 by 4 milliradians (mrad), although some lasers have larger and more uneven divergence, such as 3 by 10 mrad.

Unstable resonators generate beams of similar size but much smaller divergence, typically about 0.2 by 0.2 mrad.

Stability of Beam Direction The broad divergence and multimode output of most excimer lasers indicate that some variation in beam direction might be expected, but it is typically much smaller than divergence. However, published specifications do not list this quantity.

Suitability for Use with Laser Accessories Because of their ultraviolet wavelength, excimers are not suitable for use with conventional frequency-doubling crystals. However, harmonics can be generated by passing excimer laser pulses through gases. This technique has been used to generate coherent radiation at wavelengths as short as 35.5 nm, the seventh harmonic of KrF (Bokor et al., 1983). Although this opens up the extreme ultraviolet to laboratory study, at this writing applications remain in the research stage.

Raman interactions in molecular gases such as hydrogen can produce smaller frequency shifts when a tightly focused excimer laser beam is passed through a suitable gas cell. The Raman shift can be toward either longer or shorter wavelengths, but generally the longer wavelengths are generated more efficiently. The wavelengths and relative output powers produced by a commercial Raman shifter are shown in Fig. 13-4; qualitatively similar shifts would be produced by other such devices.

Q switching of excimer lasers is impractical because of the high gain. Both active and passive modelocking have been demonstrated in the laboratory, but they are not used in practice.

Operating Requirements

Input Power Small excimer-laser power supplies draw on the order of 10 A from 110-V, single-phase sources. Higher-power models require 208- or 380-V, three-phase service, and draw currents of 20 to 30 A. Specifications

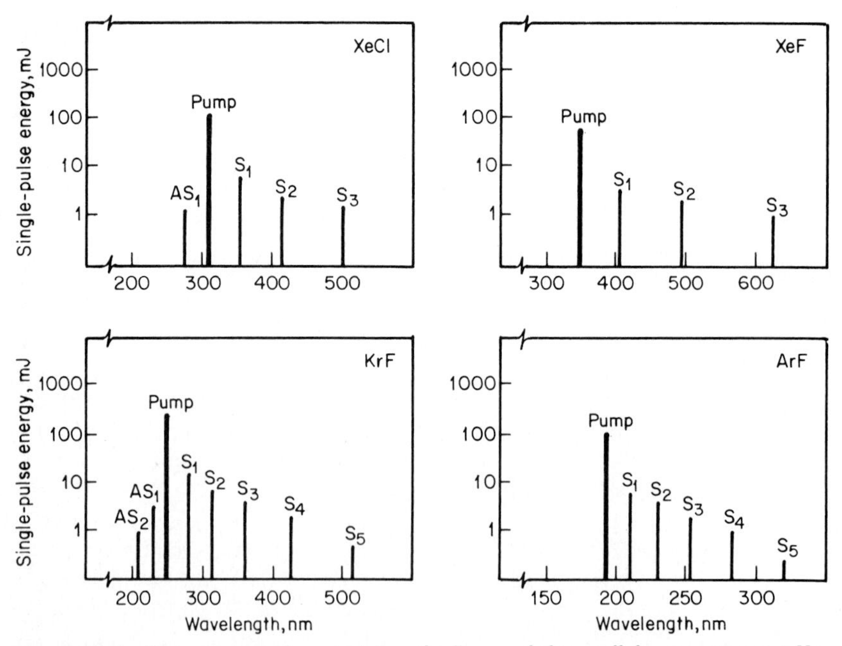

Figure 13-4 Typical output characteristics of a Raman-shifting cell for operation at 1-Hz repetition rate and optimum hydrogen pressure. S_n indicates Stokes-shifted line *n*; AS_n indicates anti-Stokes-shifted line *n*. The pumping laser is identified in the upper right corner of each small graph. *(Courtesy of Lambda Physik.)*

for some of the highest-power lasers do not mention input power requirements. All discharge-driven excimer lasers require high internal voltages, typically tens of kilovolts.

Cooling Many excimer lasers require tap-water cooling at rates of 2 to 8 liters per minute (L/min). Others, operating at lower pulse energies or repetition rates, are air-cooled.

Consumables The prime consumable in excimer lasers is the laser gas, which can account for a large part of operating costs. The gas deteriorates both during laser operation and with time, and eventually must be pumped out of the laser cavity and replaced. The number of shots possible varies among laser gases, as shown in Table 13-3 for one commercial laser. Gas lifetime also differs among different laser models.

Generally excimer lasers operate from three separate gas supplies: the inert buffer gas (typically helium or neon), the active rare gas (argon, krypton, or xenon), and the halogen being used. Halogens may be supplied in compounds such as nitrogen trifluoride and hydrogen chloride; when molecular fluorine is used, its concentration is normally reduced to 5 percent by adding helium, thus improving its storage and handling characteristics. Premixed single gas supplies may also be used. Many laser makers indicate the cost per gas fill—typically for xenon chloride—in their specification sheets; typical values are $6 to $30, obviously dependent on volume of the laser cavity. XeCl is a good choice as a baseline, both because of its popularity, and because Xe is the most expensive of the rare gases.

Although gas lifetime is normally specified as a number of laser shots, it is hard to achieve this lifetime unless the laser operates continuously at the specified maximum repetition rate. The reason is that the halogen continues to engage in chemical reactions even when the laser is off, although not as fast as when it is operating. This effect is shown in Fig. 13-5.

Gas lifetime can be extended by purifying it to remove impurities and periodically adding halogen to replace what has been consumed in the laser. Results of such operation are shown in Fig. 13-6.

Required Accessories A vacuum pump is needed to remove spent laser gas from the laser; this may or may not be included with the laser system. The spent gas cannot be exhausted directly to the air because it contains toxic and corrosive halogens, which must be removed by filtering or pumping into a storage container.

Vacuum systems which exclude air are needed for the propagation of vacuum-ultraviolet beams—the 157-nm output of F_2 and the 193-nm output of ArF. The ArF wavelength can be transmitted short distances through helium as well as through vacuum.

A few excimer lasers are sold as "bare" heads without internal power supplies. These require around 40,000 V to drive their discharges.

Lasers with spark-gap switches require a supply of dry air.

TABLE 13-3 Excimer Gas Lifetimes for a Multigas Laser without Gas Cleaning and Replenishment*

Laser or gas	F_2	ArF	KrCl	KrF	XeCl	XeF	N_2^+	F
Wavelength, nm	157	193	222	249	308	350	428	624–780
							—	—
Pulses	4×10^4	6×10^5	5×10^5	2×10^6	2×10^7	3×10^6	$>10^6$	2.5×10^5
	—						—	—
	—						—	—

* Lifetimes are the number of laser pulses required for the output to drop to half power.

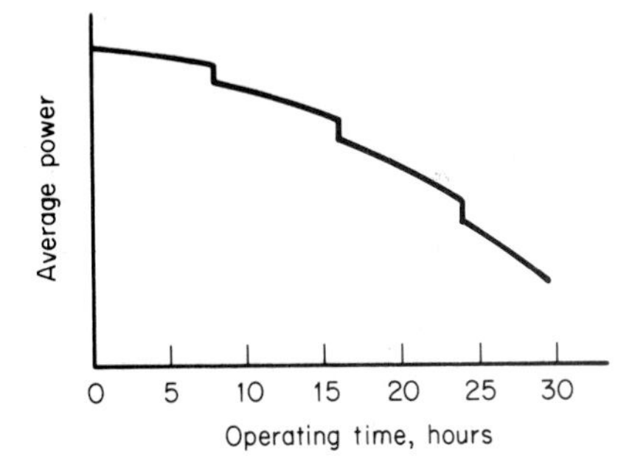

Figure 13-5 Average power fall-off for an excimer laser operated only during an 8-h workday but left off overnight, plotted against *operating* (not total) hours. The scale of the overnight drop is roughly that for KrF and XeF; there would be a smaller drop with XeCl, but a larger one with ArF. *(Courtesy of Questek Inc.)*

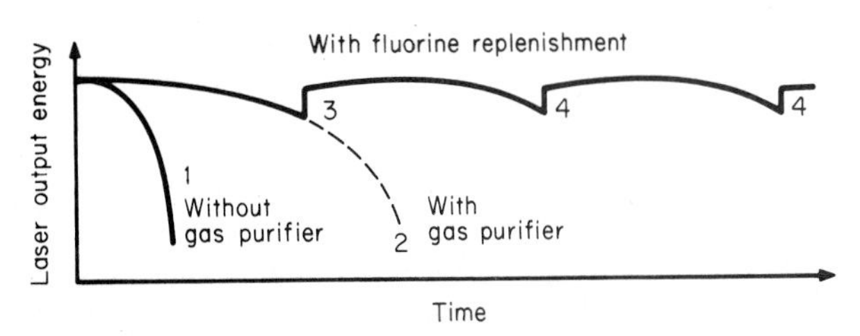

Figure 13-6 How use of gas purifier affects KrF output energy. Use of the gas purifier extends operating life by removing contaminants from the laser gas. Periodic replenishment of fluorine can sustain laser output over longer intervals. *(Based on data from Oxford Lasers Ltd.)*

Operating Conditions and Temperature Commercial excimer lasers are designed for the normal laboratory environment and work well there.

Proper operation of excimer lasers requires passivation of the gas cavity to limit the reactions involving the halogens in the gas mixture. These passification requirements are most stringent when switching laser gases, to make sure that the old gas has been thoroughly flushed out of the cavity.

Mechanical Considerations Typical dimensions for a small excimer laser are 1.4 by 0.7 by 0.35 m, with weight about 150 kg. Those packages contain both laser head and power supply. Many higher-power lasers have heads about the same size, but have the power supply in a separate package. The highest-power commercial excimers are somewhat larger.

Safety There are three potential hazards from excimer lasers: high voltages, ultraviolet light, and the laser gases themselves.

Discharge-driven excimer lasers require the application of tens of kilovolts across the laser gas. These high voltages are stored and switched within the laser and power supply, and care should be taken that all stored voltages have been dissipated before poking around inside the laser.

Short, powerful ultraviolet pulses are hazardous to the eye, and ultraviolet exposure can also harm the skin. Pulses with wavelength longer than about 315 nm can cause retinal damage, although exposure limits are not well quantified. Shorter ultraviolet wavelengths do not reach the retina but are absorbed in the cornea, where they can cause a painful but temporary sunburnlike effect called "snow blindness" or "welder's flash." Ultraviolet light also can present dangers to the lens of the eye. Skin exposure can lead to the same effects as overexposure to the sun's ultraviolet rays (Sliney and Wolbarsht, 1980).

Although the rare gases used in excimer lasers are innocuous, the halogens are not. Federal standards limit exposure to fluorine in air to 0.1 part per million (ppm), and chlorine exposure to 1 ppm (Weast, 1981). Even NF_3, a fluorine donor sometimes used to avoid the hazards of F_2, is somewhat toxic, with federal standards limiting exposure to 10 ppm. Pure F_2 is so corrosive that it presents potential fire and explosion hazards; to avoid these dangers, the F_2 used in excimer lasers is normally diluted to 5 percent concentration by adding helium, and the diluted gas supplied in cylinders.

Special Considerations The high-voltage discharges in excimer lasers generate electromagnetic interference (EMI) which can affect surrounding equipment. Laser makers are aware of this problem and generally shield their lasers to minimize EMI that could disrupt the operation of other equipment. Shielding is successful enough that some recent models include a microprocessor control system in the same case as the laser—though obviously well shielded from EMI effects. Be wary, however, of personal computers, some of which have little or no EMI shielding and may be very sensitive.

Operation of pulsed excimers can generate an audible "ping" at each pulse. The noise is more at the nuisance level than an actual disruption.

Reliability and Maintenance

Lifetime Excimer laser makers generally do not indicate a projected lifetime for their products (other than the "lifetime" of a gas fill). This may change as some companies develop models intended for industrial applications. However, as long as excimer lasers are designed for operation with multiple gases, which expose the inside of the laser to different conditions and degradation mechanisms, it will be hard to assign a precise lifetime to the lasers. The lifetime of the operating gas is short enough that it is considered a consumable.

The output windows attached to the laser cavity are limited in life by corrosion and contaminants from the laser gas. The damage depends on the laser gas and operating conditions. ArF tends to corrode output windows, while XeCl leaves a fine dust. In many cases, it is possible to remove and clean the windows.

High-voltage switching components have a reputation for reliability problems. There have been great improvements in comparatively inexpensive spark-gap switches over the past decade, but they still require adjustment and cleaning after tens of millions of shots, and suffer from ± 20 ns jitter. Thyratrons are more expensive but can operate much longer without attention. Other approaches include rail-gap and solid-state switching.

Maintenance and Adjustments Needed Periodic replacement of the laser gas is the most obvious maintenance needed in excimer lasers. If the active gas is being changed, the laser cavity must be passivated to remove contaminants left from the previous mixture, for example, to get fluorine residues removed before switching to XeCl.

If spark-gap switches are used, they must be cleaned and adjusted after tens of millions of laser shots—which at 60-Hz repetition rate can take only a few days of continuous operation.

Output windows require periodic cleaning to remove contaminants and corrosion on the surfaces exposed to the laser gas. Cleaning the interior optics is not a simple job because it involves breaking gas seals that prevent leakage around the windows, then re-forming the seals after cleaning. Once air has been allowed into the laser cavity, whether during cleaning or by operator error, the laser cavity must be repassivated with the gas mixture to be used in it.

Mechanical Durability Excimer lasers are not designed to resist heavy shocks and should be given the usual care accorded to laboratory equipment.

Failure Modes and Causes The corrosive halogens used in excimer lasers still present some problems, although the use of nickel electrodes and Teflon-coated cavities prevents the severe corrosion problems that ate away early excimer lasers. Nonetheless, exposed electrodes and windows still suffer from corrosion and deposition of contaminants. The degradation of windows and other optics can be accelerated by the powerful ultraviolet radiation produced by excimer lasers. High-voltage switching electronics are subject to their own failure mechanisms.

One of the commonest problems is operator error, accidentally allowing laboratory air into the inside of the laser cavity. Reactions of air with the laser gas or contaminants within the laser cavity can cause problems and require repassification of the inside of the laser with fresh laser gas, or possibly cleaning of the cavity and optics.

Possible Repairs Failures in high-voltage switching electronics or in optics require replacement of the damaged components. Corrosion and contamination problems require cleaning or replacement of the affected components.

Commercial Devices

Standard Configurations The typical commercial excimer laser is an all-purpose laboratory device, intended for use with several laser gases emitting at wavelengths from 157 nm in the vacuum ultraviolet (F_2) to 624 to 780 nm in the red and near-infrared (F). A single set of electrodes is generally used for all gases. One set of optics generally suffices for operation at 200 to 800 nm, but special optics are needed for the 157-nm F_2 line or high-power output on the 193-nm ArF line. Some models can produce 10.6-micrometer (μm) output when used with carbon dioxide, and suitable optics.

There are various approaches to the pulse energy–average power–repetition rate trade-off. Some lasers deliver pulses of a joule or more each at repetition rates of 1 Hz or less; electron-beam pumping can produce higher-energy pulses at lower repetition rates. Other lasers can produce much higher average powers by producing less energetic pulses at higher repetition rates.

Commercial excimer lasers almost invariably rely on discharge excitation. This reflects the economic facts of life: discharge excitation is simpler, more compact, and cheaper than electron-beam excitation.

Options Several options are common on excimer lasers:

- Electrodes designed for use with specific gas mixes, offering better performance for the specified mix than general-purpose electrodes.
- Electrodes designed for high repetition rates for use with gases such as XeCl which can sustain high repetition rates.
- Choice of spark-gap or thyratron switching. Spark gaps are less expensive, but their 20-ns jitter is comparable to the pulse length and can make synchronization impossible. Thyratrons have jitter of 1 to 2 ns.
- Unstable-resonator optics.
- Oscillator-amplifier configurations, or injection-locking of a "slave" oscillator to a master oscillator.
- MgF_2 or CaF_2 optics for wavelengths in the vacuum-ultraviolet where standard quartz optics are not transparent enough.
- Gas purification systems which remove impurities to extend the useful life of a gas mix. Periodic replenishment of the halogen component in addition to gas purification can extend lifetime even further.
- Line-narrowing and tuning optics.
- Polarizing optics.

Special Notes Most excimer lasers allow internal or external triggering of pulses. Repetition-rate limitations are often imposed by the design of the charging electronics.

Contamination of the inside of an excimer laser by air can seriously degrade performance and require a thorough cleaning. The commonest cause of such contamination is operator error in handling the laser gases, so users should be cautious.

Excimer lasers are gradually emerging from the laboratory stage to real-world applications in areas such as semiconductor processing. This change will be accompanied by an evolution of designs to meet the needs of users who know and care little about laser physics. One result will probably be an increasing emphasis on single-gas operation, most likely XeCl because of its long operating life and reasonably high power.

Pricing Current (1985) prices for excimer lasers run from around $15,000 to well over $50,000 for standard models. The highest-power lasers, and some customized models, can cost hundreds of thousands of dollars.

Suppliers When excimer lasers first came on the market, there were only a handful of suppliers. The market is becoming increasingly crowded, with a number of recent entries, and this influx of new companies should stimulate competition in both price and design.

Applications

Most excimer lasers have been used in scientific research in a variety of fields. Major areas include

- Pumping of tunable dye lasers
- Nonlinear spectroscopy
- Ultraviolet spectroscopy
- Photoexcitation and photochemistry
- Isotope separation and enrichment

A number of potential industrial applications are in various stages of development:

- Materials processing
- Patterning of semiconductor integrated circuits
- Deposition of patterns
- Polymer etching and semiconductor mask fabrication
- Laser marking
- Semiconductor device annealing
- Surface treatment of metals, ceramics, and plastics
- Photochemistry
- Atmospheric monitoring

Also being studied are potential medical applications, most involving use of short excimer pulses to ablate surface tissue, such as to make incisions in the cornea for radial keratotomy.

Although some of the semiconductor applications have hovered near the threshold of commercial use, none has found industrial applications on a significant scale as of this writing. Technical difficulties account for some of the problem, but many potential users are also worried that excimer lasers may not be ready for use as practical tools. Proving that excimer lasers can do that job is the next big challenge for excimer developers.

BIBLIOGRAPHY

J. Bokor et al.: "Generation of 35.5-nm coherent radiation," *Optics Letters 8*(4):217–219, April 1983.

Ralph Burnham and Erhard J. Schmitschek: "High-power blue-green lasers," *Laser Focus 17*(5):54–66, June 1981.

E. V. George (ed.): *1981 Laser Program Annual Report*, Lawrence Livermore National Laboratory, Livermore, Calif., 1982, pp. 7-82–7-94.

C. Breck Hitz, "The laser marketplace—1985," *Lasers & Applications 4*(1):47–56, January 1985.

D. C. Hogan et al., "Pulse length limitations in XeCl* lasers," paper MA2-1 in *Technical Digest Topical Meeting on Excimer Lasers*, January 10–12, 1983 (published by Optical Society of America, Washington).

Charles K. Rhodes (ed.): *Excimer Lasers*, Springer-Verlag, Berlin & New York, 1979. Collection of review articles.
David Sliney and Myron Wolbarsht: *Safety with Lasers and Other Optical Sources*, Plenum, New York, 1980. Comprehensive handbook on optical safety.
Technical Digest Topical Meeting on Excimer Lasers, January 10–12, 1983, Optical Society of America, Washington, D.C.
Technical Digest Topical Meeting on Excimer Lasers, September 11–13, 1979, Optical Society of America, Washington, D.C.
Thomas Varghese and S. E. Moody: "Subnanosecond excimer pulse generation using saturable absorbers," paper MC5 in *Technical Digest Topical Meeting on Excimer Lasers*, January 10–12, 1983, Optical Society of America, Washington, D.C.
Robert C. Weast (ed.): *CRC Handbook of Chemistry and Physics*, 62d ed., CRC Press, Boca Raton, Fla., 1981, pp. D-101–D-105.
J. F. Young et al.: "Microwave excitation of excimer lasers," *Laser Focus* April 1982, pp. 63–67.

CHAPTER

14

nitrogen lasers

First demonstrated over two decades ago (Heard, 1963), the 337-nanometer (nm) molecular nitrogen laser has been available commercially since 1972. One of the first ultraviolet lasers on the market, it played a key role in the spread of pulsed dye lasers into research laboratories. It served as a pump source in many pioneering experiments in dye laser spectroscopy and laser chemistry. The nitrogen laser is simple enough that many researchers have built their own units, but versatile commercial models also are available and are used in many laboratories.

The main advantages of the nitrogen laser are low cost, simplicity of operation, and high repetition rate. Atmospheric-pressure nitrogen lasers can produce nanosecond or subnanosecond pulses in the near-ultraviolet without the need for complex modelocking schemes. The 337-nm N_2 wavelength is an excellent pump source for dyes emitting over the 360- to 950-nm range. Shorter ultraviolet pulses can be produced by frequency doubling the output of nitrogen-pumped dyes.

Relatively low efficiency has proved the main disadvantage of the nitrogen laser. This limits N_2 laser pulse energy to about 10 millijoules (mJ), and restricts average power to a few hundred milliwatts in practical designs. Because of these limitations, more powerful neodymium-YAG and excimer lasers have replaced nitrogen lasers in applications such as nonlinear optics and laser chemistry. However, N_2 lasers are still widely used in laboratories that require lower power or that have limited budgets.

Internal Workings

The active medium in a nitrogen laser is nitrogen gas, usually pure, at an operating pressure between 20 torr (2700 Pa) and atmospheric pressure. In most models the gas flows through the laser and is exhausted to the atmosphere, with a vacuum pump required if—as is common—operation is below atmospheric pressure. Very fast transverse excitation allows operation at atmospheric pressure, producing nanosecond to subnanosecond pulses. Nitrogen lasers also can be sealed off at various pressures for operation at low repetition rates.

The nitrogen molecules are excited by a fast high-voltage discharge, populating the upper laser level shown in Fig. 14-1. This upper laser level is an excited electronic state with a lifetime of about 40 nanoseconds (ns), emitting a photon at 337.1 nm when dropping to the lower laser level. The laser transition is a vibronic one, involving changes both in electronic and vibrational energy levels. Careful examination of the neutral N_2 spectrum at high resolution shows that the 337.1-nm laser transition includes many discrete lines, arising because the initial and final states have a multitude of vibrational sublevels (Davis and Rhodes, 1982). By laser standards, N_2 lasers are broadband, but compared with ultraviolet lamps N_2 laser output can be considered monochromatic.

Kinetics of the nitrogen laser are in many ways unfavorable. The decay time of the lower level is about 10 microseconds (μs), much longer than the lifetime of the upper laser level. The lower laser level decays to a metastable

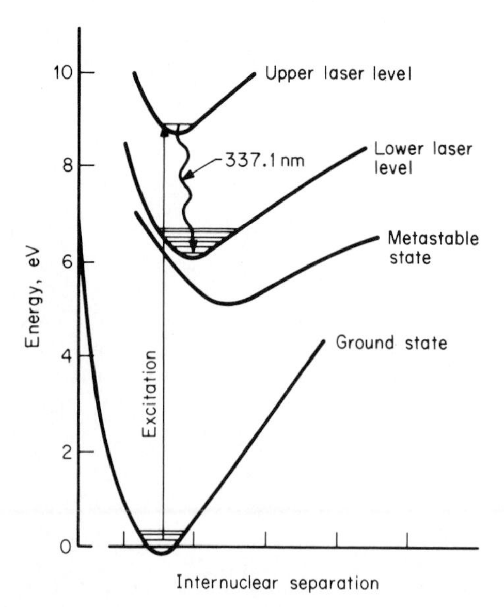

Figure 14-1 Energy levels in neutral nitrogen molecules involved in 337.1-nm laser transition. The broad curves represent electronic energy levels, with the lines within the potential wells representing vibrational energy levels for molecules in those electronic states. The laser transition is a vibronic one in which both electronic and vibrational energy states change.

level with a lifetime of many seconds (Verdeyen, 1981). Such slow decays of lower-lying levels would make a population inversion impossible in many materials. However, those problems are offset by the extremely high efficiency with which a fast electric discharge can populate the upper N_2 level. The result is a transitory population inversion, which generates laser pulses that last a matter of nanoseconds or less, then self-terminate because of ground-state bottlenecking.

Internal Structure

The rapid population and quick decay of the upper laser level combine to produce very high gains—50 dB/m or more are typical during the brief pulses. Such high gains eliminate the need for a high-quality laser resonator. Virtually all nitrogen lasers can operate in superradiant mode without any cavity mirrors. Output power can be more than doubled by use of a simple cavity with two flat mirrors—the back mirror 100 percent reflective and the front mirror 4 percent reflective—which also greatly reduces the beam divergence.

The comparative simplicity of nitrogen-laser construction is illustrated by the fact that *Scientific American* once featured a home-made version in its "Amateur Scientist" column (Strong, 1974). However, design and construction are far from trivial. From a practical standpoint, the biggest challenge in building a nitrogen laser lies in the pulse-forming network and switch, which must deliver a pulse of 15 to 40 kV with risetime under 10 ns. Both the fast-pulse circuitry and the channel through which the discharge is applied to the gas are important factors in laser performance. Channels with rectangular cross section are the usual choice, with the discharge applied transversely to the laser axis rather than along the same direction that the beam is produced. Fast rise time, essential for successful laser operation, depends on factors including electrode and discharge configurations. These design considerations are similar to those involved in design of rare-gas-halide excimer lasers, described in Chap. 13, and some multigas excimer lasers can also produce laser action from N_2 at 337 nm.

The high gain of the N_2 laser permits fairly short cavities, while the short pulse length does set an upper limit. In practice, most nitrogen laser cavities are 15 to 50 cm long.

Beam Characteristics

Wavelength and Output Power Standard commercial nitrogen lasers emit at 337.1 nm, where neutral molecular nitrogen has many closely spaced lines. N_2 can also emit at additional lines at 357.6 nm and in the near-infrared (Davis and Rhodes, 1982), but emission on those lines is suppressed in standard models. The singly ionized species, N_2^+, has a laser line at 428 nm which is obtainable from some commercial multigas excimer lasers, but that line is not produced by standard N_2 lasers and is rarely used because of its low power.

Commercial N_2 lasers are rated to produce peak powers of 1 kW to 2 MW at 337.1 nm. Pulse energies range from under 10 μJ to about 9 mJ. The low pulse energy levels reflect the gas's very limited energy-storage capacity, which originates with the short upper-state lifetime.

Efficiency Overall efficiency of a pulsed nitrogen laser is typically 0.11 percent or less, measuring input power at the wall plug. Operation at lower laser powers or the use of vacuum pumps (which draw electrical power) can cause overall efficiency to be lower. This low overall efficiency has limited nitrogen lasers to applications that require moderate power and energy, such as pumping dye lasers to produce visible pulses of 1 mJ or less, or frequency-doubled tunable pulses of no more than 50 μJ. Higher powers and energies require excimer lasers or harmonics of Nd–YAG.

Temporal Characteristics As mentioned above, internal kinetics limit nitrogen lasers to pulsed operation. The output pulse length depends both on gas pressure and on characteristics of the excitation discharge circuit. In commercial nitrogen lasers, pulse lengths range from about 300 picoseconds (ps) (at atmospheric pressure) to about 10 ns [at about 20 torr (2700 Pa)]. Many inexpensive commercial models operate at low repetition rates in the 10- to 20-Hz range; other models can operate at a couple of hundred pulses per second, and repetition rates of 1000 Hz have been achieved in the laboratory (Measures, 1984).

Spectral Bandwidth Rarely important in N_2 laser applications, spectral bandwidth is not specified for many models. In practice, laser emission is spread over multiple lines in the vibronic band at 337.1 nm, leading to a typical spectral bandwidth of 0.1 nm.

Amplitude Noise Amplitude noise in the nitrogen laser arises from pulse-to-pulse variations in output power and energy. High-performance thyratron-switched nitrogen lasers can stabilize output power to within a few percent. Makers of lower-power inexpensive lasers normally do not specify pulse-to-pulse variations, and users should expect fluctuations to be larger than those of most expensive models.

Beam Quality, Polarization, and Modes Like other lasers with high gain and without true resonators, nitrogen lasers tend to have poor beam quality. Normally output is unpolarized and in multiple transverse and longitudinal modes. Because most discharge cavities are rectangular in cross section, output beams are typically oval or rectangular in cross section, a beam profile that can be helpful in transverse pumping of dye lasers.

Coherence Length If spectral bandwidth of a nitrogen laser is assumed to be around 0.1 nm, the coherence length can be estimated at on the order of 1 mm. Spatial coherence is also small, avoiding diffraction irregularities in focused beams.

Beam Diameter and Divergence Low-power nitrogen lasers can have beams as small as a millimeter or two in diameter. For high-power models, the beam size can range upward to 6 by 32 mm, with the difference in dimensions due to the transverse pumping geometry.

Specified full-angle divergences for most models range between about 1 by 2 milliradians (mrad) to 6 by 14 mrad. Note that some manufacturers may specify divergence as a half-angle value, which must be doubled to get full-angle divergence. There is one exception to the pattern of high divergence, a beam divergence of 0.35 by 0.6 mrad—close to the diffraction limit—specified by one manufacturer, which uses a two-stage master-oscillator power-amplifier laser configuration.

Suitability for Use with Laser Accessories Nitrogen lasers cannot be *Q*-switched because of their small energy-storage capacity and because photons stay too short a time in the resonator, less than 5 ns. The short photon lifetime also makes it impossible to build up the longitudinal modes essential for modelocking. The 337-nm wavelength is too short for frequency doubling in nonlinear crystals, although the short pulses with high peak power would otherwise be attractive.

Operating Requirements

Input Power Small commercial N_2 lasers which do not require vacuum pumps can operate from 12-V batteries or draw only modest currents from a 110-V ac line. The homemade nitrogen laser described in *Scientific American* operated from a 6-V dry cell (Strong, 1974). High-power nitrogen lasers equipped with vacuum pumps may draw a total of 20 A at 110 V, or draw a smaller amperage at 220 V. Users should note that power requirements for external vacuum pumps not supplied with the laser will not be included on manufacturers' data sheets.

Cooling Forced-air cooling is standard on nitrogen lasers, although at least one manufacturer has offered water cooling as an option for operation at high repetition rates.

Consumables Most nitrogen lasers require a continuous supply of nitrogen gas, which is exhausted into the air. Exhaust is through a vacuum pump if operation is at pressures below 1 atm. Gas consumption rates range from 0.1 to 40 L/min, depending on laser design, operating power, and repetition rate. Standard-purity nitrogen is adequate because laser operation is not extremely sensitive to impurities.

An alternative to a continuous flow of nitrogen is a modular design in which the laser gas is sealed in a closed laser cavity which is packaged in a module with switching electronics. This module can be replaced as required, typically after tens of millions of pulses, and in that sense is a consumable.

Required Accessories Nitrogen lasers without sealed cavities require a supply of nitrogen gas to pass through the laser. In the majority of models which have gas flowing through the laser cavity at subatmospheric pressures, a vacuum pump is also needed. The vacuum pump is included in some models, but many require an external pump. Designs without internal pumps let users save money by using a pump already in their laboratory. An optional vacuum pump can be purchased from the laser supplier if one is needed.

Operating Conditions and Temperature Nitrogen lasers are designed for room-temperature operation in standard laboratory environments. Lower-power, compact models are designed to operate under a somewhat broader range of conditions, with some able to run from batteries. Some manufacturers have shown interest in industrial applications, but they have yet to start inserting specified operating conditions for a broad range of environments in their data sheets.

Mechanical Considerations Nitrogen lasers in the 50-kW range have been made in self-contained units, with internal power supply and sealed gas cavity, that weigh just 5 lbs (2.3 kg) and measure 9.75 by 4.5 by 2.5 in (24.8 by 11.4 by 6.4 cm). However, versions in the megawatt range are much bigger. One of the largest models has a 95-kg head measuring 1.27 by 1.06 by 0.26 m, plus a separate 23-kg vacuum pump, a 55-kg power supply, and a gas supply.

Safety Most nitrogen lasers fall under class IIIb of the federal laser product performance classification code. The 337.1-nm output of N_2 lasers falls in the UV-A region, where absorption in the lens of the eye provides some protection for the retina, but not enough to prevent high-power pulses from causing retinal damage in some circumstances. Corneal damage is also possible (Sliney and Wolbarsht, 1980). Eye response and damage mechanisms in the near-ultraviolet have yet to be fully quantified, so caution is the best policy.

The high voltages needed to drive a nitrogen laser can pose a serious shock hazard. Users should take special care that all components have been discharged before touching the power supply.

Special Considerations In many applications an external source triggers pulses from a nitrogen laser, which are then measured or studied in some way, either by themselves, in interactions with other materials, or indirectly (as when the nitrogen laser pumps a dye laser to generate a pulse of tunable light). Users involved with such applications must consider two specifications of pulse timing: the delay between the trigger pulse and the laser pulse, and the jitter or margin of error in firing the laser pulse. These two quantities differ widely among models, tending to be smallest in the most costly lasers and larger in less-expensive low-power models. Delay is often many times the pulse length, and in some cases jitter, too, can be longer than pulse duration, making it critical for users to pay attention to pulse timing when making measurements. External synchronization with subnanosecond time resolution is possible by splitting off a small part of the laser output and monitoring it with a low-cost fast photodiode.

Reliability and Maintenance

Lifetime As for most types of laboratory instruments, the lifetimes of nitrogen lasers are not normally specified. In most nitrogen lasers, the shortest-lived components are high-voltage switches, which can last tens or hundreds of millions of pulses. Thryatron switches are the most reliable, but also the most costly. In a few cases, the switching components are packaged in a disposable module along with a sealed nitrogen gas cell, which has comparable lifetime.

Maintenance and Adjustments Needed In certain modular designs, the entire laser tube and switching electronics module is replaced periodically. In most N_2 lasers, optics should be cleaned to prevent potential damage to dirty areas exposed to high near-ultraviolet powers. Thyratron replacement intervals average more than 2 years for the more advanced designs; spark gaps in simple units may have to be cleaned weekly.

Mechanical Durability Most nitrogen lasers are designed to function as laboratory equipment. Although few components are inherently fragile, generally no special effort has been made to ruggedize them.

Failure Modes, Causes, and Possible Repairs Failure of high-voltage switches is the commonest problem with nitrogen lasers; the solution is simply replacement of the affected components. Damage to output windows is a possibility, although not as likely as at the higher powers and shorter wavelengths of rare-gas-halide excimer lasers. In modular models, the components considered most likely to fail are concentrated in a single box, which is designed for ready replacement.

Commercial Devices

Standard Configurations Most commercial nitrogen lasers fall into one of two groups: high-power types or inexpensive models. The first class tends to be large and costly, with their main applications in pumping of pulsed dye lasers with reasonably high output power. Some such models resemble rare-gas-halide excimer lasers, and in fact some "multigas" excimers can operate with a nitrogen gas fill.

The other approach to nitrogen lasers has been compact versions for limited budgets. In many cases these, too, are used for pumping dye lasers, but the emphasis is on a simple design that can produce usable power levels for laboratory experiments. The modular design approach has been pioneered for these low-cost lasers, with developers hoping that it would encourage at least some applications outside the laboratory.

Compact low-cost versions usually come with laser head and power supply in a single package. In the more expensive models, the laser head may be separate from the power supply, and an external vacuum pump may also be needed for laser operation. Except in modular versions with sealed laser gas cells, a gas supply is necessary. A sampling of commercial nitrogen lasers is shown in Fig. 14-2.

Options Most options for nitrogen lasers deal with gas supply and handling. Some models come with vacuum pumps, while others allow users to supply their own. Related options include gas-flow monitoring and output-power monitors. In some models, adjustment of pulse length is optional.

Special Notes Because the major application of nitrogen lasers has been in pumping pulsed dye lasers, some models may be sold as a package with a dye laser made by the same company. Often the two lasers are designed

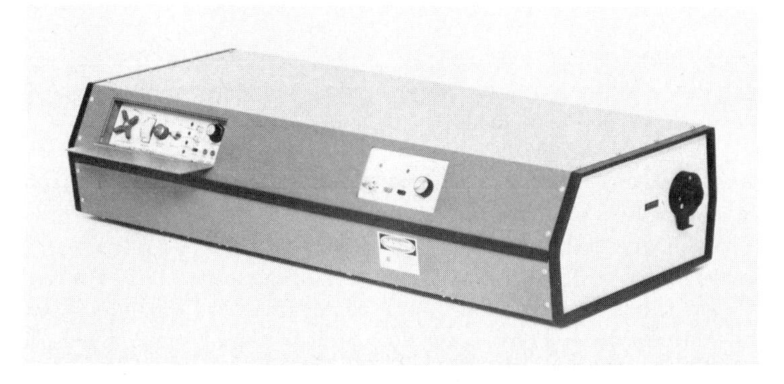

Figure 14-2 A sampling of commercial nitrogen lasers showing opposite ends of the power spectrum. *(Top)* A 1.2-m long model delivers pulses of several millijoules. *(Courtesy of Cooper LaserSonics.)* *(Bottom)* A modular sealed laser produces 0.1-mJ pulses. *(Courtesy of Laser Science Inc.)*

specifically for use with each other, including provisions for attaching them together.

Pricing In 1985, quoted prices of complete nitrogen lasers range from $1600 to tens of thousands of dollars. The least-expensive laser can be bought for $1000 without vacuum pump and gas-regulation equipment. The price range reflects a steep variation in laser performance.

Suppliers A number of companies supply nitrogen lasers, with most offering more than one model. Most suppliers specialize in either high-power or low-cost nitrogen lasers rather than offering both types. Makers of multigas excimer lasers that can emit at the 337.1-nm N_2 wavelength when filled with nitrogen are not really in the nitrogen laser market, as evidenced by data sheets that make no mention of nitrogen except in a table of wavelength and power specifications.

Applications

The major application of nitrogen lasers for many years has been pumping pulsed dye lasers. The short, high-power pulses in the near-ultraviolet are well-matched to the characteristics of many visible-wavelength laser dyes. Compact versions can be pump sources for inexpensive dye lasers that put tunable lasers within reach of limited budgets.

Like other ultraviolet sources, nitrogen lasers can be used to study and measure fluorescence effects. Their short pulse length makes them valuable for time-resolved measurements. So far most fluorescence measurements have been made in the laboratory. However, the availability of inexpensive nitrogen lasers could stimulate their use in measurement instruments, and some instrument makers are reportedly testing nitrogen lasers for such applications. Although even low-cost nitrogen lasers are much more expensive than incoherent ultraviolet lamps, their better wavelength control, shorter pulses, and more strongly directional output appear valuable in system applications. The high efficiency of N_2's 337-nm emission in stimulating fluorescence effects also could lead to new applications in fields such as analytical fluorometry. Small nitrogen lasers, or N_2-pumped lasers, can be adapted easily to replace the standard lamps on fluorescence microscopes.

Battery-powered nitrogen lasers could be important in opening up field measurement applications, particularly with the nitrogen laser pumping a dye laser. Uses could include remote sensing with a portable tunable laser.

The high peak power of a focused beam has led to nitrogen-laser applications in cell surgery and microcutting. Nitrogen lasers also can be used as ultraviolet sources for a variety of other laboratory purposes.

BIBLIOGRAPHY

Robert S. Davis and Charles K. Rhodes: "Electronic transition lasers," in Marvin J. Weber (ed.), *CRC Handbook of Laser Science and Technology*, vol. 2, *Gas Lasers*, CRC Press, Boca Raton, Fla., 1982, pp. 273–312.

H. G. Heard: "Ultraviolet gas laser at room temperature," *Nature 200*:667, November 16, 1963.

Raymond M. Measures: *Laser Remote Sensing; Fundamentals and Applications,* Wiley-Interscience, New York, 1984, pp. 188–189.

David Sliney and Myron Wolbarsht: *Safety with Lasers and Other Optical Sources,* Plenum Press, New York, 1980.

C. L. Strong: "The amateur scientist [column]: An unusual kind of gas laser that puts out pulses in the ultraviolet," *Scientific American 230*(6):122–127, June 1974.

Orazio Svelto: *Principles of Lasers,* 2d ed., Plenum Press, New York, 1982.

Joseph T. Verdeyen: *Laser Electronics,* Prentice-Hall, Englewood Cliffs, N.J., 1981, pp. 382–383.

CHAPTER

15

far-infrared gas lasers

The "far-infrared" is the part of the electromagnetic spectrum loosely defined as extending from about 10 micrometers (μm) to 1 millimeter (mm). Neither endpoint is firmly or formally defined, and the longer-wavelength part of that region, beyond about 40 μm, is sometimes called the *submillimeter* region.

The haziness of the definition reflects the fact that the far-infrared had been little explored until recent years, when far-infrared laser sources became available. In these lasers, molecular gases emit at wavelengths of 10 μm to about 2 mm, on vibrational-rotational transitions at the short-wavelength end, and on purely rotational transitions at the longer wavelengths. Laser action has been demonstrated experimentally on well over 1500 far-infrared lines. Not all of those lines have been reported from commercial lasers, partly because manufacturers have used only a limited number of gases.

In practice, two commercial lasers which emit at the short-wavelength end of the far-infrared are not usually labeled "far-infrared lasers." These are the carbon dioxide laser (Chap. 10) and the nitrous oxide (N_2O) laser (Chap. 16), both of which can be excited electrically to emit in the 10-μm region. Most other far-infrared lasers require optical excitation with either a CO_2 or N_2O laser, although a handful emitting at longer wavelengths can be excited electrically.

Commercial far-infrared lasers emit at wavelengths from about 28 μm to 2 mm. A single laser cavity can house many different gases, and typically each gas can emit at a number of separate lines. So far most of the applications of far-infrared lasers have been in research, such as diagnostics of fusion plasmas, submillimeter astronomy, and studies of semiconductor materials. Potential applications in nondestructive testing and measurement are in development.

Common materials behave much differently in the far-infrared than at shorter infrared wavelengths and in the visible. Air effectively blocks long-distance transmission between about 30 μm and roughly 500 μm, and even at longer and shorter wavelengths transmission is limited. Most common infrared window materials are opaque at wavelengths longer than about 40 μm, and few transmission curves extend beyond 50 or 100 μm. Silicon, germanium, crystalline quartz, and diamond are among the few solids transparent beyond 50 μm. Plastics such as polyethylene also can be useful in the far-infrared. Fine metal meshes can serve as far-infrared reflectors because at long enough wavelengths their surfaces appear continuous. These differences in material response help make far-infrared lasers valuable for some applications but also make it important that users avoid unwarranted assumptions about far-infrared optics.

Internal Workings

Active Media Many different molecular gases can be active media in far-infrared lasers, if they possess a permanent dipole moment. The most important molecules in commercial far-infrared lasers are alcohols and other carbon compounds. Minor changes to the molecules, such as isotopic substitution of deuterium (hydrogen 2) for normal hydrogen can give new laser lines. A sampling of major lines of commercial far-infrared lasers is given in Table 15-1. Tabulations of lines seen experimentally are much more extensive (see, for example, Knight, 1982a and 1982b; Button et al., 1984).

In commercial far-infrared lasers, the gases are put into the laser tube at 30 to 300 millitorr (4 to 40 Pa). The gas may be flowing, or sealed into the tube. A sealed tube can be pumped out and a new gas inserted when the user wants another set of laser lines or when the gas has degraded.

Not all far-infrared laser transitions demonstrated in the laboratory can be used in commercial far-infrared lasers. The most important group of these lines is the family of vibrational-rotational lines between about 10 and 25 μm, which some specialists call the mid-infrared, produced by molecules such as ammonia and carbon tetrafluoride (see Harrison and Gupta, 1983). Special optics are needed because the emission lines are close to the standard 10-μm pump wavelength, so these molecules could not be expected to lase well in a device designed for operation throughout the far-infrared.

Energy Transfer Optical pumping is preferred because it permits precise selection of the initial excited state, important in producing a population

TABLE 15-1 Major Far-infrared Laser Lines Available From Commercial Lasers

Wavelength, μm	Gas	Wavelength, μm	Gas
41.0	CD_3OD	255	CD_3OD
46.7	CH_3OD	375	$C_2H_2F_2$
57.0	CH_3OD	433	HCOOH
70.6	CH_3OH	460	CD_3I
96.5	CH_3OH	496.1	CH_3F
118.8	CH_3OH	570.5	CH_3OH
148.5	CH_3NH_2	699.5	CH_3OH
163.0	CH_3OH	764.1	$C_2H_2F_2$
184	CD_3OD	890.0	$C_2H_2F_2$
198.0	CH_3NH_2	1020.0	$C_2H_2F_2$
229.1	CD_3OD	1222.0	$C^{13}H_3F$

Source: From tabulations by Apollo Lasers, Inc., and MPB Technologies, Inc.

inversion and laser emission in the far-infrared. Pumping is normally with an external CO_2 or N_2O laser, tuned to emit a narrow frequency band on a single line in the 10-μm region. This narrow-band emission excites a specific vibrational state of the far-infrared laser molecule, generating a population inversion on a rotational or vibrational-rotational transition.

The processes involved in far-infrared laser action can be seen by studying Fig. 15-1, which shows some of the energy levels in a polar molecule. Two vibrational levels are shown, the low-lying or ground state v_0 and the adjacent excited level v_1. Each vibrational state is associated with a large number of rotational levels (indicated by a J number). Far-infrared laser emission normally takes place between adjacent rotational levels of the same vibrational state.

It is hard to produce a selective population inversion on such a transition because the separation of rotational levels is less than thermal energy kT. Most excitation techniques tend to populate all the levels, but optical pumping can provide selective excitation if the pump frequency is within about 50 megahertz (MHz) of a molecular absorption frequency. The normal pump sources are carbon dioxide lasers and others (including isotopically substituted gases) emitting in the 10-μm range, with grating tuning of the output wavelength needed to precisely match emission and absorption frequencies. The selective excitation raises the far-infrared laser species to an excited vibrational level, and the laser transition occurs when the molecule drops from one rotational state of that level to a lower one.

Only a few far-infrared lasers can be excited by electric discharges. Two important examples are water vapor and hydrogen cyanide (HCN). These discharge-excited lasers are important for generating high output powers.

The complex structure of molecular energy levels lets many molecules emit on multiple lines even after they have been excited to a specific state. As described below, the cavity is tuned to select a single wavelength. In many molecules, selection of the initial excited state limits the range of emission wavelengths.

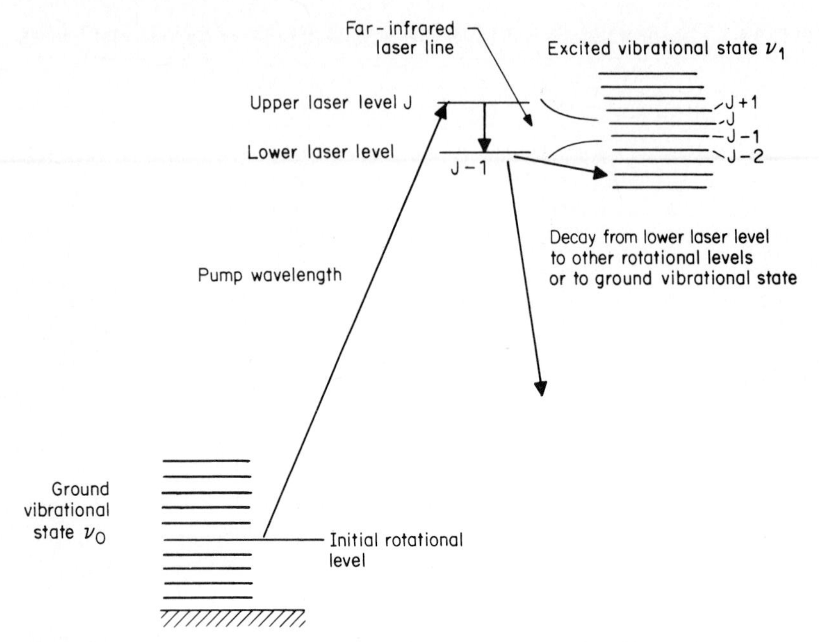

Figure 15-1 Representative energy-level diagram for a far-infrared laser, showing how optical pumping raises molecule to an excited vibrational state, where laser action in the far-infrared takes place on a transition between two vibrational levels. The laser levels are pulled out of the series of other rotational levels for clarity; their spacing actually is similar to that of other levels.

Internal Structure

The most common form for commercial far-infrared lasers is a metal or dielectric waveguide, 25 to 50 mm in diameter and 1 to 3 m long. Dielectric waveguides produce linearly polarized beams, but they suffer from excessive propagation losses if the wavelength exceeds about 2 percent of the waveguide diameter. Metal waveguides are usable throughout the far-infrared and allow lower pumping thresholds, but they tend to produce higher-order laser modes with mixed polarization. (For certain applications polarization can be important.)

The pumping geometry, shown in Fig. 15-2, is longitudinal. The pump wavelength enters the waveguide tube at one end, and the emission wavelength exits at the other end. This approach imposes different requirements on the optical elements at each end of the laser. At the pump end, the vacuum window must transmit 10-μm pump radiation; zinc selenide is a common choice because it is also transparent in the visible, aiding alignment. The mirror at that end of the cavity must transmit the pump beam but let little of the circulating power out of the cavity. The usual approach is to make a hole 1 or 2 mm in diameter in the center of the mirror, which admits a tightly focused pump beam, but represents only a tiny fraction of the mirror area.

At the output end, the vacuum window should transmit well throughout most of the far-infrared—a criterion that limits selection to materials such as pure silicon and special grades of crystalline quartz, which absorb strongly at 10 μm. The cavity output coupler is often a mirror with a central hole of 3 to 10 mm diameter, small enough to avoid excessive cavity losses. Other possibilities include partly transparent meshes, and hybrid mesh or hybrid hole designs.

Both mirrors must reflect at the 10-μm pump wavelength and throughout the far-infrared region. Reflection at the pump wavelength is important because the laser gas is tenuous and only weakly absorbing at the pump wavelength, making long path lengths essential for efficient energy transfer.

The two mirrors together with the waveguide tube define the resonant cavity. One mirror is fixed, but the other must be linearly movable with a micrometer or piezoelectric positioner. Adjustability of cavity length over a distance small compared to the tube length (typically a few millimeters) is needed to match cavity length to an integral multiple of half the emission wavelength. In shorter-wavelength lasers, the gain curve of the medium normally spans two or more longitudinal cavity modes, but in the far-infrared, the gain curve is much narrower than the cavity mode spacing. Thus the cavity length must be adjusted until a longitudinal cavity mode falls within the gain curve. This mechanism also helps select one line from the multiple lines which can be emitted by a single gas, because in general the cavity length will equal an integral number of half-wavelengths only for a single line.

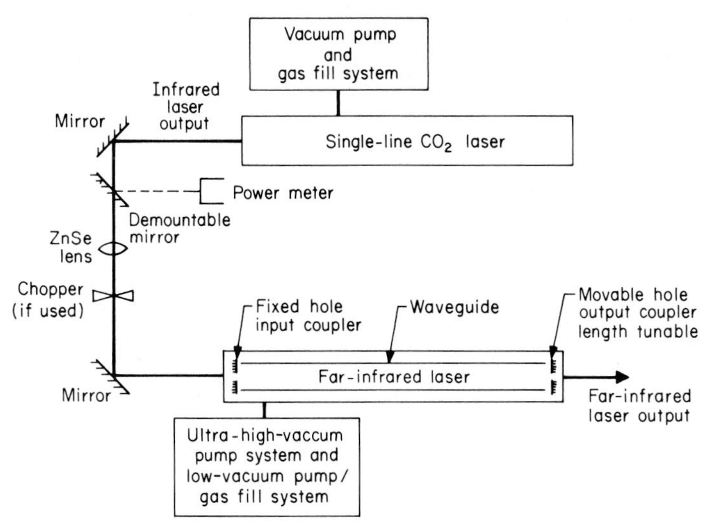

Figure 15-2 Typical arrangement for pumping a far-infrared laser optically. *(Adapted from diagram by MPB Technologies, Inc.)*

The laser resonator is housed in a vacuum-tight housing with facilities for vacuum pumping and gas filling. Typically an outer glass tube surrounds the waveguide, with laser mirrors and vacuum windows mounted on either end. The gas may be sealed in the housing or may flow through the laser.

Variations and Types Covered

The range of far-infrared lasers demonstrated in the laboratory is significantly broader than the range of commercial models. Commercial models are built as general-purpose instruments for use with many gases and at many wavelengths. Experimental versions are often custom-built for use with a specific gas or in a specific wavelength region, and thus can be optimized for performance under those conditions. With the demand for far-infrared lasers limited, manufacturers simply cannot justify design of different models for each laser line or material—and researchers seeking to study the far-infrared could not justify buying half a dozen or more different instruments to cover that spectral region. However, some commercial models can be optimized for a specific wavelength range by changing the output coupler and/or waveguide as required.

The distribution of lines from commercial far-infrared lasers is adequate for most purposes, but it is hard to find many lines between 10 and about 40 μm. In part this reflects the scarcity of known lines in part of this region, but it also is related to the practical problems when pump and emission wavelengths are closely spaced. The problems include the need for different optics at such short wavelengths, and the difficulty in separating pump and output lines, mentioned earlier.

This chapter concentrates on commercial far-infrared lasers and their characteristics. Information on experimental lasers can be obtained from references cited in tabulations of far-infrared laser lines (e.g., Knight, 1982a and 1982b; Button et al., 1984) or from recent issues of scholarly journals. Emphasis here is on waveguide lasers, in which the radiation is guided by a waveguide structure and does not obey the laws of free-space propagation. Such waveguide lasers tend to be more compact and to have much higher quantum efficiency than conventional open-resonator lasers.

Beam Characteristics

Wavelength and Output Power Table 15-1 lists some of the important wavelengths available from commercial far-infrared lasers. Output powers range from under 1 milliwatt (mW) to more than 100 mW continuous wave when pumping with single-line powers of 10 to 100 W. Peak powers with a pulsed pump laser are about five times the continuous-wave level, and peak powers about a thousand times the continuous-wave level are possible with a *Q*-switched pump laser. Megawatt pulses in the far-infrared can be produced by pumping with a transversely excited atmospheric (TEA) CO_2 laser.

Efficiency Maximum theoretical quantum efficiency of a far-infrared laser is given by

$$\eta_{max} = \frac{1}{2}\frac{\nu_{FIR}}{\nu_{pump}}$$

where ν_{FIR} is the far-infrared frequency and ν_{pump} is the pump laser frequency. Typical far-infrared lasers have quantum efficiencies of a few percent of this maximum value, but efficiencies to about 30 percent have been demonstrated experimentally. Overall energy efficiency is much lower because the emitted photons have much lower energy than the pump photons—1 to 25 percent for 1-mm to 40-μm emission pumped by a 10-μm laser.

Temporal Characteristics Depending on the pump-beam characteristics, output from far-infrared lasers can be steady continuous-wave, chopped continuous-wave, or pulsed. Pulsed or chopped output is needed when using pyroelectric detectors, which detect changes in power rather than steady power. The pump beam need not be chopped to generate a chopped output in the far-infrared. Instead, part or all of a steady continuous-wave far-infrared output beam can be chopped. This approach is valuable for applications that require a continuous-wave beam, because it allows monitoring of a portion of the far-infrared beam that is split from the main beam, then chopped.

Spectral Bandwidth Bandwidth of a far-infrared laser with tuned cavity can be on the order of a few megahertz, with long-term (tens of seconds) frequency jitter as low as a few tens of kilohertz. This corresponds to long-term frequency stability of better than 1 part in 10^8, and stabilities of 4 parts in 10^{12} have been demonstrated experimentally over 50-millisecond (ms) intervals. Normally, far-infrared lasers emit only a single wavelength in a single longitudinal mode, but it is at least theoretically possible to have simultaneous oscillation at two different wavelengths.

Amplitude Stability If input pump power and wavelength are stable, the amplitude of a far-infrared laser can remain stable to within a few percent for at least a 30-minute interval. Frequency stabilization of the *pump* laser is important to stability of far-infrared amplitude, because if the pump wavelength drifts away from the molecular absorption line, the pump-beam energy will not be absorbed, and far-infrared output will be reduced or nonexistent. Stabilization systems can be used with far-infrared lasers.

Beam Quality, Polarization, and Modes Waveguide far-infrared lasers operate in a single longitudinal mode. Dielectric waveguides can operate in the lowest-loss EH_{11} waveguide mode, which couples well into the gaussian free-space TEM_{00} mode. Dielectric waveguides produce linearly polarized beams, but cylindrical metal waveguides tend to produce beams of mixed polarization.

Specifications normally do not give beam divergence, which is large by laser standards because of the long wavelength, and which varies with wavelength. In the far infrared, even a diffraction limited beam diverges rapidly. For a 10-mm output aperture, the diffraction limit ($1.2\ \lambda/D$, where λ is wavelength and D is output aperture) is 120 milliradians (mrad) at 1-mm wavelength and about 5 mrad at 40 μm. Diffraction-limited beam divergence is even larger if the output-coupling aperture is smaller. Divergence can be reduced by using whole-area couplers which couple radiation out of the waveguide across its entire cross-sectional area.

Operating Requirements

Input Power Far-infrared lasers require optical pumping with a few watts to 100 W in the 10-μm region. The pump wavelength must be selected to match a molecular absorption line of the laser gas. Electricity is not needed per se to produce the laser beam, but it is required to operate controls, vacuum pumps, stabilization circuitry, and the pump laser.

Cooling Because the far-infrared laser tubes themselves have to dissipate excess pump power of no more than 100 W, they are air-cooled. Some laboratory results indicate that output power in the far-infrared can be increased by water cooling the waveguide. In any case, many pump lasers used to drive far-infrared lasers require water cooling.

Consumables The gases used in far-infrared lasers are consumables. They degrade gradually with lifetimes measured in hours to weeks, and must be changed to obtain different wavelengths from the laser. However, overall gas consumption is small because the pressure in the tube is very low.

Required Accessories A single-line pump laser and a high-vacuum pump (to change gases) are necessary for operation of a far-infrared laser. The usual pump laser is a conventional or waveguide CO_2 laser which can be tuned to single emission lines and tuned off the peaks of those lines to precisely match absorption lines. Isotopic variants of CO_2, such as $^{13}CO_2$, may be substituted in the laser to obtain additional pump lines (Davis et al., 1981; Wood, et al., 1980a). Single-line N_2O lasers also may be used to provide additional pump lines in the 10-μm region.

Operating Conditions and Temperature Far-infrared lasers are designed for normal laboratory conditions. The dependence of output power on cavity length makes the laser sensitive to thermal gradients, but the use of Invar rods to stabilize the cavity length minimizes this problem. Although no special warm-up is needed for the far-infrared laser, some warm-up may be needed to assure stable single-line output from the pump laser.

Mechanical Considerations Far-infrared lasers are small in cross section, but their cavities are 1 to 3 m long, making them cumbersome; weights are 50 kg and up. Bench space is required for the pump laser and any stabilization equipment used.

Safety The hazards of far-infrared wavelengths longer than the 10-μm output of CO_2 lasers have not been well-quantified. The best guideline is to use the same safety precautions and eye protection that would be used if the pump laser were being used by itself.

Reliability and Maintenance

Far-infrared lasers typically carry a 1-year warranty and can operate in most places where standard CO_2 lasers can function.

Particular attention must be paid to cleanliness and stability of the laser tubes. Cleanliness of the laser tube is important because operating gas pressures are low, and some of the gases are reactive. A major concern is ways in which impurities can react with some of the laser gases. For example, hydrogen-exchange reactions between water and deuterated methanol can convert the deuterated compound to normal methanol, producing a new set of possible emission lines. To limit contamination problems, the laser tube must be pumped down to pressures on the order of 10^{-5} torr (10^{-3} Pa) when changing gases.

A few gases used in far-infrared lasers can react with materials that may be used in the laser or vacuum pump. For example, formic acid can attack copper gaskets in the laser, forming compounds that must be cleaned out of the tube.

Thermal and mechanical stability are vital because of the need to match cavity length to oscillating wavelength. Changes in the distance between the mirrors can weaken oscillation at the desired wavelength. To provide the required stability, the laser tube is normally mounted in a housing with thick Invar rods running along its length.

Optical alignment of the flat mirrors so they are parallel with respect to each other is not as critical as for shorter-wavelength lasers, although it is still important. Minor misalignments can reduce output power and cause the laser to oscillate in higher-order modes. If ZnSe optics are used in the laser, the mirrors can be aligned visually with a helium-neon laser. However, some far-infrared lasers are designed for only factory alignment.

Commercial Devices

Standard Configurations Commercial far-infrared lasers are built for general-purpose laboratory use throughout the far-infrared. The laser heads themselves are similar in general size and appearance—1 to 3 m long, with much smaller cross section, in a housing designed for use on an optical table. The heads may be sold alone, or as part of a complete system including pump laser, stabilization equipment, and controls. The pump laser normally fits on the optical bench, along with the output-monitoring equipment used to stabilize output wavelength of the pump laser. Power supply for the pump laser, operator controls, and other auxiliary equipment may be mounted on an equipment rack or separate console. A sampling of commercial far-infrared lasers and systems is shown in Fig. 15-3.

Options In some senses, everything beyond the basic far-infrared laser head is optional. Typical options listed by manufacturers include:

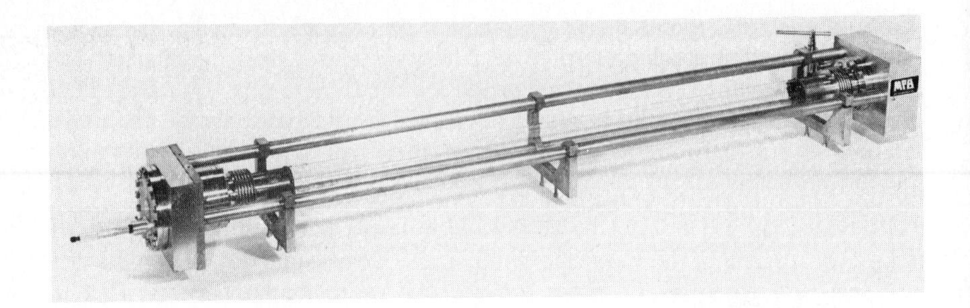

Figure 15-3 Two commercial far-infrared lasers showing *(top)* one simple structure *(Courtesy of MPB Technologies Inc.)*, and *(bottom)* a system including computer controls. *(Courtesy of Apollo Lasers Inc.)*

- Output-stabilization systems which control pump wavelength
- Selection of output coupler and waveguide configuration
- Special couplers for wavelengths below about 50 μm
- Remote tuning of the far-infrared cavity length
- Detection systems
- Vacuum pumps and gas-fill systems
- Two-wavelength or dual-cavity options for heterodyning or interferometry
- Systems to stabilize cavity length
- Choice of pump laser integrated with far-infrared laser
- Types of pump laser, including CO_2, isotopically substituted CO_2, or N_2O

Pricing As of this writing, prices of far-infrared lasers run from slightly over $10,000 to more than $70,000. The lowest prices are for laser heads alone, the highest for complete systems with many accessories.

Suppliers A handful of companies inside and outside of the United States make far-infrared lasers. All of them also make pump lasers, and in general far-infrared lasers are made specifically to mate with pump lasers from the same company. Marketing emphasis does vary, with some suppliers stressing laser heads and pump modules, while others emphasize complete systems.

Makers of far-infrared lasers also can supply optics and detectors for the far-infrared. Outside of pyroelectric detectors, and front-surfaced mirrors, few optical components used at wavelengths beyond 50 μm are used at much shorter wavelengths, and such specialized far-infrared components are hard to obtain from suppliers of conventional visible and infrared optics.

Applications

Most applications of far-infrared lasers are in research. This largely reflects the fact that far-infrared lasers have opened this part of the spectrum for serious study, and there has been little time to develop practical applications. Major research applications—in addition to study of the far-infrared in its own right—concentrate on measurement. They include

- Diagnostic measurements of plasmas, particularly for research in nuclear fusion
- Studies of semiconductor materials such as germanium and gallium arsenide
- Molecular spectroscopy
- Atmospheric spectroscopy
- Heterodyne sources for far-infrared astronomy
- Far-infrared imaging

Some new efforts could lead to practical applications in industrial measurement and testing. One example is nondestructive testing of high-voltage cables to search for internal breaks (Cantor et al., 1981). Another is measurement of the water content of paper (Boulay et al., 1984).

BIBLIOGRAPHY

R. Boulay, R. Gagnon, and B. Rochette: "Paper sheet moisture measurements in the far infrared," *International Journal of Infrared & Submillimeter Waves*, September 1984.

Kenneth J. Button (ed.): *Infrared and Millimeter Waves*, vol. 1, *Sources of Radiation*, Academic Press, New York, 1979.

Kenneth J. Button (ed.): *Infrared and Millimeter Waves*, vol. 7, *Coherent Sources and Applications Part II*, Academic Press, New York, 1983.

Kenneth J. Button, M. Inguscio, and F. Strumia (eds.): *Reviews of Infrared and Millimeter Waves*, vol. 2, *Optically Pumped Far-Infrared Lasers*, Plenum, New York, 1984.

Arnold J. Cantor et al.: "Application of submillimeter wave lasers to high-voltage cable inspection," *IEEE Journal of Quantum Electronics QE-17*(4):477–489, April 1981.

Paul D. Coleman: "Far infrared lasers, introduction," in Marvin J. Weber (ed.), *CRC Handbook of Laser Science & Technology*, vol. 2, *Gas Lasers*, CRC Press, Boca Raton, Fla., 1982, pp. 411–419.

B. W. Davis et al.: "New FIR laser lines from an optically pumped far-infrared laser with isotopic $^{13}C^{16}O_2$ pumping," *Optics Communications 37*(4):303–305, May 15, 1981.

R. G. Harrison and P. K. Gupta: "Optically pumped mid-infrared molecular gas lasers," in Kenneth J. Button (ed.), *Infrared and Millimeter Waves*, vol. 7, *Coherent Sources and Applications Part II*, Academic Press, New York, 1983.

Dean Hodges, "Review of advances in optically pumped far-infrared lasers," *Infrared Physics 18*:375–384, 1978.

D. J. E. Knight: "Tables of CW gas laser emissions," in Marvin J. Weber (ed.), *CRC Handbook of Laser Science & Technology*, vol. 2, *Gas Lasers*, CRC Press, Boca Raton, Fla., 1982a.

D. J. E. Knight: "Ordered list of far-infrared laser lines," NPL Report QU 45, National Physical Laboratory, Teddington, Middlesex, UK 1982b. Revised periodically.

Mary S. Tobin: "A review of optically pumped NMMW [near-millimeter-wave] lasers," *Proceedings of the IEEE 73*(1):61–85, January 1985.

R. A. Wood et al.: "Application of an isotopically enriched $^{13}C^{16}O_2$ laser to an optically pumped far-infrared laser," *Optics Letters 5*(4):153–154, April 1980.

R. A. Wood et al.: "Operating characteristics of an optically pumped waveguide FIR laser," *Optics Communications 35*(1):105–108, October 1980.

CHAPTER

16

other commercial gas lasers

The commercially important gas lasers described in Chaps. 7 to 15 are only a few of the gas lasers that have been demonstrated in the laboratory. Most of the gas lasers listed in extensive tabulations based on research literature (for example, Weber, 1982; Beck et al., 1980) are of purely academic interest. However, a few fall into an intermediate category, lasers which are available commercially, but only on a small scale, often with little marketing support.

This chapter covers five lasers that fall into that category:

- Carbon monoxide lasers at 5 to 6 micrometers (μm)
- Nitrous oxide (N_2O) lasers at 10 to 11 μm
- Xenon helium lasers at 2 to 4 μm
- Xenon lasers near 0.5 μm
- Iodine lasers at 1.3 μm

Not covered are a number of types that are minor members of families described in other chapters. Examples include the helium-selenium laser, a member of the metal vapor ion laser family described in Chap. 9, and the molecular and atomic fluorine lasers which are described with rare-gas-halide excimers in Chap. 13. Also not covered are isotopic variations, in which rare isotopes are substituted for common ones in order to obtain different wavelengths from gas molecules, particularly carbon dioxide.

Carbon Monoxide Lasers

Similar in many ways to the CO_2 laser, the CO laser was discovered in 1964 at Bell Labs by C. Kumar N. Patel (Patel and Kerl, 1964), who earlier had demonstrated the first CO_2 laser. Like CO_2, CO normally is excited with an electric discharge and emits on many vibrational-rotational lines. CO emission is at shorter wavelengths, with most lines between 5 and 6 μm.

The carbon monoxide laser has a potential for high power and high efficiency that led to study of it as part of the laser weapon program. By 1975, discharge efficiency as high as 63 percent had been measured, and researchers were talking about theoretical efficiency levels in the 80 percent range (Bhaumik, 1976). However, two serious practical problems limited interest in CO lasers. The strongest CO emission lines coincide with bands of high atmospheric absorption, leading to atmospheric transmission problems too serious for use in ground-based laser weapons. In addition, the gain of CO lasers drops off rapidly with temperature, and highly efficient operation is possible only at cryogenic temperatures.

Although these problems damped CO laser development, the CO laser remains the best source in the 5- to 6-μm region. It has found a variety of applications in spectroscopy and also has been used in studies of optical bistability (Bowden et al., 1981).

A couple of companies offer continuous-wave CO_2 lasers designed so minor modifications can allow their use as line-tunable CO lasers. Modifications for CO operation include substitution of optics transparent at 5 to 6 μm, addition of cooling equipment, and insertion of the proper gas mix. When cooled to −10 to −15°C, such CO lasers can emit on a few dozen lines between about 5.2 and 6.2 μm. Single-line powers of about 3 W are possible on a few lines near 5.3 and 5.4 μm, but power drops with increasing wavelength, to about 100 milliwatts (mW) for the longest wavelengths. Efficiency and output power can be doubled by cooling the lasers to about −70°C. Multiline continuous-wave output is also possible, and high powers have been demonstrated in the laboratory. However, most current applications require single-line output.

Some transversely excited atmospheric (TEA) CO_2 lasers also can be modified for use with CO, producing multipeaked pulses lasting a few microseconds, as shown in Fig. 16-1. The output is at multiple wavelengths, with complex energy-coupling effects causing the emission wavelength to shift during the pulse.

CO lasers operate with a flowing gas mixture that contains CO, nitrogen, helium, usually xenon (which enhances output power), and sometimes air. In continuous-wave models, total pressure is around 35 torr (4500 Pa); higher pressures may be used in TEA lasers. Because it includes the odorless but deadly carbon monoxide, the exhaust gas from the laser is hazardous. CO concentrations of 50 parts per million are considered unhealthy, and higher

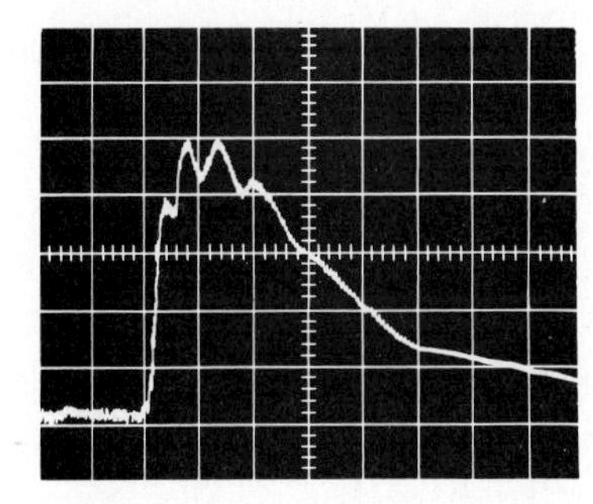

Figure 16-1 Typical pulse from transversely excited atmospheric pressure CO laser, with scale 500 ns/per division. *(Courtesy of Lumonics Inc.)*

concentrations can cause unconsciousness and death. Thus care must be taken to direct exhaust from the vacuum pump in the usual flowing-gas configuration *out* of the workspace and into the atmosphere.

Nitrous Oxide

Nitrous oxide (N_2O) is a linear triatomic molecule similar in structure to CO_2. The nitrous oxide laser was discovered shortly after the CO laser (Patel, 1965). Like CO_2, the N_2O laser can emit on up to 100 rotational sublevels of a vibrational transition near 10 μm. The 10.65-μm center wavelength of the N_2O vibrational transition is slightly longer than the 10.4-μm center wavelength of the corresponding CO_2 transition. The N_2O laser lines in Fig. 16-2 lie between 10.3 and 11.1 μm, filling in some of the holes in the CO_2 line spectrum. As with CO_2 and CO, excitation is by passing an electrical discharge through the gas.

Only minor modifications are needed to operate an N_2O laser in some CO_2 laser tubes. In line-tunable continuous-wave lasers, single-line N_2O powers can reach 10 W; in pulsed TEA lasers pulse powers can reach 30 W. These output powers are much lower than available from CO_2, about one-seventh the power level from continuous-wave line-tunable models, and a smaller fraction of the peak output of TEA pulsed lasers. Measurements indicate gain of the N_2O laser is only about a fifth that of CO_2 (Djeu et al., 1968a and b).

Because CO_2 lasers can produce much higher output power in the same wavelength region, there is little call for N_2O lasers except in applications which require precise wavelengths not available from CO_2. There are several such applications, including lidar experiments in the atmosphere (Oppenheim and Melman, 1970), spectroscopy, frequency mixing, clock applications (Thomas et al., 1980), and pumping of far-infrared lasers (see Chap. 15).

The active medium in a typical continuous-wave N_2O laser is a mixture of about 12% N_2O, 16% N_2, 9% CO, and 63% He, at a total pressure around 10 torr (1300 Pa). Normally this mixture flows through the laser and is exhausted through a vacuum pump. As with the CO laser, this exhaust gas is both potentially lethal and odorless, so care must be taken to vent the spent gas outside.

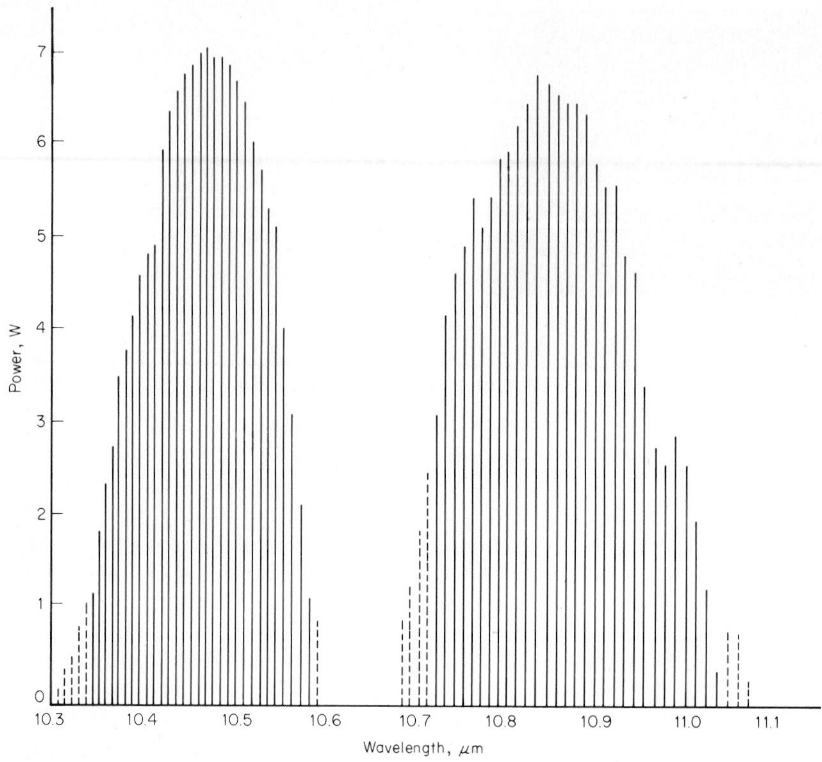

Figure 16-2 Line spectrum of N_2O laser for continuous-wave and chopped operation. The dashed lines can only be produced when laser output is chopped; all other lines indicate output power available on that line in continuous-wave operation. The gas mix is N_2O, 12%; N_2, 16%; CO, 9%; He, 63%. *(Courtesy of Apollo Lasers Inc.)*

Xenon-Helium

Xenon has a number of laser lines between 2 and 4 μm in the near-infrared. The 2.026-μm line was among the early lasers to be demonstrated (Patel et al., 1962). The stronger 3.506-μm line was reported less than a year later (Bridges, 1963). Other lines are weaker and were discovered later. The laser emission can come from pure xenon, but addition of helium at pressures tens to hundreds of times that of the xenon raises output power significantly.

Gain is high on the 3.506-μm line, about 2.5 per centimeter in a 0.75 mm bore, and this has been called a high-power laser in pulsed operation (Davis, 1982). However, it suffers from power saturation that limits continuous-wave output and has attracted only minimal commercial development or interest.

Only a single company makes commercial Xe-He lasers, which can operate continuous-wave, chopped, or pulsed at rates to 1 kilohertz (kHz). Maximum continuous-wave power is about 10 mW, with up to about 30 mW peak power obtainable in pulses. Powers are highest if the gas flows through the laser, but sealed operation is possible also.

The major lines do lie in windows of good atmospheric transmission, but so far applications have been limited to research.

Visible Xenon

Triply ionized xenon [Xe^{3+} or Xe (IV)] has several blue-green laser lines. Unlike the rare gas argon and krypton ion lasers described in Chap. 8, these xenon lines are pulsed. Experimental peak powers have reached several tens of kilowatts, but commercial versions deliver peak powers on the order of 1 kW. The main emission lines are at 526.0, 535.3, and 539.5 nm, with a few weaker lines in the blue-green. Experimental xenon lasers have emitted on lines in the blue and ultraviolet as well (Bridges, 1982).

The pulses of 0.1 to 0.4 microsecond (μs) can be repeated at rates to around 200 Hz and have peak powers that are high by ion laser standards. The combination of that high power with the short wavelength has made pulsed xenon lasers desirable for applications requiring the focusing of moderate powers onto small spots. Examples include trimming of thin-film resistors in microelectronic circuits, and making the masks used in semiconductor fabrication. Other potential applications include holography, studies of turbulence, and research requiring such short, moderate-power blue-green pulses. So far most pulsed visible xenon lasers have been incorporated in electronics manufacturing equipment. The lasers normally are made by the company producing the complete system and generally are not marketed as individual devices.

Iodine Lasers

The 1.315-μm atomic iodine laser exists in two basic variations: the chemical oxygen-iodine laser mentioned in Chap. 11, and the optically pumped photodissociation iodine laser. As mentioned in Chap. 11, chemical oxygen-iodine lasers are being developed because their potential high powers makes them laser weapon candidates. Few details on that program are reported in public (Hecht, 1984).

The major force pushing work on the older photodissociation iodine laser is interest in its potential for laser-induced nuclear fusion (Hohla et al., 1977; Witte et al., 1977; Brederlow et al., 1983). Flashlamp pumping of large iodine lasers can produce nanosecond pulses with peak powers in the terawatt range, as demonstrated in the Asterix III laser at the Max Planck Institute for Quantum Optics in Garching. On a smaller scale, flashlamp pumping is used in a commercial model that recently came on the market (Bannister and King, 1984).

The photodissociation iodine laser relies on ultraviolet light to break up iodine-containing molecules, producing atomic iodine in an excited state, as shown in Fig. 16-3. The excited iodine can be stimulated to emit a photon at 1.315 μm. After the laser pulse, much of the iodine recombines with the molecular fragments, and it is often possible to reconstitute the starting compound.

Normally, iodine laser operation is pulsed. Recently continuous-wave operation for several weeks at powers to 38 mW has been demonstrated in a longitudinally flowing gas mixture, pumped by light from a high-pressure mercury arc lamp (Schlie and Rathge, 1984).

The commonest iodine compound used, and the standard type in commercial iodine lasers, is C_3F_7I. In some simple commercial models, ampoules of the iodine compound are inserted into the laser, and the vapor produced exhausted by a vacuum pump. This allows up to a thousand shots to be produced at repetition rates of no more than one every 10 s, to allow for removal of the spent gas. The other approach used commercially is recycling of the laser gas to reconstitute the iodine compound in a sealed laser tube. This approach allows up to a million shots to be produced at repetition rates of 1 to 10 Hz.

Commercial iodine lasers can operate in long-pulse (fixed-*Q*), *Q*-switched, or modelocked modes. Fixed-*Q* operation can produce 3-joule (J) pulses lasting 2.5 to 4 μs. *Q*-switching can generate 1-J, 15- to 30-ns pulses because of iodine's good energy-storage capacity. Modelocking can produce pulses of 2 ns to 500 picoseconds (ps), and addition of buffer gas to broaden the natural linewidth can allow generation of pulses as short as 100 ps. Short *Q*-switched and modelocked pulses can be used for generation of second, third, and fourth harmonics at 658, 438, and 329 nm, respectively.

Oscillator-amplifier configurations are possible with iodine lasers, an approach studied for fusion applications. Continuing efforts to improve iodine laser performance are concentrating on developing pump lamps better matched to the iodine molecule's absorption, and studying new iodine molecules. One long-term goal is an iodine laser that could be pumped efficiently by sunlight.

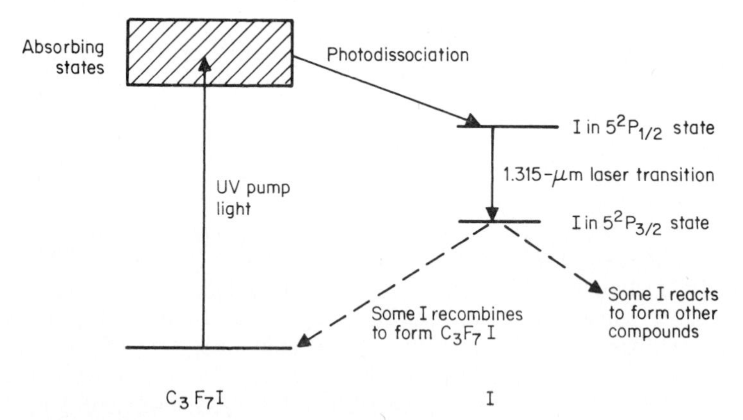

Figure 16-3 Transitions involved in iodine photodissociation laser. Ultraviolet excitation raises the starting molecule, C_3F_7I, to an excited state in which it quickly photodissociates to free excited I, which is the upper laser level. After the I drops to the lower level, it recombines with other molecular fragments to form C_3F_7I or other compounds.

Laser fusion and laser weapons are typical of iodine laser applications in that they are in the research-and-development stages. One emerging application for small iodine lasers is study of fast chemical reactions in aqueous solutions (Bannister et al., 1984), taking advantage of the strong water absorption at the iodine wavelength. Because of that absorption, a 1-J laser pulse can heat a sample of a milliliter or less by several degrees in nanoseconds. Small iodine lasers also can be used to study pulse propagation in optical fibers or in the atmosphere, and in development of larger iodine lasers. Research is in progress on the potential for medical and industrial applications.

BIBLIOGRAPHY

J. J. Bannister et al.: "The iodine laser and fast reactions," *Chemistry in Britain,* March 1984, pp. 227–233.

J. J. Bannister and T. A. King: "The atomic iodine photodissociation laser," *Laser Focus 20*(8):88–97, August 1984.

R. Beck, W. Englisch, and K. Gurs: *Table of Laser Lines in Gases & Vapors,* 3d ed., Springer-Verlag, Berlin and New York, 1980.

Mani L. Bhaumik: "High-efficiency electric-discharge CO lasers," in E. R. Pike (ed.), *High-Power Gas Lasers 1975,* Institute of Physics, Bristol and London, 1976, pp. 243–267.

Charles M. Bowden, Mikael Ciftan, and Hermann R. Robl: *Optical Bistability,* Plenum Press, New York, 1981.

G. B. Brederlow, E. Fille, and K. J. Witte: *The High Power Iodine Laser,* Springer-Verlag, Berlin and New York, 1983.

W. B. Bridges: "High optical gain at 3.5 μm in pure xenon," *Applied Physics Letters 3:*45–47, 1963.

W. B. Bridges: "Ionized gas lasers," in Marvin J. Weber (ed.): *CRC Handbook of Laser Science & Technology,* vol. 2, *Gas Lasers,* CRC Press, Boca Raton, Fla., 1982, pp. 171–269.

Christopher C. Davis: "Neutral gas lasers," in Marvin J. Weber (ed.): *CRC Handbook of Laser Science & Technology,* vol. 2, *Gas Lasers,* CRC Press, Boca Raton, Fla., 1982.

Nicholas Djeu, Tehman Kan, and George J. Wolga: "Gain distribution, population densities, and rotational temperature for the (00°1)–(10°0) rotation-vibration CO_2-N_2-He transitions in a flowing laser," *IEEE Journal of Quantum Electronics QE-4:*5, May 1968a, pp. 256–260.

Nicholas Djeu, Tehman Kan, and George J. Wolga: "Laser parameters for the 10.8-μ N_2O molecular laser," *IEEE Journal of Quantum Electronics, QE-4:*11, November 1968b, pp. 783–785.

Jeff Hecht: *Beam Weapons: The Next Arms Race,* Plenum Press, New York, 1984.

K. Hohla et al.: "Prospects of the high-power iodine laser," in Helmut J. Schwartz and Heinrich Hora (eds.), *Laser Interaction & Related Plasma Phenomena,* vol. 4A, Plenum, New York, 1977, pp. 97–113.

U. P. Oppenheim and P. Melman: "Spectroscopic studies with tunable N_2O laser," *Journal of the Optical Society of America 60:*3, March 1970, pp. 332–334.

C. K. N. Patel: "CW laser action in N_2O (N_2-N_2O) system," *Applied Physics Letters 6:*12–13, January 1, 1965.

C. K. N. Patel and R. J. Kerl: "Laser oscillation on $X'\Sigma^+$ vibrational rotational transitions of CO," *Applied Physics Letters 5:*81, 1964.

C. K. N. Patel, W. L. Faust, and R. A. McFarlane: "High-gain gaseous (Xe-He) optical maser," *Applied Physics Letters 1:*84–85, 1962.

L. A. Schlie and R. D. Rathge: "Long operating time CW atomic iodine probe laser at 1.315 μm," *IEEE Journal of Quantum Electronics QE-20* (10):1187–1196, October 1984.

J. E. Thomas et al.: "Stable CO_2 and N_2O laser design," *Review of Scientific Instruments 51:*2, February 1980, pp. 240–243.

Marvin J. Weber (ed.): *CRC Handbook of Laser Science & Technology* vol. 2, *Gas Lasers,* CRC Press, Boca Raton, Fla., 1982.

K. Witte et al.: "Terawatt iodine laser," in Helmut J. Schwartz and Heinrich Hora (eds.), *Laser Interaction and Related Plasma Phenomena,* vol. 4A, Plenum, New York, 1977.

CHAPTER

17

dye lasers

The organic dye laser has found many applications in scientific research because of its unusual flexibility. Its output wavelength can be tuned from the near-ultraviolet into the near-infrared; the use of frequency-doubling crystals can extend emission further into the ultraviolet. Dye lasers can be adjusted to operate over an extremely narrow spectral bandwidth, producing ultrapure light for studies of optical properties of materials in very narrow wavelength regions. They also can produce ultrashort pulses, with durations much shorter than a picosecond.

This versatility comes at a cost in complexity. Individual dyes can emit light at a broader range of wavelengths than other lasers, but a range that is still narrow compared with the entire visible spectrum. Tuning wavelength across the entire visible range requires several changes of dye. The dyes degrade, and because they must be dissolved in a solvent, require a complex liquid-handling system. They require excitation with a separate source of intense light, another laser or a flashlamp, which adds to the cost and complexity, although the choice of pump sources does give some added flexibility. Complex optics are needed to produce either ultranarrow-linewidth output or picosecond pulses, and the two cannot be produced simultaneously. Designers of dye lasers have made great strides to build systems which can be used readily by biologists, chemists, and others whose specialties are far from laser technology, but dye lasers are still more scientific instruments than industrial tools.

Dye lasers have made vital contributions to spectroscopy. It was not long after the first demonstration of a dye laser, by Peter P. Sorokin and J. R. Lankard at the IBM Watson Research Center in Yorktown Heights, N.Y. (Sorokin and Lankard, 1966), that optics were developed to adjust wavelength and produce narrow-bandwidth output. Powerful techniques were soon developed to probe energy levels and physical processes within atoms and molecules. Arthur L. Schawlow of Stanford University, who made major contributions to this research, was cited for his contributions to laser spectroscopy when he shared in the 1981 Nobel prize in physics.

Internal Workings

Active Medium The active medium in a dye laser is a *fluorescent* organic compound (a dye) dissolved in a liquid solvent. Intense illumination by light from a separate source—another laser or a flashlamp—excites the dye molecules, producing a population inversion. The dye then produces stimulated emission, generating a laser beam. The process is called optical pumping, and inherently involves some energy losses, so the output wavelength is longer than the absorbed wavelength.

Dyes are large molecules containing multiple ring structures, and they have complex spectra. Most important dyes fall into a number of families with chemically similar structures. Members of these families differ in the end groups attached to their outer edges, and these chemically superficial differences lead to important differences in characteristics such as laser emission wavelength, tuning range, absorption wavelengths, and tolerance of operating conditions. Listings of dyes in which laser action has been demonstrated go on for many pages (Steppel, 1982; Maeda, 1984). Many of these dyes are offered commercially for laser use.

Solvents are also an important consideration, and many are in use. Although a few dyes are water-soluble, most require organic solvents such as methanol and dimethyl sulfoxide. In certain cases, alcohols may be mixed with water. Solvents can influence both the optical and degradation characteristics of dyes.

Dye degradation is an important practical issue. The complex molecules tend to be decomposed by intense pump light. Degradation becomes increasingly severe with increasing pump intensity and with shorter pump wavelength. Some reactions are photochemical, some thermal, and some a combination. The operating lifetimes of dyes depend on their chemical structure as well as pump conditions. Typically lifetimes are specified in watt-hours of pump power per liter, with values ranging from a few watt-hours to over a thousand. Under favorable operating conditions, some of the stablest dyes can last more than a year.

Efforts have been made to develop dye lasers in which the dye is carried in a medium other than a liquid. Laser action has been demonstrated with dyes in the vapor phase and embedded in a solid plastic host. Both approaches have some conceptual attractions, but they suffer some serious practical drawbacks and have yet to find practical applications.

Energy Transfer The energy that drives a dye laser comes from an external light source. Absorption of a photon raises the dye molecule to a highly excited state, nonradiative processes drop it to the upper level of the laser transition, then after the laser light is emitted nonradiative processes remove molecules from the lower level of the laser transition. The actual energy-level structure is quite complex, and even the diagram in Fig. 17-1 is oversimplified.

Dyes can both absorb and emit light over a range of wavelengths because interactions among electronic, vibrational, and rotational energy levels create a continuum of levels in some energy ranges. The electronic transitions are those between groups of states in Fig. 17-1, and their energies normally correspond to wavelengths in or near the visible region. Superimposed upon the electronic transitions are vibrational transitions of the molecule, represented in the diagram by darker lines within each group of levels. The smallest energy shifts are caused by rotational transitions, represented by light lines. Although the levels are shown separated for clarity, in practice they are spaced closely enough to form a continuum when subjected to normal line-broadening mechanisms.

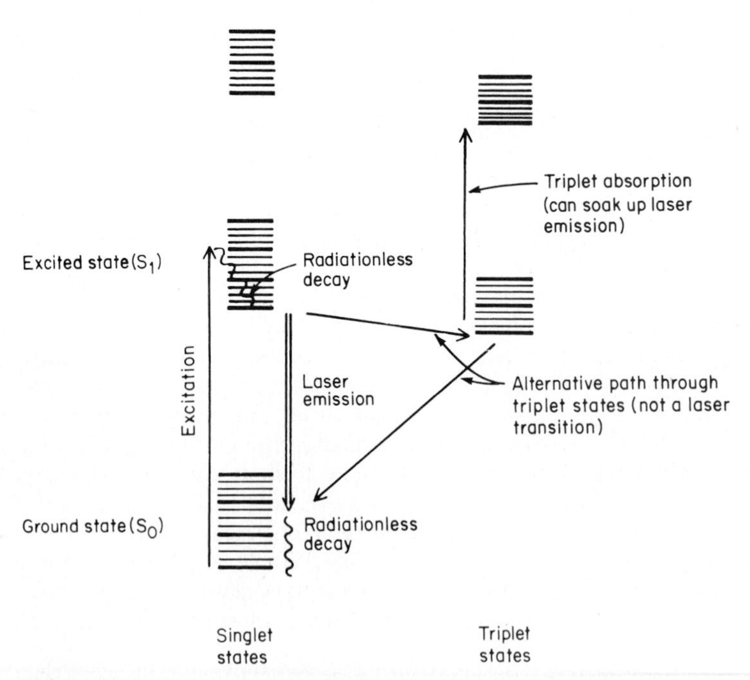

Figure 17-1 Typical energy-level structure in a laser dye. The electronic energy levels (groups of lines) contain vibrational (dark lines) and rotational (light lines) substates.

It is this continuum of energy levels that lets dyes absorb and emit light at what by laser standards is a broad range of wavelengths, as shown in Fig. 17-2. Initial excitation to the first excited electronic level is followed by nonradiative relaxation to the bottom of that group of energy levels, where a population inversion accumulates. Energy storage in the upper laser level is small because that state's lifetime is only a few nanoseconds. Stimulated emission occurs when a molecule makes a transition from that upper level to a level in the lower band. Spontaneous emission can occur in a wide range of wavelengths in that band, with the highest probability of emission at central wavelengths. With an untuned resonant cavity, amplification is strongest at the peak laser wavelengths, where the difference between emission and absorption is greatest. Insertion of wavelength-tuning optics allows only a limited range of wavelengths to oscillate within the cavity, so amplification occurs at these wavelengths.

In practice, the picture is complicated by the parallel set of electronic energy bands shown in Fig. 17-1, known as *triplet* states, while the ones from which laser emission occurs are known as *singlets.* These states arise because electrons in dye molecules are bonded together in pairs. In the ground state, the two electrons have different spins and occupy the same energy level. If the electrons maintain opposite spins as one ascends the energy-level ladder, they stay in singlet levels. If the spin of the higher-energy electron reverses to become parallel to that of the lower-energy electron, they enter the triplet states, which have slightly lower energy than the corresponding singlet levels. Because the triplet states have lower energy, the molecules tend to drop into them. That creates problems because those levels have such long lifetimes that they trap molecules, which thus cannot take part in laser emission, but which can cause some absorption. Pump pulses in the 10-ns range or shorter—

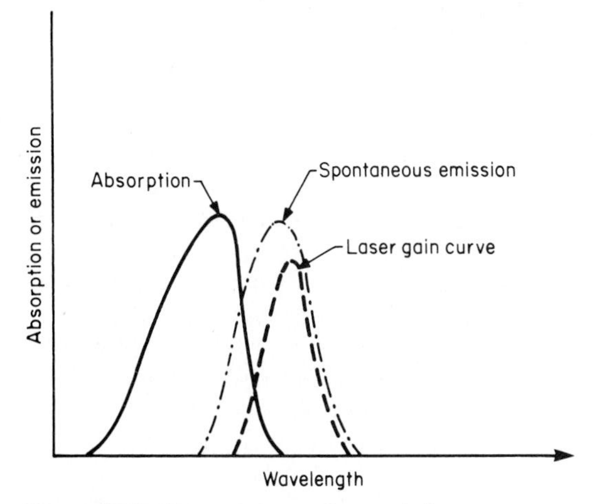

Figure 17-2 Representative shapes of absorption, spontaneous emission (fluorescence), and laser emission curves for a laser dye.

from excimer, frequency-shifted neodymium-YAG, or nitrogen lasers—are too short to produce triplets. Triplet production can limit tuning ranges of dyes with longer-duration pumping, and can prevent many from operating in the continuous-wave mode, despite the use of rapidly flowing dye jets to control triplet production by moving the dye quickly out of the excitation region. Much has been written on how triplet production limits dye performance (Steppel, 1982), but the details are rarely relevant to laser users.

Internal Structure

Although all dye lasers share the same active medium, they differ widely in internal structure. The differences reflect both the choice of pump source and the intended applications of the laser.

The pump source dictates much of the design of a dye laser because of the energy-transfer kinetics of the dye. The few-nanosecond lifetime of the upper laser level makes the pulsed dye laser a high-gain, high-loss system. The high pump powers needed to pass laser threshold are a problem for continuous-wave dye pumping; they require extremely tight focusing of the beam, and combined with other considerations make the continuous-wave dye laser a low-gain, low-loss system. Rapid dye flow is essential in continuous-wave dye lasers to prevent the accumulation of triplet states from quenching emission, a concern that led to design for the dye jet for such systems.

Configurations also differ. Flashlamp-pumped dye lasers must contain the light source and plug directly into a source of electrical energy. Laser-pumped dye lasers obtain their pump energy from an external laser and are designed to accommodate the characteristics of the pump source.

The choice of cavity optics and cavity structure depends on the intended application. For many types of spectroscopy, line-narrowing optics are needed, but short-pulse research requires broadband emission and a modelocked pump laser. Both applications and pump sources can vary widely, so the descriptions that follow cover major types of dye lasers, not all possible configurations, and overlook details of some custom-built laboratory systems which have set performance records.

Optics Wavelength-selective optics are used in most dye lasers. In the common configurations shown in Fig. 17-3, a wavelength-dispersive element (a prism or diffraction grating) is inserted inside the laser cavity. The dispersive element is aligned so light at one wavelength is reflected back along the axis of the laser cavity, while other wavelengths are dispersed. This assures that laser oscillation will take place only at the selected wavelength (assuming it is within the dye's gain bandwidth). In the simple examples shown, turning the grating or rear cavity mirror changes the resonant wavelength. The range of oscillating wavelengths can be further restricted by inserting etalons or other wavelength-selective elements into the laser cavity.

An alternative tuning approach is use of a tunable wavelength-selective element. One such element is a tuning wedge, which forms a Fabry-Perot cavity and limits oscillation wavelength to a narrow range by interference

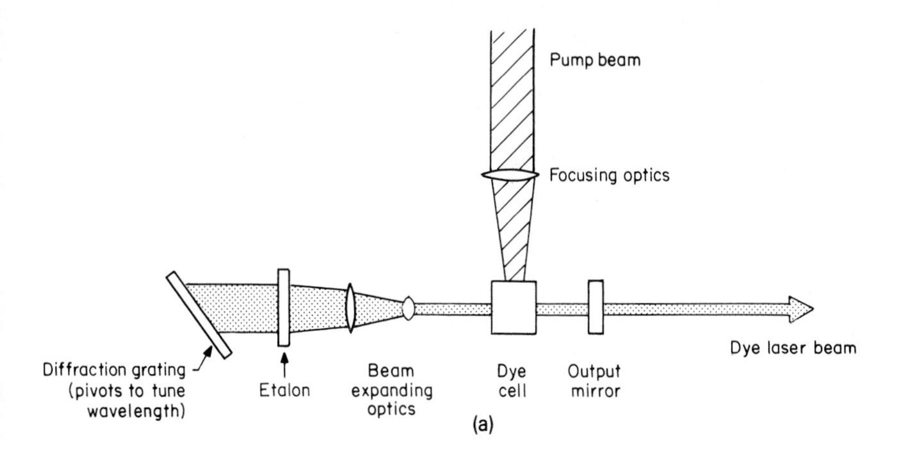

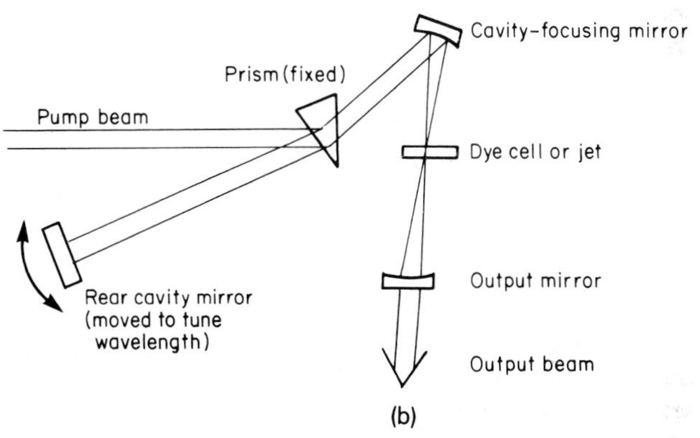

Figure 17-3 Typical cavity designs for dye lasers with dispersive wavelength-tuning elements. *(a)* Grating-tuned dye with pulsed pumped laser; *(b)* prism tuning of a continuous-wave dye laser.

effects. Vertical motion of the wedge changes the spacing of the resonant elements, thus tuning the cavity. Another such element is the birefringent tuner, which changes the polarization of the light to an extent dependent on wavelength. If the cavity is arranged to permit oscillation only of linearly polarized light, oscillation is possible only at the wavelength for which the birefringent tuner produces linearly polarized light.

Optics are also used to focus the pump light into the dye in most cases. An exception is the coaxial flashlamp, shown in Fig. 17-4, in which dye solution flows through a bore surrounded by an annular-cross-section flashlamp; the flashlamp plasma is opaque to the radiation it produces, so only light from the inner surface of the lamp reaches the dye solution. Other flashlamp-pumped dye lasers use a linear flashlamp with a cavity that reflects the pump light onto a cylindrical tube holding the dye solution.

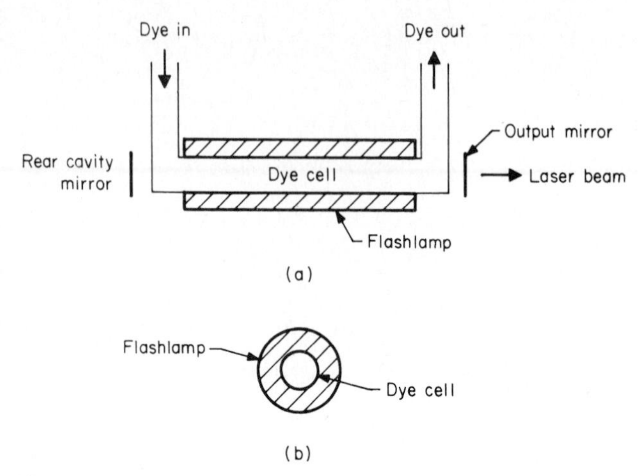

Figure 17-4 Design of a coaxial flashlamp dye laser. *(a)* Side view; *(b)* cross section.

Laser pumping can be done at different angles to the dye laser axis. The pump beam can be directed at a small angle to the axis, passing around the cavity mirrors or in some cases deflected around them by wavelength-dispersive tuning optics. Such longitudinal pumping can offer high efficiency, important for low-gain continuous-wave dye lasers. Alternatively, the pump beam can be directed transversely to the dye-laser axis, an approach which simplifies alignment; the sacrifice in efficiency is an acceptable trade-off for high-gain pulsed dye lasers. Transverse pumping can be used with dye lasers in an oscillator-amplifier configuration by splitting a single pump beam so part is directed into the oscillator and the remainder drives the amplifier. Special focusing optics are not needed with high-peak-power pulsed pump lasers, but they are essential for continuous-wave pumps because of the high pump intensity requirement to reach laser threshold. Typically continuous pump lasers are focused to narrow waists a few micrometers across in the dye jet.

Dye lasers with the ring cavity configuration shown in Fig. 17-5 and described below generally contain another optical element, which serves to restrict oscillation to a single direction. The device contains two polarization-rotating components, one which rotates polarization in the same direction regardless of the way light travels, the other which rotates polarization in a direction dependent on the beam direction. The rotation cancels out for light going in one direction, but not for light going in the opposite direction, which then cannot oscillate in the laser cavity because it suffers high losses. The result is a traveling wave in the ring laser cavity which can extract more single-frequency power than the standing wave formed in a conventional laser cavity.

Cavity Length and Configuration Many different cavity configurations are used with dye lasers. The choice depends on the pump source, the intended applications, and the nature of the dye. In most dye lasers the dye cell accounts for only a small part of the cavity length. (Flashlamp-pumped dye lasers are an exception, with the dye contained in a tube aligned along the laser axis.) Except for low-power models, the dye solution flows constantly through the dye cell. The lasers are designed so the dyes can be interchanged to shift to other wavelengths.

In continuous-wave dye lasers, the dye is usually in the form of a free-flowing jet, with the pump beam focused on the narrowest part of the jet. This approach avoids optical damage to solid windows that could be caused by the tight focusing of the continuous pump beam.

Pulsed-dye-laser cavities are designed to be small, because beam quality is improved if the beam can make several round trips during the pump pulse. Minimum length is limited by the need to house intracavity tuning elements. Typical cavity lengths thus may be no more than a few centimeters. Standard dye laser cavities containing tuning elements are often folded, with the two resonator mirrors not directly aligned with each other along the laser axis, as shown in Fig. 17-3. The folding in some cases can be at a sharper angle, or the cavity could include more than one folding element. In some designs, the wavelength-selecting element serves as one of the cavity mirrors. From the outside, the differences are not apparent; the packaged dye laser seems short.

There are exceptions to this pattern. Dye lasers designed for synchronous pumping with a modelocked laser must have matching cavity length or a length that is an integral multiple of that of the pump laser, so the cavity round-trip times match as required for modelocked pulse generation. Pump lasers used for such pumping typically have cavity length of 1 to 2 m, so the dye laser cavity is usually stretched to that length. This can be done with a linear extension or by a multiply-folded delay line.

An unusual cavity configuration with some attractive characteristics for continuous-wave operation is the ring dye laser, with mirrors that direct the beam in a triangular or figure-eight pattern such as shown in Fig. 17-5. The beam emerges through one of the mirrors which is partially transparent, and the beam oscillates in only one direction around the ring when a direction-selecting element is used. External size is similar to that of conventional continuous-wave dye lasers. Despite its complexity, the ring design is attractive because its traveling-wave oscillation is at a far narrower range of wavelengths than a conventional standing-wave dye laser. It can therefore produce more power at a single frequency.

Many high-power pulsed lasers are long because they use the oscillator-amplifier configuration shown in Fig. 17-6. In this design, the oscillator cavity itself is short, but the length of the dye laser is extended by a separate amplifier stage. The designs often deliberately leave extra room for the insertion of accessories. For pulsed dye lasers, the oscillator-amplifier configuration is a more efficient way to generate high powers than the construction of large oscillators, and operational difficulties are minimal. Often a single pulsed laser pumps both the oscillator and amplifier, as in Fig. 17-6.

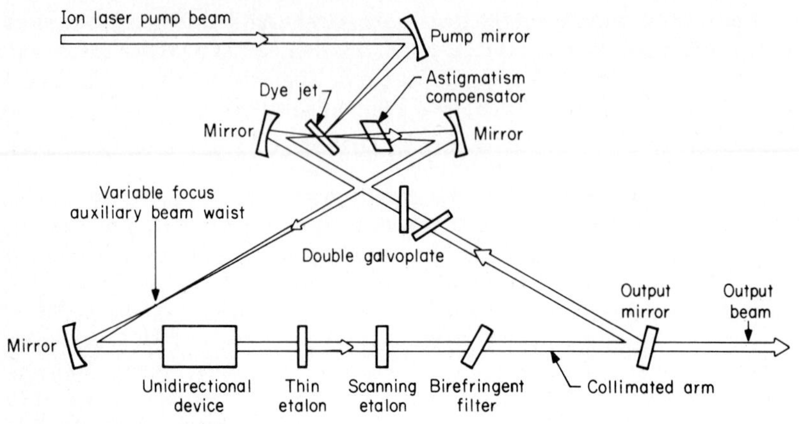

Figure 17-5 Configuration of a ring dye laser with a traveling wave inside the laser cavity. *(Based on drawing from Spectra-Physics Inc.)*

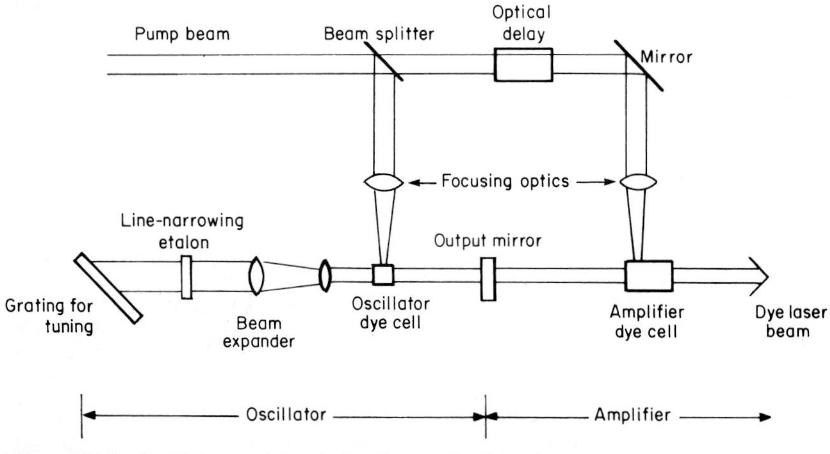

Figure 17-6 Oscillator-amplifier design for a pulsed dye laser.

Inherent Trade-offs

Three characteristics figure high on the wish lists of many dye laser users: extremely narrow linewidth, high average power, and ultrashort pulses. An individual dye laser can offer one of these features, but not all three. The inability to combine narrow linewidth and short pulses is fundamental because the minimum attainable linewidth and shortest possible pulse duration are related according to the formula:

$$\text{Pulse length} = \frac{0.441}{\text{bandwidth}}$$

(Svelto, 1982). This is an expression of a Fourier transform relationship between time and frequency domains; it is a fundamental physical limitation.

The problem of combining high average power with either of the other goals is more practical. Continuous-wave operation is needed to produce extremely narrow bandwidth, but it yields comparatively feeble output. Ultrashort pulses contain little energy because of their length (although peak powers are reasonable), so the average power they can deliver is limited.

There are a number of trade-offs inherent in the choice of dyes and pump sources. Ultraviolet photons deliver more energy to the dye, but also destroy the dye faster, so shorter-wavelength dyes tend to be shorter-lived, and shorter-wavelength lasers tend to degrade dyes faster. Pump-light sources are a major factor in costs of a dye laser system, and they involve complex trade-offs. Flashlamp pumping offers low capital costs, but beam quality is limited, and coaxial flashlamps are expensive to replace. Neodymium-YAG and excimer laser pumps are more expensive to buy but can be less costly to operate. Nitrogen lasers can be made small and inexpensive, but they are low in power and efficiency. Ion laser pumps are expensive and offer a different set of characteristics for dye lasers. Copper vapor pump lasers are more efficient and generally more powerful than argon, but the technology is still being perfected, and output is repetitively pulsed rather than continuous. Those trade-offs are complex and make it important to carefully define needs when checking available products.

Variations and Types Covered

As indicated above, there are considerable variations among dye lasers, with several distinct types on the market. Because these variations have been described above, they will not be covered here. The rest of this chapter will concentrate on the major types that are commercially available, with general remarks covering most types unless otherwise noted.

Beam Characteristics

Wavelength and Output Power The wavelength and output power of dye lasers depend on the choice of dye and pump source, as well as on design of the laser. Fundamental-frequency output wavelengths range from about 330 to beyond 1250 nm, with the shortest and longest wavelengths attainable only with pulsed pump lasers; wavelengths as short as about 216 nm can be produced by frequency-doubling the dye laser beam. For continuous pumping, maximum output is on the order of several watts; maximum average power for modelocked operation is somewhat less. For pulsed operation, peak powers can range from kilowatts to megawatts, with hundreds of megawatts possible in amplified picosecond pulses. Average power can range from below a watt to tens of watts. However, those are specified values for the best of dyes; powers can be much lower with inefficient dyes or when operating away from wavelengths of peak dye efficiency.

Each dye covers only a limited range of wavelengths. The tuning ranges of individual dyes vary somewhat depending on the pump-light source and on the solvent used, as indicated in Table 17-1, which lists tuning ranges of three dyes offered by one major vendor. Conversion efficiency is highest near the center of the wavelength band, dropping off toward the edges.

Wavelength coverage of individual lasers can be limited by the choice of cavity optics. Typically the cavity mirrors are designed to have low losses over a range of wavelengths much wider than those covered by any one dye, but still not as broad as the range covered by all available dyes. Extending tuning to other wavelengths requires use of an alternative set of optics. The use of single-frequency optics, to narrow linewidth to extremely small values, can restrict oscillation to only part of the wavelength region covered by a single dye.

Output power depends on the choice of pump laser and the design of the dye laser as well as the choice of dye. Powers available from some representative types of dye lasers are listed in Table 17-2. Note that these specifications are not intended to be exhaustive, either of the characteristics of dye lasers or of the models available on the market. Specifications generally

TABLE 17-1 Characteristics of Selected Laser Dyes

Dye	Peak wavelength, nm	Tuning range, nm	Pump source	Solvent
p-Terphenyl	340	323–364	KrF (249 nm)	Cyclohexane
	340	333–348	Nd–YAG (266 nm)	Cyclohexane
	341	334–347	XeCl (308 nm)	Cyclohexane
	341	335–355	Flashlamp	Dimethylformamide
	341	335–349	XeCl (308 nm)	*p*-Dioxane
Stilbene 420	424	410–454	XeCl (308 nm)	Ethanol/H_2O, 9:1
	424	411–436	Nd–YAG (355 nm)	Methanol
	425	400–460	N_2 (337 nm)	Ethanol/H_2O, 1:4
	425	405–467	XeCl (308 nm)	Ethanol
	425	408–453	N_2 (337 nm)	Methanol
	432	406–448	Ar (uv)	Ethylene glycol/ methanol, 9:1
	445	421–468	N_2 (337 nm)	H_2O
	449	420–470	Ar (uv)	Ethylene glycol
Rhodamine 590 (Rhodamine 6G)	560	548–580	Nd–YAG (532 nm)	Methanol
	572	564–600	Copper vapor	Ethanol
	579	568–605	N_2 (337 nm)	Ethanol
	583	566–610	XeCl (308 nm)	Methanol
	587	565–615	Flashlamp	Methanol
	590	570–650	Ar (458, 514 nm)	Ethylene glycol
	596	577–614	Flashlamp	Methanol/H_2O, 1:3
	602	560–654	Kr (blue-green)	Methanol/ethylene glycol mixture
	610	585–633	Flashlamp	4% ammonyx LO in H_2O

Source: Exciton Chemical Corporation.

give maximum output power for the dye Rhodamine 6G, a particularly efficient dye with peak output near 600 nm which can be used in all types of dye lasers. Other dyes will generally give lower powers, as will Rhodamine 6G away from its peak wavelength.

Dye laser power is proportional to pumping power above laser threshold. Output power depends on how well the excitation wavelength matches the absorption bands of the dye. This makes the choice of pump laser a vital factor in determining dye laser performance, so in studying specifications it is essential to understand what pump-light source was used.

Efficiency The efficiency of laser-pumped dye lasers is measured as an output-input ratio, the percentage of input energy that is converted to output energy. Conversion efficiencies well over 50 percent have been obtained in favorable laboratory conditions for inherently efficient processes such as pumping Rhodamine-6G with Nd–YAG's 532-nm second harmonic (Tarasov, 1983, p. 32). In normal operation, dye laser conversion efficiencies are lower. Typical specifications give conversion efficiencies of a few percent to a few tens of percent. Actual efficiency could be much lower for operation of an inefficient dye near the edges of its tuning range. Some literature also gives quantum efficiency, which measures the percentage of input photons that generate output photons. The quantum efficiency is inevitably higher than the energy-conversion efficiency because output photons have less energy than input photons. Another common scale, particularly for continuous-wave lasers, is slope efficiency, the ratio of extra output to extra input above laser threshold.

For flashlamp-pumped dye lasers, efficiency is given as the "wall-plug" efficiency—the laser output divided by electrical input to the laser. Typical specified values are on the order of 1 percent for pumping Rhodamine 6G, and lower for less-efficient dyes. Note, however, that this figure gives overall efficiency in converting electrical energy to dye laser energy, not just the optical conversion efficiency that is measured for laser-pumped dye lasers. It takes into account losses in the flashlamps as well as in the dye-pumping process. However, it does not consider operation of liquid pumps for cooling water and dye solution, and other power demands that can reduce overall efficiency to the 0.2 to 0.4 percent range.

TABLE 17-2 Power Levels and Pulse Energies Typical of Commercial Dye Lasers Pumped by Various Sources

Pump source	Pulse energy, mJ	Continuous-wave or average power, W
Excimer or Nd–YAG	5–75	0.01–10
N_2	0.02–1	0.003–0.3
Copper vapor	0.4–1	Over 1 W
Flashlamp	100–10,000	1–50
Argon ion	—	0.1–3 W

Note: This is not an exhaustive listing.

If pump laser efficiency were factored into calculations of overall efficiency of laser-pumped dye lasers, the resulting values generally would be below those of flashlamp-pumped dye lasers. This is because the pump lasers themselves are inefficient in converting electrical energy to light.

Temporal Characteristics of Output Dye lasers can produce continuous or pulsed output, depending on the pump source. Dye laser emission is possible only while the pump source is delivering light to the dye, although emission can be restricted to shorter pulses by modelockers or cavity dumpers. Different pump sources can produce distinct pulse patterns

- Flashlamp pumping produces pulses on the order of a microsecond long, or up to 500 μs long in some cases, with repetition rates ranging from a few shots per minute for high-energy types to a few hundred pulses per second for lower-energy models.
- Pumping with a pulsed excimer or neodymium laser produces pulses on the order of 10 ns long, with repetition rates from tens to a couple of hundred pulses per second.
- Pumping with a pulsed nitrogen laser produces pulses about 0.5 to 7 ns long, with repetition rate to a few hundred hertz.
- Pumping with a continuous ion laser produces a continuous beam. Output can be cavity-dumped to generate pulses on the order of 20 picoseconds (ps) at repetition rates in the 1-MHz range.
- Pumping with a copper vapor laser generates pulses of tens of nanoseconds repeated at the several-kilohertz repetition rate of the pump laser.
- Synchronous pumping with a modelocked ion or YAG laser can generate trains of picosecond pulses at 100- to 200-MHz repetition rates. Cavity dumping or external pulse selection techniques can reduce repetition rate for modelocked pulses.

Spectral Bandwidth In principle, spectral bandwidth of a dye laser depends on the dye's gain bandwidth and on the cavity optics, but in practice the cavity optics are usually the dominant factor. Most dyes have a gain bandwidth over 10 nm, but normal gain-narrowing processes limit the linewidth of laser emission to a few nanometers in untuned cavities.

In most pulsed dye lasers, standard wavelength-selecting optics can readily reduce bandwidth to a few thousandths of a nanometer. An intracavity etalon can compress bandwidth further to about 0.001 nm, equivalent at visible wavelengths to about 1 gigahertz (GHz) in frequency or about 0.025 inverse centimeter in wavenumber. Further line narrowing is difficult because of the inherent instabilities of pulsed operation.

Commercial dye laser systems with sophisticated line-narrowing optics and active control systems can achieve bandwidths around 500 kHz. Much narrower linewidths, on the order of a hertz, have been reported in laboratory experiments, but require very elaborate stabilization.

Production of picosecond pulses requires linewidth of about 10^{12} Hz, equivalent to roughly 1 nm at visible wavelengths. By using nearly the entire gain bandwidth of the dye, it is possible to produce pulses shorter than a tenth of a picosecond.

Amplitude Noise If wavelength is held constant, and operation is well above threshold, amplitude noise in a dye laser depends mainly on the nature of the pump source and the noise it contributes. For pulsed-laser-pumped dyes, amplitude noise in the pump laser should be used as a guideline. For flashlamp pumping, typical pulse-to-pulse variations run a few percent. Operation at high repetition rates can cause inhomogeneity problems in the medium in pulsed lasers. Intensity stabilization of continuous-wave dye lasers makes it possible to restrict variations in output power to less than 1 percent for an hour of operation.

However, changing the output wavelength causes independent changes in output power because the gain of laser dyes depends strongly on wavelength. Output power can vary by a large factor while tuning across a dye's gain bandwidth. Accessories can help to stabilize output power near the middle of a dye's operating range, but sharp variations are likely near the edges of the tuning range.

Beam Quality, Polarization, and Modes Like many other characteristics of dye lasers, the beam quality, polarization, and emission modes depend both on the pump source and on the optical cavity of the dye laser. In general, beam quality tends to be best with continuous-wave ion laser pumps, and worst with flashlamp pumping. Beams from other dye lasers are typically two to three times the gaussian diffraction limit. Quality of flashlamp-pumped beams depends entirely on the cavity and tends to be limited because the main goal is generally high power.

Continuous-wave dye lasers typically are specified as producing a TEM_{00} output beam, but for other types of pumping the output beam mode is typically unspecified. Stable-resonator cavities are typical, but unstable resonators are optional on many flashlamp-pumped dye lasers, where they can improve beam quality.

Grating tuning, common in dye lasers, or the use of any intracavity component at Brewster's angle, produces linearly polarized output, with the angle and degree of polarization depending on the design. Intra- or extracavity polarizers can be added if higher degrees of polarization are needed.

Coherence Length Coherence length of dye lasers is highly variable because it depends on spectral bandwidth. A rough value can be calculated from the formula.

$$\text{Coherence length} = \frac{\text{speed of light}}{\text{frequency bandwidth}}$$

(Wilson & Hawkes, 1983, p. 257).

Beam Diameter and Divergence Typical values of beam diameter and divergence for different types of dye lasers are:

- Flashlamp-pumped: 3–15 millimeter diameter, 1.5–2 milliradian divergence, with standard optics
- Pulsed-laser pumped: 1–5 mm diameter, 0.5–3 mrad divergence
- Ion-laser pumped: 0.5–1 mm diameter, 1.5–2 mrad divergence

Stability of Beam Direction This quantity is not normally specified for dye lasers. In practice, it is of little importance because most dye laser beams are directed over short distances in a laboratory.

Suitability for Use with Laser Accessories Dye lasers are routinely used with many accessories. The most important are harmonic generators, frequency mixers, modelockers, and cavity dumpers.

Second-harmonic generators are often used to shift the visible output of pulsed dye lasers into the ultraviolet, obtaining shorter wavelengths than dyes can directly produce. The result is a continuously tunable source of pulsed ultraviolet light. Harmonic generators can be used inside the cavities of continuous-wave dye lasers, but output power of continuous dye lasers is generally insufficient for reasonably efficient extracavity harmonic generation.

Frequency mixers are sometimes used to extend dye tuning range. For example, the fundamental frequency of a dye laser might be mixed with a neodymium-YAG harmonic in a nonlinear crystal to produce ultraviolet output. The sum-frequency wavelength could be tuned by tuning the dye laser wavelength.

Dye lasers are well suited for modelocking because of the broad gain bandwidth of dyes. Typically, modelocked pulses are generated from continuous-wave dye lasers by synchronous pumping with an actively modelocked ion or YAG laser; repetition rate of the modelocked pulses is determined by the cavity round-trip time. Dye lasers pumped by pulsed lasers can be passively modelocked. Picosecond pulses can be produced by commercial products. Special laboratory systems using passive modelocking have generated pulses as short as 55 femtoseconds (fs) (0.055 ps), which other techniques can further shorten to as little as 8 fs (Knox et al., 1985).

Argon- or krypton-laser-pumped dye lasers can be cavity-dumped independently of, or together with, synchronous pumping. Cavity dumping of a continuously pumped dye laser produces pulses on the order of a cavity round-trip time, 15 ns or longer. Simultaneous cavity dumping and synchronous pumping produce pulses with peak power about 30 times higher than that of ordinary synchronously pumped pulses and also reduces repetition rates to levels reasonable for many applications.

Q switching is not done with dye lasers. The upper laser level has a lifetime of only a few nanoseconds, too short for storage of energy in the gain medium.

Operating Requirements

Input Power Dye lasers which derive their optical power from an external pump laser have modest electrical power requirements because the electrical power is not used to generate the laser beam. However, power can be required for optical and electronic controls, dye-solution pumps, and other equipment.

Some data sheets for low-power dye lasers make no mention of electrical power requirements, and the laser may not have to be plugged in. However, higher-power dye lasers can need up to 5 A at 110 V (or half that much current at 220 V) to operate fluid pumps and control equipment. All pump lasers need electrical power.

Flashlamp-pumped dye lasers *do* need significant electrical power to drive the flashlamp that pumps the dye. Power requirements run from around 15 A at 110 V to 25 A at 220 V, in all cases single-phase current.

Cooling Flashlamp-pumped dye lasers typically remove excess heat by flowing water through the system at rates to 2 gal/min (8 L/min). The water cools both the dye solution and the flashlamp.

In laser-pumped dye systems, the dye flow system usually serves a dual purpose, cooling the solution as well as keeping it flowing through the dye cell to avoid degradation. Some dye flow systems include heat exchangers to transfer heat from the dye solution to slowly flowing water or some other medium. A few dye lasers which operate at low average power use sealed dye cuvettes without a liquid flow system.

Consumables The prime consumable in a dye laser is the dye itself. The complex organic molecules are vulnerable both to thermal and photochemical degradation. The severity of the effects increases as the pump wavelength becomes shorter and as light intensities increase; short-wavelength ultraviolet light is particularly damaging. Stated lifetimes range from around an hour to over 1000 hours, and they vary considerably among dyes and depending on operating conditions. Under favorable conditions, the most durable dyes can last over a year.

Dyes are offered either as premixed solutions or as compounds which can be added to solvents. Makers of dye lasers typically offer dyes to their customers, but the dyes are actually made by independent companies which specialize in their production. Often addition of fresh dye to a solvent can restore normal operation, although breakdown products can also build up to detrimental levels.

Required Accessories The wide variation in commercial offerings makes the idea of "required accessories" a hazy one for dye lasers. A pump laser is required for all but flashlamp-pump dye lasers. Other than that, however, the accessories needed depend on the applications. Line-narrowing and tuning optics are needed if the application requires narrow-band or tunable output. Modelocking is needed to generate ultrashort pulses. Dye flow systems are needed for moderate- and high-power operation to avoid rapid dye degradation. However, these are usually considered options or are supplied with the laser.

Operating Conditions and Temperature Dye lasers are designed to function in a normal laboratory environment at room temperature, but could encounter problems if bounced about in the back of a truck or operated at extreme temperatures. Solvent evaporation is a potential problem at elevated temperatures because many common dye solvents have lower boiling points than water. For example, methanol boils at 65°C.

Mechanical Considerations Dimensions and weight vary significantly among the different types of dye lasers.

Although a few flashlamp-pumped models are packaged as integrated units, typically there are three modules, a laser head, an electronics module or power supply, and a dye circulator. Dye heads weigh 15 to 100 kg and are 0.5 to 1.3 m long. The power supply is normally more compact and boxy, weighing 20 to 60 kg, while the dye circulator is also boxy and weighs 7 to 70 kg without dye solution.

By avoiding the need for internal flashlamps and power supplies, laser-pumped dye lasers can be much more compact. The smallest model is one packaged together with a compact nitrogen pump laser in a 7-lb (3-kg) package 12¼ in (31 cm) long. Many dye laser oscillator heads (without pump lasers) are similar in size, but there are some important exceptions. If a dye laser is synchronously pumped with a modelocked YAG or ion laser, the laser cavity must be extended to match the length of the pump-laser cavity, which is usually 1 to 2 m. This can be done by mounting a linear tube or adding a folded multipass delay line.

Another exception is the high-power pulsed dye laser that is pumped by a pulsed laser. Such lasers often incorporate separate oscillator and amplifier stages and include room for many accessories. As a result, the complete package may be over a meter long and half a meter or more wide. However, most of that volume is empty space, and the device itself may weigh less than 50 kg.

Safety Dye lasers present four potential types of hazard: the laser beam itself, electric shock, toxic dyes and solvents, and explosion and fire hazards.

The optical hazards from dye laser beams are similar to those from other lasers operating at the same wavelengths (Sliney and Wolbarsht, 1980). The principal danger is eye damage, because the visible, near-infrared, and ultraviolet light from dye lasers generally can penetrate the eye. High-power pulses are particularly dangerous because a single pulse can do serious damage, particularly in the ultraviolet. One special hazard of the dye laser is a direct result of its tunability: because the emitted wavelength can change, there are no safety goggles that can block just the narrow range of wavelengths emitted by a dye laser. It is also desirable to block any stray pump light, or any of the fundamental frequency light left after frequency doubling. To meet these requirements, goggles must block a comparatively broad chunk of the visible spectrum. Unfortunately, this is like wearing sunglasses inside, and it can obstruct vision, presenting other hazards as well as frustrating some wearers enough that they abandon the goggles. Also, if the entire visible spectrum is being covered, users will have to change goggles at least once during the experiment for proper eye protection.

Electrical hazards are presented by the high voltages needed to fire flashlamps and to operate pump lasers. Note that special care should be taken if safety goggles obstruct vision, so no high-voltage points are exposed where they might be contacted inadvertently by half-blinded goggle wearers.

Both laser dyes and solvents can be toxic (Kues and Lutty, 1975). Although no one is likely to swallow laser dyes, spills can be absorbed through the skin. Some organic solvents enhance penetration effects of dangerous dyes. Solvent vapors are a theoretical hazard, but fortunately most common solvents such as alcohols are toxic only at high concentrations (Weast, 1981).

Most organic solvents are highly flammable, both in liquid and vapor form, and present potential fire hazards. The main dangers are sparks, elevated temperatures, and flashlamp explosions. Fires have been triggered when an electrical discharge passed through solvents, and when solvents have contacted hot objects such as arc-lamp tubes. Flashlamps operated above their rated power levels can explode and trigger fires (Grant and Hawley, 1975).

Reliability and Maintenance

Lifetime As mentioned earlier, laser dye solutions have limited lifetimes, depending on the dye and operating conditions. Factors that can decrease dye lifetime are high pump energy and short pumping wavelength.

The rated lifetimes of dye laser flashlamps range from around 100,000 shots to several million, decreasing as pulse energy increases. Estimated lifetimes are sometimes given in terms of total number of joules delivered by the flashlamp, typically 10^7 to 10^8 joules (J). Linear flashlamps can be replaced readily, but replacement of coaxial flashlamps can be more complex because the internal tube contains the dye medium.

Except for dyes and flashlamps, lifetime figures are generally not given for dye lasers, which are designed for laboratory use, not heavy-duty use.

Maintenance and Adjustments Needed Several types of maintenance may be needed for dye lasers

- Periodic replacement of dye solutions or replenishment of dye.
- Cleaning of dye flow system and filters.
- Cleaning and adjustment of optics. In general, the more complex the optical system, the more frequently adjustment will be needed.
- Replacement of flashlamps in flashlamp-pumped lasers.
- Checking for optical damage, particularly to dye cells, and realigning optics to avoid any damaged regions or replacing severely damaged components.
- Maintaining alignment of the pump laser with the dye cell.
- Maintaining adjustment of tuning optics.
- Periodic maintenance of liquid flow pumps and electronic components.

Specific requirements depend on the specific dye laser used.

Mechanical Durability Dye lasers are scientific instruments built for use in the controlled environment of a laboratory. Specifications do not indicate shock resistance, but rough handling could damage the liquid flow system and knock the optics out of alignment.

Failure Modes and Causes Dye lasers can fail to meet a variety of specifications. If the pump source or the laser dye is not at fault, the problem is probably in the optics. The cavity could have drifted out of alignment, key components could be out of adjustment, or optical damage could have occurred at key points such as dye-cell walls. Note that continuous-wave dye lasers generally have flowing dye jets rather than dye cells to avoid the possibility of optical damage there.

Users should be aware that flashlamps and pump lasers can degrade in ways that reduce output power to lower levels than specified, and possibly to levels below the threshold for laser action in the dye. Proper diagnosis of the problem may require measuring power of the pump source.

Possible Repairs Other than fixing the light source or replacing the dye solution, the main repairs are optical realignment and replacement or repair of damaged optics. If damage has occurred at a small area, such as on the surface of a dye cell, it may be possible to move the component so the damaged area is not in the optical path. Some high-power pulsed dye lasers have dye cells designed so damaged areas can be moved out of the pump laser focal spot.

Commercial Devices

Standard Configurations There are many dye laser configurations on the market, each with its own advantages and advocates. Many models are designed for use with a specific type of pump source, but some can accept different types of pulsed lasers. The major configurations offered commercially are

- Coaxial flashlamp-pumped, in which the dye cell is in the central bore of a coaxial flashtube, as shown in Fig. 17-4.
- Linear flashlamp pumped, in which the dye cell is in the same reflective cavity as a linear flash tube.
- Nitrogen laser pumped, sometimes in a compact, integrated package.
- Excimer laser pumped, functionally meaning pumping with the 308-nm output of xenon chloride because that gas mixture offers the best combination of long excimer gas lifetime and limited degradation of the laser dye.
- Neodymium-YAG pumped, with the 532-nm second harmonic, the 355-nm third harmonic, or for some infrared dyes with YAG's fundamental 1.06-μm wavelength in the near-infrared.
- Pulsed laser pumped, in a version designed to accept YAG, excimer, or nitrogen pumping.
- Copper vapor pumped, similar to other types pumped with pulsed lasers, but designed to operate at higher repetition rates and average powers, with pump lines in the green and yellow.
- Continuous wave, pumped with an ion laser (usually argon, but sometimes krypton), using a linear, folded, or ring cavity.

- Synchronously pumped with a modelocked ion or frequency-doubled YAG laser. The doubled YAG is becoming increasingly popular as a continuous-wave source of modelocked pulses for synchronous modelocking.

Dye lasers pumped with either a pulsed laser or a flashlamp are also offered in an oscillator-amplifier configuration that boosts output power above what is readily obtainable from an oscillator alone. In laser-pumped versions, a single pulse from the pump laser is split and used to pump both oscillator and amplifier stages. In flashlamp pumping, separate lamps are used for oscillator and amplifier stages, with the pulse-generating electronics synchronized.

The dye laser may be packaged together with the pump laser, but more often the two are sold separately. The many companies which make both dye and pump lasers typically design their dye lasers specifically to mate with their pump lasers, although pump lasers are similar enough that this does not prevent the use of other models.

A few alternatives have been demonstrated, but they are not available commercially. The latest to reach the market is pumping with the copper vapor laser described in Chap. 12. This technique was developed to obtain high average powers from dye lasers for experiments in uranium isotope enrichment at the Lawrence Livermore National Laboratory (Lawrence Livermore National Laboratory, 1982). The recent commercialization of the copper vapor laser is leading to smaller-scale applications.

It should be noted that all the configurations on the market reflect efforts to meet requirements set by laboratory users of dye lasers. In practice, there are four principal design goals, all to some extent mutually contradictory:

- Extremely narrow spectral bandwidth (requiring continuous-wave pumping with an ion laser).
- Extremely short pulses (dictating modelocking and a broad spectral bandwidth).
- High power (either peak or average) operation and moderately narrow bandwidth.
- Low cost.

Options Dye lasers are scientific instruments designed to meet a wide variety of user requirements. To do so, manufacturers offer many options, beyond the simple choice of dye and pump source. Major options are listed below, but users should note that some of these options may be standard features on certain models:

- Etalons, to narrow spectral bandwidth
- Wavelength-tuning optics (standard on many models), including tunable birefringent filter, tuning wedges, prisms, and gratings
- Optics with broad spectral range, to avoid the need for interchange of optics while tuning wavelength
- Automatic frequency scanning across a range of wavelengths
- Active stabilization of output frequency

- Raman shifters
- Frequency-doubling crystals to produce ultraviolet second harmonics of visible-wavelength tunable output
- Frequency-summing arrangements, to add the tunable dye laser output frequency to that of a fixed-frequency laser
- Modelockers
- Polarizing optics
- Cavity dumpers
- Cavity-extending optics to match length of the dye laser cavity with that of a pump laser for synchronous pumping
- Optical output monitors
- Amplifiers for pulsed lasers
- Unstable resonator optics (for pulsed lasers)
- Dye filters, to limit particle size in the dye solution to less than the wavelength of light and thus avoid excess scattering
- Dye circulators

Special Notes The relative merits of different pump sources for pulsed dye lasers have been argued at great length, generally by companies which make dye lasers and/or pump sources. Although there were some early expectations that excimer and YAG pumping might force the nitrogen laser off the market, the various approaches have instead found different market niches. In practice each approach has its own practical advantages:

- Flashlamp pumping: low capital cost, high average power
- Nitrogen laser pumping: amenable to compact inexpensive designs for low-power operation; low operating costs; higher peak power than flashlamp pumping; able to produce subnanosecond pulses without modelocking; simple, broad tuning range; simple gas handling; repetition rate above 100 Hz possible
- Excimer laser pumping: high peak powers in 10-ns pulses, high repetition rates (over 100 Hz), high average power, broad tuning range
- Nd–YAG pumping: good beam quality, high peak powers, flexibility because of possible choice of harmonics, no gas-handling requirements

In many cases, a key factor is the choice of dye, because some dyes operate more efficiently with some pump sources than with others. The considerations affecting a choice are complex, and there are few easy guidelines. The best course for users is to compare their requirements and their budgets with the specifications and prices of a variety of commercial dye lasers.

Pricing Pricing of dye lasers can be confusing because of the many configurations available. Flashlamp-pumped dye lasers are sold with internal pump lamps, and a *few* laser-pumped dye lasers are sold as systems incorporating pump lasers. However, the attractively low prices of some sophisticated dye lasers may only indicate that the expensive pump lasers are sold separately. Note also that the standard optics configurations vary widely; some models come complete with highly sophisticated optics, while others come equipped only with mirrors needed to produce broadband output. The 1985 prices listed below cover the typical range:

- Flashlamp-pumped dye: $4600 to $58,000
- Pulsed-laser pumped (without pump laser): $1100 to $43,000
- Pulsed laser with pump laser (usually compact versions packaged with a nitrogen laser to match low budgets): $3500 to $15,000
- Ion-pumped conventional cavity (without pump): $8800 to $32,000
- Ion-pumped ring cavity (without pump) $15,000 to $58,000

Suppliers The dye laser market is quite competitive, with most major variations offered by at least two companies. Competition is most intense in models pumped by pulsed lasers, with about a dozen suppliers; smaller numbers make models pumped by ion lasers or flashlamps. Virtually all companies specialize in only one type of dye laser; for example, makers of flashlamp-pumped dye lasers do not offer models pumped by ion or pulsed lasers. The market choices often reflect company specializations in pump lasers, and in fact a number of suppliers added dye lasers to an existing line of lasers that were being used to pump dye lasers from other companies.

Although several companies have been making dye lasers for years, there are also some newcomers to the market. The pace of change is fast, so users would be wise to consult a current industry directory when they start shopping. There are enough suppliers to make pricing competitive.

Applications

About 80 percent of the dye lasers sold in 1985 were used in research and development (*Lasers & Applications,* 1986). Applications in medical therapy are growing rapidly and dye lasers also are used in measurement, inspection, and entertainment.

The major scientific applications of dye lasers are in spectroscopy and measurement, both in the time and frequency (or wavelength) domains. The ability to tune dye laser wavelength and to limit emission to a narrow spectral bandwidth makes dye lasers extremely useful in studying the absorption and emission of light by various materials. Although other light sources can be used in spectroscopy, dye lasers offer higher light intensities in narrower spectral bands. They have been particularly useful in the study of atomic and molecular physics, and in stimulating fluorescence.

Ultrashort picosecond pulses from dye lasers have been invaluable in studies of time response, such as measurements of excited-state lifetimes and decay rates. Ultrashort pulses have been used in research on semiconductor properties, chemical-reaction kinetics, photosynthesis, and biomolecular processes. The scope of these time and frequency domain studies is far beyond the realm of this book (see, for example, Arecchi et al., 1983).

In medicine, dye lasers are used to treat cancer and eye and skin disorders. Dye lasers also are used in diagnostic and laboratory work, including cell sorting. Research areas include ophthalmology, dermatology, cancer treatment, and cell surgery.

In entertainment, dye lasers are used to generate varicolored beams for laser light shows and displays. Although dye lasers offer a unique way to scan the entire visible spectrum, their practical display applications are limited by their high cost and operational complexity.

Complexity and high cost have discouraged applications of dye lasers in industry, where reliability and low cost are paramount concerns. Fortunately, few industrial applications require light at a specific narrow range of wavelengths. The only major commercial-scale process using a dye laser under development is isotope enrichment. Two parallel programs are being conducted at the Lawrence Livermore National Laboratory in California. In one, large copper vapor lasers pump high-power dye lasers, which are tuned to the precise wavelength that will ionize uranium 235, but not the commoner uranium 238. The ions are then collected to produce material enriched in uranium 235 for use as nuclear fuel. A separate, highly classified program uses technology that presumably is similar to remove undesired isotopes from plutonium produced in nuclear reactors; after processing the plutonium would be used in nuclear bombs. At this writing, neither process is operating on a commercial scale.

Medical dye laser applications are making the transition from research to practical use, and other applications may follow.

BIBLIOGRAPHY

F. T. Arecchi, F. Strumia and H. Walther: *Advances in Laser Spectroscopy*, Plenum, New York, 1983.

Atomic Vapor Laser Isotope Separation, Lawrence Livermore National Laboratory, Livermore, Calif., 1982.

W. W. Duley: *Laser Processing and Analysis of Materials*, Plenum, New York, 1983.

W. B. Grant and J. G. Hawley: "Prevention of fire damage due to exploding dye laser flashlamps," *Applied Optics 14*(6):1257–1258, June 1975.

W. H. Knox, et al,: "Optical pulse compression to 8 fs at a 5-kHz repetition rate," *Applied Physics Letters 46*(12):1120–1121, June 15, 1985.

Henry A. Kues and Gerard A. Lutty: "Dyes can be deadly," *Laser Focus 11*(5):59–61, May 1975.

"The Laser Marketplace—1986," *Lasers & Applications 5*:1, January 1986, pp. 45–46.

Mitsuo Maeda: *Laser Dyes: Properties of Organic Compounds for Dye Lasers*, Academic Press, Orlando, 1984.

O. G. Peterson: "Dye lasers," in C. Tang, *Methods of Experimental Physics*, vol. 15, part A, Academic Press, New York, 1979.

F. P. Schafer (ed.): *Dye Lasers, 2d ed.*, Springer-Verlag, Berlin & New York, 1977.

C. V. Shank, E. P. Ippen, and S. L. Shapiro: *Picosecond Phenomena*, Springer-Verlag, Berlin & New York, 1978.

David Sliney and Myron Wolbarsht: *Safety with Lasers & Other Optical Sources*, Plenum, New York, 1980.

Peter P. Sorokin and J. R. Lankard: "Stimulated emission from an organic dye, chloroaluminum phlalocyonine," *IBM Journal of Research & Development 10*:162,1966.

Richard Steppel: "Organic dye lasers," in Marvin Weber (ed.), *CRC Handbook of Laser Science & Technology*, vol. 1, *Lasers & Masers*, CRC Press, Boca Raton, Fla., 1982.

Orazio Svelto: *Principles of Lasers*, 2d ed., Plenum, New York, 1982, p. 191.

L. V. Tarasov: *Laser Physics*, Mir, Moscow, 1983, p. 32. English edition.

Robert C. Weast (ed.): *CRC Handbook of Chemistry and Physics*, 62d ed., CRC Press, Boca Raton, Fla., 1981.

J. Wilson and J. F. B. Hawkes: *Optoelectronics: An Introduction*, Prentice-Hall International, Englewood Cliffs, N.J., 1983.

CHAPTER

18

near-infrared semiconductor diode lasers

Semiconductor diode lasers are more a family of devices than a single type. Practically speaking, there are two major categories: diode lasers fabricated from compounds formed from columns III and V of the periodic table, such as gallium arsenide, which emit in the near-infrared (or sometimes just into the red), and diode lasers made from "lead salt" compounds which emit at longer infrared wavelengths from around 2.7 to 30 micrometers (μm). This chapter describes only the near-infrared III–V diode lasers; lead salt types are covered in Chap. 19.

Many more diode lasers are sold than any other commercial type, although because the unit price is low there are several other laser types which lead diode lasers in dollar value of sales (*Lasers & Applications,* 1986). The technology actually goes back over two decades to 1962, when three research groups independently and nearly simultaneously demonstrated gallium arsenide diode lasers (Hall, 1976). However, it took a decade to transform a fragile device requiring cryogenic cooling into one capable of the seemingly simple task of emitting a continuous beam at room temperature. The pace of development has accelerated greatly over the past decade, and diode lasers have become one of the fastest-moving laser technologies. The technology has been stimulated both by the spread of fiber-optic communications (which require a semiconductor light source) and by the need for compact, inexpensive light sources for a variety of information-handling applications. The result has been a great increase both in the quality of diode lasers on the market and in the demand for them.

Description

As the name implies, the diode laser is really a semiconductor diode which emits a coherent laser beam. The diode laser (or laser diode) is similar in structure to the light-emitting diode or LED, and in fact diode lasers generally can function as incoherent LEDs—albeit inefficient ones. LEDs, however, lack the structural elements and reflective faces they would need to function as diode lasers.

The laser diode itself is a block of semiconductor material containing a *pn* junction. Typically, an individual laser chip is less than a millimeter in all of its dimensions, and only a 10-μm wide strip within it produces a laser beam. The internal structure is considerably more complex than that of a simple *pn* junction, particularly for sophisticated diode lasers. As described later, there are several types of diode lasers, which differ significantly in detail.

Because of their small size, commercial diode lasers are almost always packaged in a housing that simplifies handling and aims the output beam or beams (which can emerge from one or both faces of the device) in a predictable direction. Housings come in several types intended for various applications. Lasers can be mounted in arrays as well as singly; the array approach increases the output power, although at a cost in beam quality and sometimes in duty cycle. Traditionally arrays were assembled from discrete chips, but recently monolithic arrays with multiple laser stripes have come on the market.

Internal Workings

Operation of a diode laser relies on the internal physics of semiconductor electronics. As in other semiconductor devices, current in a diode laser is carried by electrons free to move within the crystal when an electric field is applied, and by holes, vacancies for valence electrons within the crystal lattice. Forward-biasing the *pn* junction lowers internal potential barriers and thus causes carrier injection from one side of the junction to the other. In this case, forward bias means that a negative voltage is applied to the *n* material and a positive voltage to the *p* material, creating a local excess of carriers. Equilibrium is restored when the excess electrons fall to lower energy states, i.e., holes. The process, shown schematically in Fig. 18-1, is called electron-hole recombination.

The electrical result of forward biasing a semiconductor diode is current flow, with the recombination of electrons and holes releasing energy at the junction. In silicon diodes, most of that energy is released as heat, but in certain other materials much of the released energy is light. In these light-emitting materials, when an electron drops from the conduction band to the

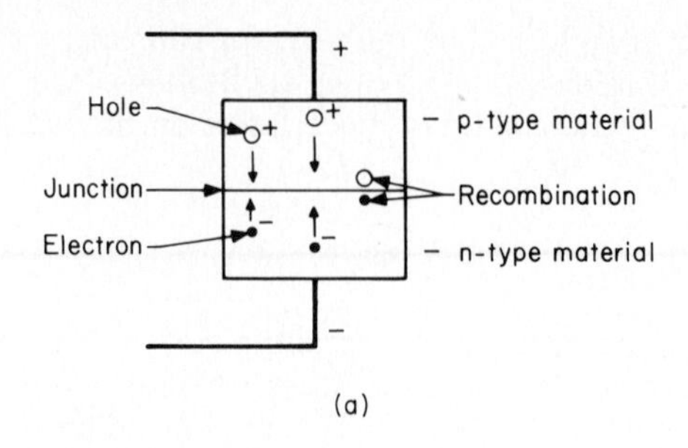

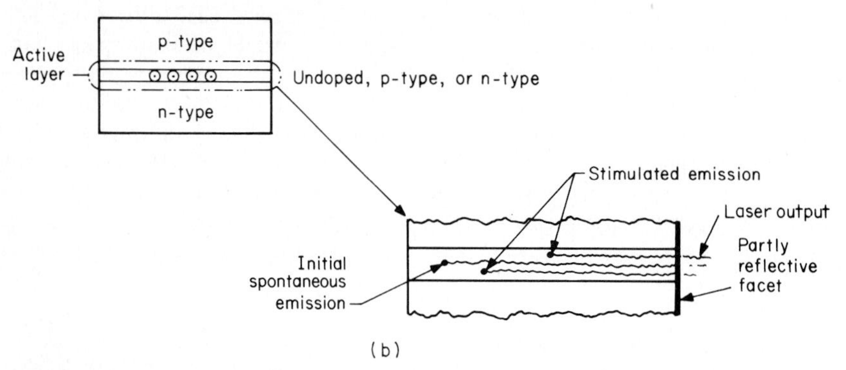

Figure 18-1 Operation of a diode laser (simplified). *(a)* Forward bias produces current flow and recombination; *(b)* stimulated emission in active layer.

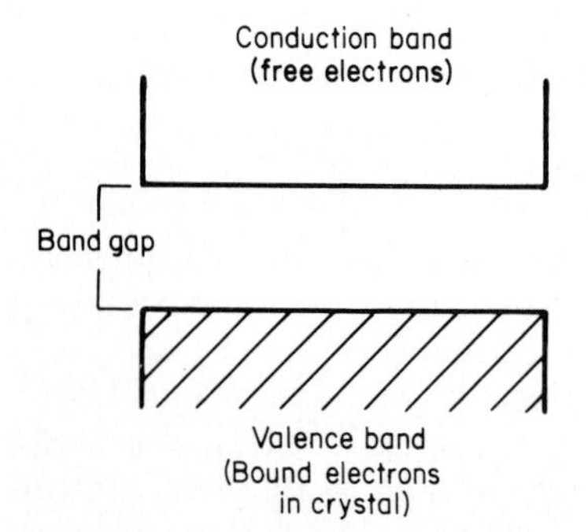

Figure 18-2 Energy-level structure in an unbiased semiconductor.

valence band during recombination, it can emit much of the energy in the form of a photon with energy roughly equivalent to the band gap—the energy difference between the two bands—as shown in Fig. 18-2.

Light emission occurs spontaneously, at random, in LEDs, but not in diode lasers, which are designed so that stimulated emission dominates. In a diode laser, recombination of the excited carriers, which remain briefly in an excited state, is stimulated by a photon of the recombination energy *before* the excited state can drop naturally to the lower energy level by spontaneous emission.

High drive currents are needed in diode lasers to produce a population inversion of excited carriers, so light will be amplified as it passes through the material. Addition of a pair of resonator mirrors, which in practice means cleaving opposite ends of the laser crystal to produce reflective facets, provides the optical feedback that is needed for the stimulated emission to overcome cavity losses and create self-sustained laser oscillation. Spontaneous emission dominates in LEDs because they have no reflective facets and operate at lower drive currents.

The energy of the photons released by the semiconductor is a function of the bandgap energy, which depends on the crystalline structure and the chemical composition of the semiconductor material. Thus the wavelengths emitted by semiconductor lasers are determined by the materials from which they are made. Various III–V semiconductor compositions can be used to produce diode lasers with output from the red end of the visible spectrum to about 1800 nanometers (nm) in the near-infrared. There are efforts to extend the wavelength range in both directions.

Only a fraction of the electrical energy dissipated by a diode laser emerges in the laser beam; a large part is turned into heat. The need to minimize and remove this heat puts constraints on diode laser operation, particularly because the operating characteristics of diode lasers degrade as temperature increases. Because of these problems, early diode lasers were limited to low-temperature and/or pulsed operation. Sophisticated structures are needed to reduce the laser's threshold current density and heat dissipation enough to permit continuous operation at room temperature.

Internal Structure

As indicated earlier, the key element in a semiconductor laser is an internal *pn* junction, which has led to the use of the term *junction* laser. (The term *injection* laser is sometimes used because of the electrical injection process which creates excess carriers.) In its simplest form, a diode consists of a block of semiconductor material in which part is doped with electron donors to form *n*-type material and the other part is doped with electron acceptors (hole generators) to produce *p*-type material.

In practice, fabrication of a diode laser starts with a wafer of semiconductor with a controlled impurity level, say, *n*-type material. Then a different impurity is diffused into selected areas of the surface of the wafer; in the example, electron acceptors are diffused into the material in sufficient quantity to overwhelm the *n*-type electron donors and create *p*-type material. The junction occurs at the diffusion-layer boundary. In practice, several layers of different composition are grown within a diode laser, with the extra layers generally serving to improve performance.

The internal structure of a diode laser is similar to that of an LED. From a structural standpoint, the most important difference is that two of the edge faces (or "facets") of the diode laser are cleaved to reflect part of the generated light back into the semiconductor. (Coatings may be added to enhance reflectivity, and often one facet is coated to make it totally reflective, so that all of the laser emission will emerge from the other facet.) The other major difference is the operation of diode lasers at much higher drive currents than LEDs. The combination of these factors leads to the great differences in the output characteristics of LEDs and diode lasers.

The recombination process at a semiconductor junction can emit photons in any direction, although the planar nature of the junction tends to concentrate LED emission somewhat. Commercial LEDs either generate beams from their edges or from their tops; in the latter case a hole is usually etched to remove the light directly from the junction. The reflective facets of a diode laser direct the emitted light back through the semiconductor junction, providing the optical feedback needed to sustain laser action. The reflected light stimulates the emission of other photons from other recombination sites—the light amplification by the stimulated emission of radiation which provides the acronym laser, as indicated in Fig. 18-1. If the net optical amplification or gain in a round trip of the cavity is higher than the losses, the result is laser oscillation.

Stimulated emission serves to increase the optical power emitted. Spontaneous emission in an LED is a rather inefficient process because the energy from many recombination sites is dissipated as heat rather than as photons. Stimulated emission can extract more of this energy as light. The stimulated emission process also tends to produce the strongest amplification of wavelengths where an LED is most likely to emit light, while suppressing emission at other wavelengths, thus causing bandwidth of a laser diode to be much narrower than that of an LED.

A resonator is not enough to produce laser action. As in any other type of laser, stimulated emission can dominate to produce a laser beam only if there is a population inversion—which in a diode laser means a very high density of electron-hole pairs. This requires a high operating current. At low current values, even a diode with a resonator will behave like an LED. At higher currents, the device becomes an amplifier or superluminescent diode but does not have enough gain to produce laser oscillation. Only when the threshold current is passed will the diode operate as a laser.

Although a single value is usually given for threshold current, the transition from LED to laser action is not instantaneous. As shown in Fig. 18-3, the threshold current level marks a rapid change in the slope of the line plotting output power vs. drive current. The slope efficiency of laser emission is so much higher than that of LED emission that low-resolution measurements may make output-vs.-current curves appear to be zero until the threshold current is reached. Practically speaking, however, the threshold current is usually an extrapolated point obtained by extending the linear part of the laser output-vs.-current curve back to zero-output line.

Types of Diode Laser Structure The variety of diode laser structures is far too extensive to cover in detail in a book like this. Indeed, the list grows longer with every volume of journals such as *Applied Physics Letters* and

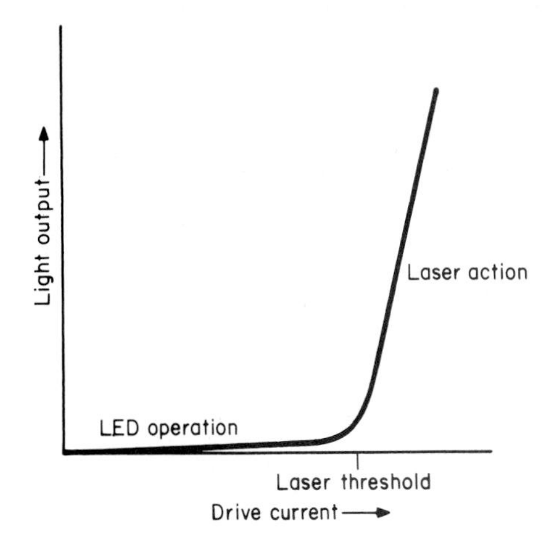

Figure 18-3 Output of a laser diode below threshold (when it operates as an LED) and above threshold (when it operates as a laser). The much steeper slope for laser operation indicates higher-efficiency emission.

Electronics Letters. For practical purposes, I will focus only on broad categories that cover the most important types and concentrate on the variations which have the most significant effects on performance. The state of the art is moving very rapidly, and any detailed description of diode laser structures would almost inevitably be obsolete by the time it could be published in book form. A sampling of major structures is shown in Fig. 18-4.

- *Homojunction* diode lasers are made entirely of a single semiconductor compound, typically gallium arsenide, with different portions of the device having different dopings. The junction layer is the interface between *n*- and *p*-doped regions of the same material. This structure was used in the first diode lasers but has since been replaced by other structures which can provide better laser characteristics.
- *Single-heterojunction* (or single-heterostructure) lasers are those in which the active layer has one boundary with a material having a different bandgap. In practice, this means that it is sandwiched between two layers of different chemical composition—typically GaAs and GaAlAs, which have different bandgaps. Most often the active layer is GaAs, and the laser's emission wavelength is at 904 nm. Single-heterojunction lasers have better properties than homojunction lasers and are widely used to produce high peak powers in pulsed operation. They are offered as single devices or as "arrays" assembled from separate chips. However, single-heterostructure lasers are relatively inefficient and have high threshold currents that make them unsuitable for continuous or high-duty-cycle operation, such as in fiber-optic communications systems.
- *Double-heterojunction* (or double-heterostructure) lasers have an active (light-emitting) layer bounded by two layers of different material—for example, GaAs sandwiched between two GaAlAs layers. These lasers have proved the most suitable for continuous operation and have found widespread use in fiber optics. They cannot produce pulses with high peak power, but they can operate continuously or with high duty cycle at powers to tens of milliwatts

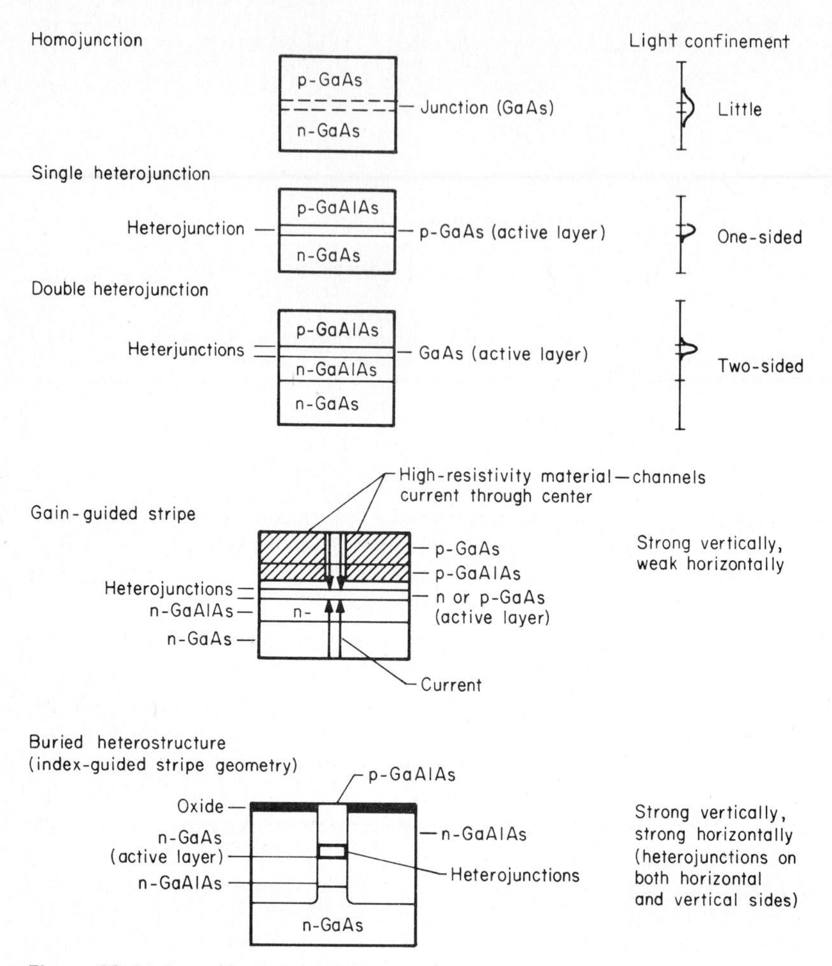

Figure 18-4 A sampling of diode-laser structures.

at room temperature. Their active layers are thinner than those of single-heterojunction lasers, so they can have threshold current densities low enough to operate continuously at room temperature.

- *Stripe-geometry* lasers are a subcategory of double-heterojunction lasers in which emission is confined to a narrow stripe along the length of the laser. Typically the stripe is 1 to 10 μm wide, compared with widths of 50 μm or more for earlier double-heterostructure diode lasers. The stripe-geometry laser has a number of attractions. Because the drive current is concentrated in a smaller stripe area, the threshold current for laser action is lower than in wider-stripe lasers. The narrower stripe also improves beam quality by limiting the number of spatial modes in which the laser can oscillate, often to a single mode. By helping to focus the beam down to a small spot, this aids information-handling applications. The small emitting area corresponding to the stripe width also is important for fiber-optic applications, because it aids in coupling to single-mode fibers.

The active-region width can be limited by two basic techniques. One is called *gain guiding* and relies on variations in the injected-carrier density and optical gain across the active region. The other is *index guiding,* in which the active region is defined by changes in the composition of the semiconductor that establish refractive-index boundaries in the plane of the active region. (These refractive-index boundaries are analogous to those perpendicular to the junction plane in a double-heterostructure laser.) Index-guided lasers tend to have better-quality beams, sharper turn on, and more linear output-current characteristics, and have found increasing uses in applications that require small spot size. However, they are vulnerable to some noise and feedback effects, problems that have left room for gain-guided lasers in some areas of fiber optics.

Single-heterojunction diode lasers have long been arranged in linear arrays or stacks to produce higher-power pulses than single lasers. This technique recently has been extended to production of monolithic linear arrays of stripe-geometry lasers, all in the same device plane (Cross et al., 1985). In commercial versions, all stripes are modulated in unison to obtain higher pulsed or continuous-wave powers than possible with single devices. Further work is in the laboratory stage. One possibility is independent modulation of separate stripes in the same array, perhaps to drive separate optical fibers. Another possibility already demonstrated is the "phased array," in which modulation of separate stripes differs in phase by an amount that causes their output to merge into a single-lobed beam that is narrow and perhaps even steerable. One of the problems with early commercial versions has been that output tends to be in a multilobed beam that is not desirable for many applications.

Also in the laboratory stage is work on lasers with active regions composed of many extremely thin layers of semiconductor with differing composition—typically two compositions in alternating layers. Versions in development include the multi-quantum-well laser and the strained-layer superlattice laser. Such refined structures offer prospects for controlling modulation characteristics and wavelength more than is possible with conventional designs.

Development is also in progress on diode laser structures that would provide the extremely narrow linewidth required for fiber-optic communication at extremely high speeds. One approach is the distributed-feedback laser, in which a grating in the semiconductor structure provides feedback that limits oscillation to a very narrow range of wavelengths (Suematsu et al., 1985). Another is the cleaved-coupled-cavity or C^3 laser (Tsang, 1985), where a single laser is split into two sections that are electrically separate but optically coupled in a way that limits oscillation to one wavelength. A third approach attaches a passive external element to a diode laser chip, again limiting oscillation to a single wavelength.

Gain in semiconductor lasers is very high, so the resonator cavities are short—typically a few hundred micrometers long. The resonator cavities are unusual in shape because of the nature of the active layer, which typically is a tenth of a micrometer to a couple of micrometers thick, and between a few and about 100 micrometers wide. Diode lasers can produce beams emerging from both facets, although normally they are packaged to emit only a single beam. The unusual resonator structure leads to some unusual beam characteristics, which are described later in this chapter.

Inherent Trade-offs

Major advances in semiconductor materials and device fabrication have greatly improved laser diode characteristics, greatly alleviating the early problem of short lifetimes. New materials and structures have expanded the range of output wavelengths and operating characteristics, but some fundamental trade-offs remain in commercial devices:

- Beyond certain limits, the lifetime of a diode laser decreases sharply with operating temperature and output power.
- Operating wavelengths much shorter than the 750- to 780-nm range come only at a cost of short laser lifetime. The problem worsens with shorter wavelength and higher output power. Some researchers are optimistic, but not all, and it is not clear how much improvement can be expected.
- Individual devices are designed either to produce high-peak-power pulses with low duty cycle, or to operate steadily at much lower power levels. Present devices are not optimized for both types of operation.
- The higher powers from diode laser arrays come at the cost of lower beam quality than single lasers. Progress is being made in the laboratory using phased control of the individual stripes in monolithic arrays and injection locking of a monolithic array to output of a single laser. However, output quality remains a limitation of commercial devices.

Variations and Types Covered

As described earlier, operating characteristics of diode lasers differ with the device structure and material composition, creating a broad range of possible types. This chapter focuses on the most important types on the market, concentrating broadly on families of devices rather than narrowly on structural details of individual devices. These major types are

- Single-heterojunction GaAs–GaAlAs lasers, packaged individually or in arrays, to produce pulses near 900 nm.
- Double-heterojunction GaAs–GaAlAs lasers, packaged individually to produce continuous output, normally between 780 and 900 nm.
- Monolithic arrays of up to 500 "multiheterojunction" GaAs–GaAlAs laser stripes. Arrays with up to about 100 stripes can operate continuous-wave, producing higher powers than single devices. Larger arrays can only operate in pulsed mode.
- Double-heterojunction InGaAsP–InP lasers, packaged individually, to generate continuous output between about 1150 and 1600 nm.

Stripe-geometry lasers have become increasingly popular because of their low threshold currents and good modulation and beam quality. Similar structures can be used in both GaAlAs and InGaAsP materials. Research continues on a multitude of new laser structures, but the details are beyond the scope of this book. Also not included here are descriptions of work on diode lasers for wavelengths shorter than about 750 nm and longer than about 1.6 μm.

Beam Characteristics

Wavelength and Output Power Wavelength of a diode laser depends primarily on the bandgap of the material in which the electrons and holes recombine. In a binary (two-element) compound such as gallium arsenide, the bandgap has only one possible value. This is not the case in semiconductor compounds in which the relative proportions of different elements can vary, including ternary (three-element) compounds such as GaAlAs and quaternary (four-element) compounds such as InGaAsP. The range of possible wavelengths is indicated schematically in Fig. 18-5.

The practical availability of diode lasers of various wavelengths is more limited than the figure might at first indicate. One reason is that some compounds lack the "direct-bandgap" energy-level structure needed for efficient production of light; these "indirect-bandgap" materials are not suitable

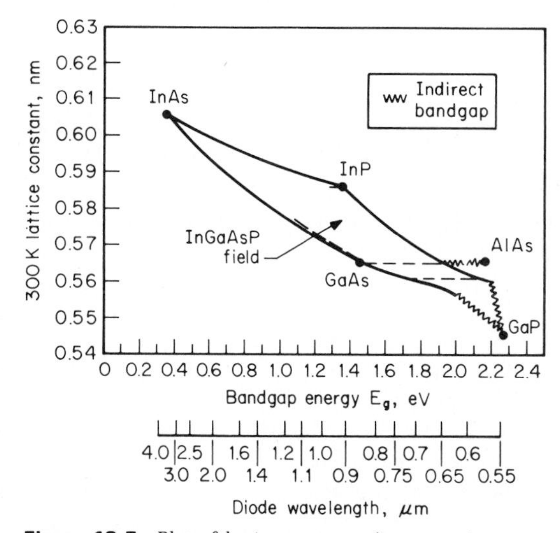

Figure 18-5 Plot of lattice constant (interatomic spacing) vs. bandgap energy and diode wavelength for InGaAsP and GaAlAs. The use of four elements in a quaternary compound maps out a two-dimensional operating region, while operation of the ternary compound GaAlAs is constrained to points along a straight line. "Indirect bandgap" semiconductors are unsuitable for laser operation because too little recombination energy is converted into light. *(Adapted from Kressel, 1982, p. 12.)*

for use as laser diodes. Another is that ternary or quaternary compounds are grown on substrates of binary compounds, and for the growth to proceed properly, the spacing of atoms in the substrate must be close to that in the compound being grown. (In practice, GaAlAs is grown on GaAs, while InGaAsP is grown on InP.) In addition, some of the compounds are simply hard to grow. The development of strained-layer superlattice structures may enhance the range of compounds that can be grown, but at present there is a limited range of diode laser materials used commercially:

- GaAs: 904 nm
- GaAlAs: about 720 to 900 nm, although short wavelengths are very short-lived. the shortest wavelengths of commercial devices are around 750 nm.
- InGaAs: around 1060 nm
- InGaAsP: generally around 1300 to 1550 nm in commercial devices, although the material is usable from around 1000 to 1700 nm

Precise control over the composition and hence output wavelength is almost impossible for ternary and particularly quaternary compounds. Thus, in practice there is a margin of error inherent in fabricating diode lasers from such compounds, and laser diodes are made to emit at a nominal wavelength rather than a precise one. If a precise wavelength is required, fabricated lasers have to be tested, and one meeting the precise specifications must be identified. Suppliers normally specify variations of ±20 to 30 nm from a nominal center wavelength.

Commercial single-heterostructure lasers are almost invariably made from GaAs–GaAlAs, except for the special-purpose models made of InGaAs to simulate neodymium range finders. Typical single-heterojunction lasers produce pulses with peak powers of one to a few tens of watts, lasting a fraction of a microsecond. Peak powers of a few hundred watts can be obtained by assembling pulsed lasers into arrays.

Double-heterojunction lasers of GaAlAs can produce continuous or average powers from one to a few tens of milliwatts, while maximum output powers of InGaAsP lasers at this writing remain somewhat lower. Rapid pulsing at a 50 percent duty cycle can permit operation at powers up to about twice the continuous-output limit, and in many applications such pulsed output is a more realistic approximation of actual operation than continuous output. Monolithic arrays of GaAlAs lasers can produce continuous-wave powers to 500 mW at room temperature. Similar devices can generate 100-microsecond (μs) pulses with peak power of several watts at low duty cycle. These figures may go up as the technology is brought to market.

Efficiency The overall efficiency of a laser diode (the fraction of the electrical drive power converted to optical output) depends on two factors, the threshold current and the slope efficiency—the percentage of each added milliwatt of electrical input converted into light. Because of the threshold effect, efficiency generally increases with output power; lasers emitting a few milliwatts typically have efficiency reaching several percent, while GaAlAs lasers producing 10 to 20 mW continuously may be more than 10 percent efficient. Increasing the operating temperature causes threshold current to rise and slope efficiency to drop, degrading overall efficiency. Because the lost energy is converted to heat, the drop in efficiency with rising temperature raises the possibility of thermal runaway of the diode. There has been much work aimed at reducing threshold currents to help avoid that problem and to increase the operating lifetime of diode lasers, and the lowest values for commercial lasers are in the 10- to 20-mA range.

Temporal Characteristics of Output At room temperature, single-heterojunction diode lasers are capable only of pulsed operation. Typical specifications call for pulses of 0.1 to 0.2 microsecond (μs) at maximum repetition rates of thousands of pulses per second. Nearly square pulses can be generated because rise times are generally under 1 ns. Arrays can produce pulses of about the same length, but heat dissipation requirements limit their repetition rates to lower levels.

Double-heterojunction lasers normally are considered continuous emitters, but their output can be modulated at high speeds by varying the drive current. Rise time, typically under 1 ns, can approach 0.1 ns for stripe-geometry lasers. Repetition rates are limited not by heat dissipation but by the need to avoid interference between successive pulses. Major concerns include delay time in responding to electrical pulses, and transient "relaxation oscillations" which can last 0.1 to 1 ns in typical diode lasers, as shown in Fig. 18-6. The fastest modulation speeds for commercial diode lasers are several gigahertz, and modulation bandwidths more than twice that have been demonstrated in the laboratory (Su et al., 1985). Data rates in commercial digital fiber-optic systems are approaching the one-gigabit-per-second range, and much higher transmission speeds have been demonstrated.

Spectral Bandwidth Spectral bandwidth of GaAlAs/GaAs diode lasers typically runs from 0.1 nm for a single-mode double-heterojunction laser to around 10 nm for a pulsed single-heterojunction laser. (LEDs typically have

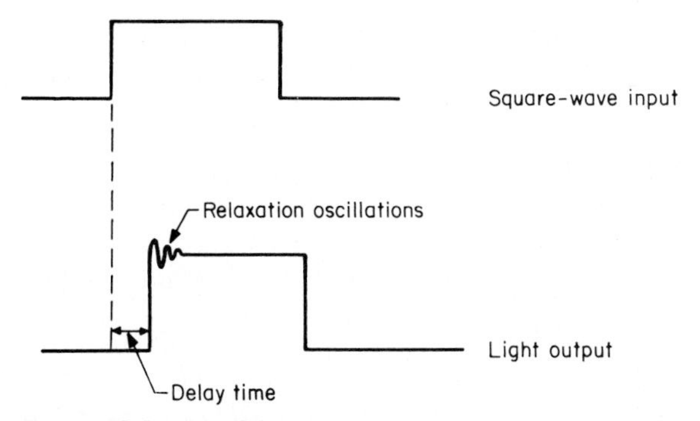

Figure 18-6 How delay time and relaxation oscillations can affect the emission of a laser diode.

a bandwidth of about 50 nm at similar wavelengths.) The bandwidth at longer wavelengths tends to be larger because it is proportional to a fraction of the wavelength. Thus a laser structure which produces 4-nm linewidth at 800 nm may have 6-nm linewidth at 1300 nm. The best single-mode lasers are rated at 0.1-nm bandwidths at either wavelength, which reflects measurement rather than device limitations. Under carefully controlled conditions, single-mode laser bandwidths can be limited to tens of megahertz (roughly equivalent to 10^{-5} nm).

Frequency Stability The center wavelength of a diode laser depends on temperature, typically shifting about 0.25 nm/°C for GaAlAs or about 0.5 nm/°C for InGaAsP. The picture is complicated by the fact that many diode lasers can oscillate simultaneously on several different modes, each with a different wavelength. As temperature changes, the laser can "mode hop" from one longitudinal mode to another, abruptly shifting wavelengths. This can happen as pulse length changes during pulsed operation, or as operating current (and output power) change during continuous operation. Increases in drive current can shift many continuous lasers from multimode operation to output strongly dominated by a single mode. The overall picture thus becomes a complicated one, in which output wavelength may gradually change, then abruptly jump to another value, as shown in Fig. 18-7. Such behavior can cause frequency instabilities in time and/or with temperature variations, as well as excess noise at mode-hopping points. If wavelength stability is critical, active temperature compensation that keeps the diode at a steady temperature can help minimize many of these problems.

There is a subtle frequency shift or "chirp" which becomes significant for direct modulation of diode lasers at high speeds. As the current pulse builds in the semiconductor, the change in electron density alters the refractive index of the material. This leads to a change in the effective length of the

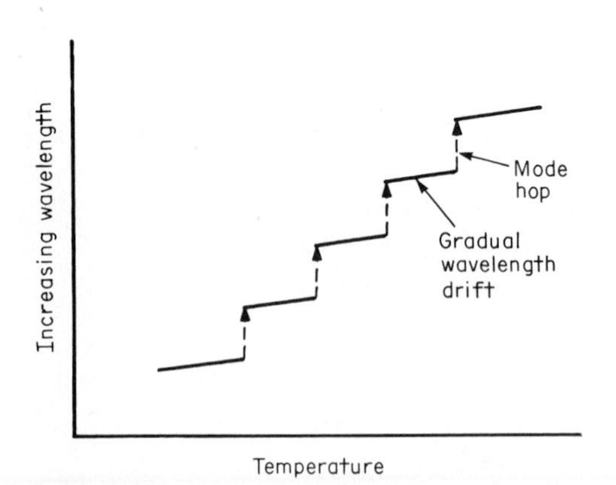

Figure 18-7 Peak wavelength of a diode laser varies slowly with temperature until emission "hops" to another longitudinal mode.

laser cavity and in the oscillation wavelength. As a result, the wavelength of laser light changes *during* the laser pulse, by a fraction of an angstrom. This can lead to problems in very high speed fiber-optic communications in fibers which have significant spectral dispersion (change in refractive index with wavelength) at the laser wavelength.

Amplitude Noise Manufacturers generally do not specify an amplitude noise for diode lasers. Data sheets typically indicate linearity by including a plot of light output vs. drive current. In practice, no special compensation is needed for most digital modulation, but fast analog modulation requires sophisticated linearization techniques.

Beam Quality, Polarization, and Modes With a cavity that is very short by laser standards, and an emitting area that is generally a rectangle much wider than it is high, a diode laser produces an oval-shaped, rapidly diverging beam. Polarization is invariably unspecified in data sheets.

The terminology used for diode laser modes can be confusing. The term *single mode* is usually used to denote single longitudinal mode, which functionally means narrow-linewidth emission. That can be very important in applications requiring linear response to modulation. However, there is also another type of single mode—single *spatial* mode, which is an indication of beam quality and can be important in applications such as coupling light into single-mode optical fibers. Beam quality also depends on the waveguiding mechanism inside the diode. Gain-guided lasers generally produce broad, astigmatic, nongaussian beams, while index-guided lasers produce nonastigmatic gaussian beams that are more desirable for many applications.

Cylindrical optics are often used to focus the oval beam from a diode laser into a more conventional circular cross section for use in optical systems. Complicated and expensive corrective optics may be needed if the diode laser beam is highly astigmatic.

Coherence Length Manufacturers rarely specify the coherence length of diode lasers. The short optical cavity is conductive to long coherence length, but the broad spectral width of multimode lasers counters that effect, leading to coherence lengths on the order of millimeters. Single-longitudinal-mode lasers may have coherence lengths on the order of 10 to 30 m.

Coherence length can be an important specification because in many diode laser applications coherence can cause undesirable effects. Examples include speckle or modal noise in multimode fiber-optic systems, or optical feedback interference from reflections in equipment such as optical disk readers. In some cases, designers have turned to LEDs to avoid coherence effects; in others, multimode lasers have been sufficient. An intermediate alternative between the LED and diode laser is the superluminescent diode, which is similar to a diode laser in structure and operates at comparably high drive-current levels, but lacks the reflective facets that are needed to generate a coherent laser beam. At present, superluminescent diodes are available commercially, but not widely.

Beam Diameter and Divergence The beam from a diode laser starts out as a long, narrow oval and diverges rapidly by laser standards. The emitting area of a single-mode GaAlAs laser is a few micrometers wide along the junction and a fraction of a micrometer thick (perpendicular to the junction). The beam spreads out at a 40° angle perpendicular to the junction and a 10° angle in the plane of the junction. Both figures are generally larger for longer-wavelength lasers. In practice, divergence varies somewhat with operating conditions and output power, actually decreasing with increasing output power. Specified values also vary from that rough average; note, however, that some figures give only half-angle divergence, while others, such as those above, give the larger full-angle figure.

Beam divergence in both directions is often somewhat larger for multimode oscillation of gain-guided double-heterojunction lasers. LEDs have even larger beam divergence. On the other hand, the divergence of pulsed single-heterojunction lasers is somewhat smaller than that of double-heterojunction lasers, particularly in the direction perpendicular to the junction (which is usually thicker than that of a double-heterojunction laser).

Study of a sampling of data sheets indicates that beam divergence can vary significantly with operating conditions and among different manufacturers. Fortunately, it is rarely a critical design parameter. In fiber-optic systems it helps to determine what fraction of a laser's output is coupled into a fiber, but in many other systems external optics collimate the diode laser's diverging beam to a much tighter, circular form.

Stability of Beam Direction Beam direction stability is rarely mentioned in data sheets, but tight accuracy is unreasonable to expect given the large beam divergence. One data sheet specifies stability as ±2°, small compared with divergence, and that is probably fairly typical.

Suitability for Use with Laser Accessories The comparably low output power from diode lasers makes them generally unsuitable for use with accessories that require high power. However, harmonic generation has been demonstrated in the laboratory, using diode lasers with high peak powers and nonlinear materials with high nonlinear coefficients (Günter et al., 1979). Such devices may become practical with development of higher power diode lasers.

The ability to directly modulate the output of diode lasers simply by modulating their drive current has limited the incentives for developing sophisticated modulation techniques. However, the frequency chirp caused by direct modulation causes enough problems for extremely high speed fiber transmission to revive interest in external modulation. Interest in Q switching and modelocking has centered on prospects for extremely high repetition rates and very short pulses. Q switching at rates to 7 GHz has been demonstrated in the laboratory, as has modelocking to produce pulses in the 10-picosecond (ps) range. At this writing, however, those capabilities remain of primarily research interest.

Operating Requirements

Input Power By laser standards, diode lasers require very modest input power. For double-heterojunction lasers, the maximum rated input voltage is typically about 2 V for devices emitting at 800 to 900 nm. It may be less for longer-wavelength lasers, which have smaller bandgaps and hence smaller voltage drops. Drive-current requirements depend on laser threshold, which is as low as 10 mA for the best commercial devices to somewhat over 100 mA for less-expensive versions. A single continuous-output laser can draw a current of a few tens of milliamperes to operate just above threshold, to a few hundred milliamperes for operation at the upper limit of its rated range. Thus the actual power requirement of a continuous-output double-heterojunction laser is typically under a watt. The low power and voltage requirements are comparable with those of semiconductor electronics and are a major attraction of diode lasers for many information-handling applications. In monolithic arrays, the lasers are electrically in parallel, so drive voltages are in the 1.5- to 2-V range. Typical threshold currents are around 250 mA. Operating currents are around 500 mA for continuous-wave operation at modest power, and around a couple of amperes for pulsed operation with peak power in the 1-W range.

Input power requirements are rather different for single-heterojunction lasers designed for pulsed operation at low duty cycles, typically 0.1 to 0.2 percent. These lasers can withstand voltages of several volts and peak currents of tens of amperes in pulses lasting a microsecond or less. Peak input power during the pulses may be hundreds of watts, with peak output power of 20 W or more. Because of the low duty cycle, the average power requirement is comparable to that of double-heterojunction lasers operating continuously.

Stacked arrays of single-heterojunction lasers produce pulses at even lower duty cycles than single pulsed lasers. Typically, the lasers are electrically in series, multiplying voltage per junction by the number of junctions in the array. When operated slightly above threshold, operating voltages are simple multiples of the number of junctions in an array; for example, 1.4 V for a single diode, 4.2 V for three, and 7.0 V for five. In pulsed operation they are normally driven far above threshold, with input currents of tens of amperes and voltages of tens of volts for arrays of five or six diodes.

Cooling Requirements Virtually all III–V diode lasers are designed for room-temperature operation and usually come in a package intended for heat-sinking to remove the excess heat produced during laser operation. Such cooling mechanisms are adequate for many applications, but active cooling may be needed if stable operation—particularly in output wavelength—is critical. For demanding applications such as high-performance fiber-optic transmitters, active thermoelectric coolers are the usual choice because they are compact and have modest power requirements. Typically, thermoelectric coolers stabilize temperature around 20°C.

Consumables Diode lasers are self-contained and do not require consumables.

Required Accessories A diode laser is a tiny chip of semiconductor material that must be packaged into a useful form for practical applications. The key elements that make up a package are

- A housing to permit easy mounting and to help carry away excess heat
- An output window or optical fiber to direct the laser output
- Electrical connections between the chip and the outside world

External modulators are almost never needed because the laser output can be modulated directly by changing the drive current. To avoid changes in output due to aging or temperature fluctuation, the laser can be packaged with a semiconductor photodetector and a feedback loop to control drive current. In this approach, the laser is coated so that most of the output emerges from one facet and a much weaker beam emerges from the rear facet. While the main beam emerges from the package to be used, the weaker beam is monitored by the packaged photodetector. A feedback loop connects the photodetector and drive circuit, so that any decrease in detected output will trigger an increase in drive current to restore the laser output to the desired level. This approach assures that the laser generates a stable output power throughout its lifetime, which is important for many applications.

Auxiliary optics are needed if a tightly focused beam is required from a diode laser. Typically such focusing optics are made and sold by optics suppliers rather than by diode laser makers, although some packages containing both semiconductor lasers and optics have begun to appear on the market.

Operating Conditions and Temperature Generally III–V diode lasers operate at or near room temperature. Single-heterojunction lasers, many of which are intended for military applications, are typically specified for operating temperatures of −55 to +60°C, with even broader ranges of storage temperature. Somewhat narrower operating ranges are normally specified for double-heterojunction lasers, which are used primarily in civilian equipment. Threshold current and degradation effects generally increase with operating temperature, so limiting the temperature can pay dividends in increased reliability and operating life, in addition to stabilizing laser output power. Care should be taken at low temperatures, because the decrease in threshold current that accompanies a drop in temperature could lead to unexpectedly high output power if the drive current was not adjusted. In some cases, the increase would be enough to raise output power beyond rated levels, potentially leading to facet damage.

The diode laser itself requires no warm-up, responding in a matter of nanoseconds or less to direct electrical modulation. The duty-cycle limitations of pulsed single-heterojunction lasers are intended to prevent excessive heating of the devices, which can cause rapid degradation and failure.

Mechanical Considerations Semiconductor lasers themselves are tiny; the chips measure a fraction of a millimeter in each dimension. For commercial diode lasers, the mechanical considerations depend on packaging. Typical packages are shown in Fig. 18-8; many are similar in size and shape to transistor housings or compact electrical connectors. Many packages are designed for mounting directly on circuit boards or heat sinks. Except for possible packaging limitations, the laser itself is generally resistant to normal mechanical shocks.

Safety Although near-infrared diode lasers are low in output power, they are not completely innocuous. Users should not be misled by the apparent feebleness of the emission from "visible-wavelength" diode lasers. The retina's response to light drops off very sharply with increasing wavelength beyond 700 nm, but it is still possible for the eye to detect faint deep-red emission from a diode laser emitting at 780 nm (which may be marketed as "visible" although it would not be useful for display purposes). However, the beam appears feeble only because the retina responds to just a tiny fraction of the light—its response to 780-nm light is on the order of 10^{-5} of that to 568-nm light—and in reality fairly intense light may be reaching the retina.

The large divergence of diode lasers offers some natural protection against inadvertent exposure to power levels dangerous to the eye. However, safety specialists warn that tightly focused beams at wavelengths as long as 1400 nm could penetrate the eyeball and be focused onto the retina to cause damage (Sliney and Wolbarsht, 1980).

Light at longer wavelengths is not transmitted through the ocular medium to the eye, so the 1550-nm lasers being developed for long-wavelength fiber-optic communications present much less hazard. Indeed, military researchers have been studying the prospects of using such lasers as eye-safe range finders.

Because of their low operating voltages, diode lasers do not present significant electrical hazards.

Special Considerations Advances in diode laser technology have been exceptionally rapid during the past few years, and new types will continue moving from the laboratory to the marketplace. The most advanced models typically appear first as "developmental" products even before some life tests have been completed. The most rapid advances are coming at long wavelengths, high continuous powers, and narrow output linewidth. Progress in moving to the shorter wavelengths desired for information-handling applications has been tempered by severe lifetime problems. Users with particularly demanding requirements should keep an eye on the state of the art.

Reliability and Maintenance

Lifetime Lifetime was a serious problem with early diode lasers, but it seems to have been tamed for standard commercial devices. Although few laser makers list a rated lifetime on their data sheets, this is more an indication of problems in testing than limited reliability. Performance is now good enough that real-time life tests are not able to cast much light on lifetime in the short interval between product development and marketing. It is possible to get an indication of laser lifetime from accelerated aging tests, in which the laser is operated well above room temperature (for example at 70°C) for

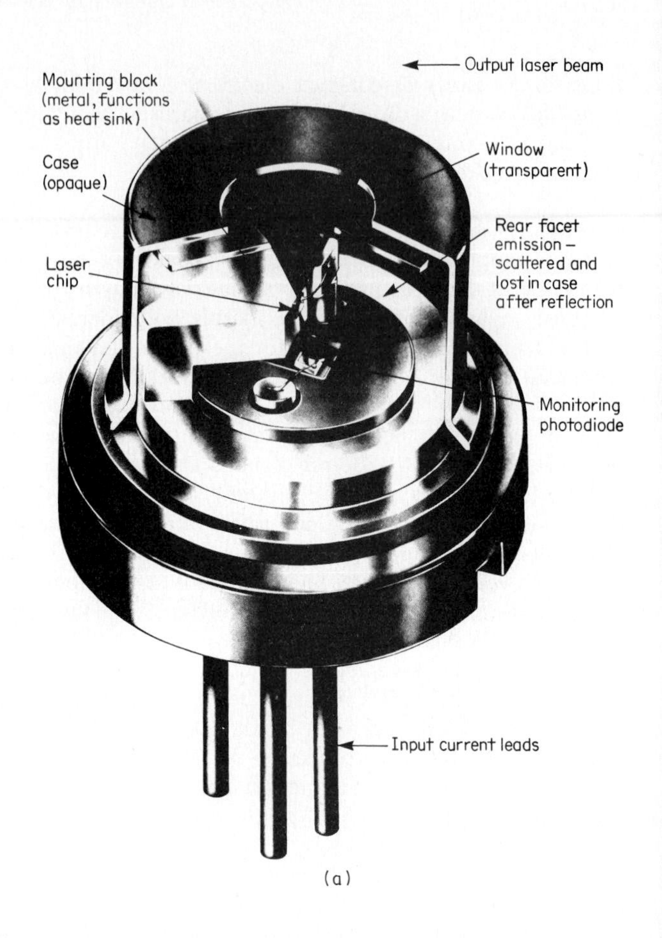
Output laser beam
Mounting block
(metal, functions
as heat sink)
Window
(transparent)
Case
(opaque)
Rear facet
emission –
scattered and
lost in case
after reflection
Laser
chip
Monitoring
photodiode
Input current leads

(a)

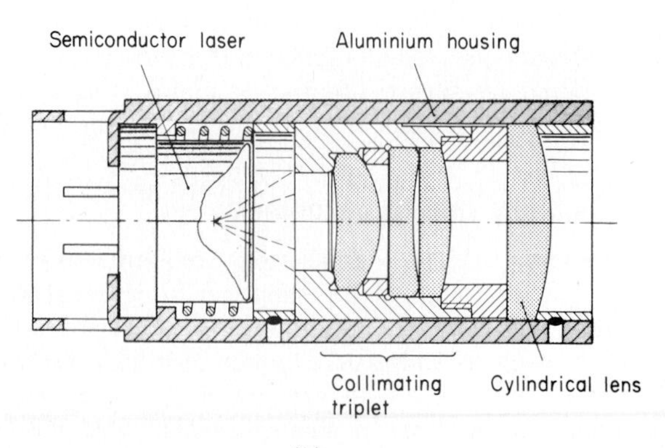
Semiconductor laser
Aluminium housing
Collimating
triplet
Cylindrical lens

(b)

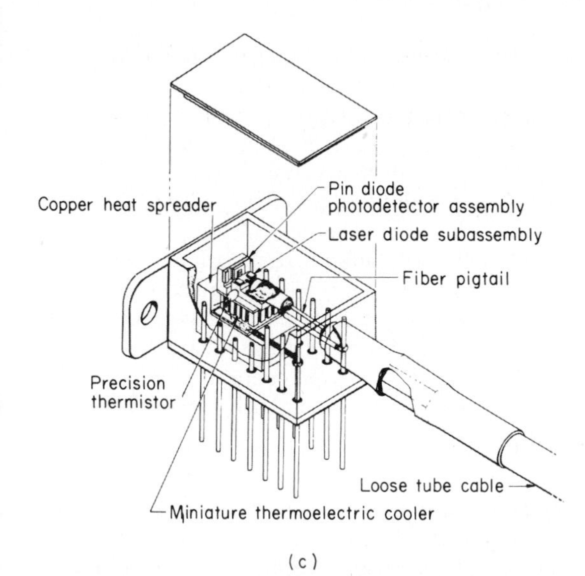

Figure 18-8 Typical packages for commercial diode lasers. *(a)* Cutaway view of a diode laser packaged in a transistor-type housing, showing output window (top), the laser mounted on a heat sink, and a photodiode which monitors output from the laser's rear facet. The diagram has been simplified by omitting internal connecting wires *(Courtesy of Amperex); (b)* light pen with packaged diode laser and collimating optics *(Courtesy of Amperex); (c)* laser output coupled to a fiber pigtail in a DIP (dual in-line package). *(Courtesy of M/A Com Laser Diode Inc.)*

long intervals. Extrapolation of high-temperature degradation rates indicates expected room-temperature lifetime reasonably well, but many manufacturers remain cautious and avoid specifying lifetimes on data sheets.

It is reasonable to assume operating lifetimes of tens or hundreds of thousands of hours for single double-heterojunction GaAlAs lasers emitting several milliwatts, but lifetimes of single-stripe lasers emitting tens of milliwatts are likely to be somewhat shorter. For multistripe monolithic arrays, median lifetime at 30°C has been extrapolated to be 4900 hours for 200-mW output and about 26,000 hours for 100-mW output (Harnagel et al., 1985). In general, lifetime decreases with increases in operating power, temperature, or threshold current of the laser. Improved technology has led to steady increases in device lifetimes in recent years, and this trend should continue.

Much of the work on diode laser lifetime has concentrated on GaAlAs lasers emitting at 780 to 900 nm. Heavier aluminum doping allows operation at shorter wavelengths, but only at a penalty in device lifetime which becomes worse as wavelength decreases, effectively limiting the shortest wavelengths available.

Different mechanisms limit the lifetimes of InGaAsP lasers emitting at longer wavelengths. Studies of 1300-nm devices developed for fiber-optic communications indicate that lifetimes are at least comparable with those of GaAlAs lasers. Less research has been done on 1550-nm lasers because of the severe problems in producing them with reasonable yields.

Relatively little information is available on lifetimes of single-heterojunction diode lasers, apparently because their major applications are not as demanding of long lifetimes as the fiber-optic communication systems which stimulated development of double-heterojunction lasers. With the high stresses of pulsed operation at high powers, and their simpler structures, single-heterojunction lasers might be expected to have shorter lifetimes. One data sheet quotes mean time to failure of a diode laser array as 5000 operating hours, which would be considered short for a continuous-output laser. Other manufacturers' literature does not indicate expected lifetimes.

Maintenance and Adjustments Needed Diode lasers themselves require no special maintenance other than operation under proper conditions. However, for many applications it may be desirable to incorporate the laser into an automatic feedback-controlled system to maintain stable output power or operating temperature. Output power can be stabilized by monitoring emission and using feedback from the monitor to control drive current; temperature control requires active compensation, such as a thermoelectric cooler. Power and temperature stabilization systems are generally designed for automatic self-adjustment.

Failure Modes and Causes There are four main failure modes for diode lasers

- Catastrophic optical damage to the laser facets, caused by excessively high laser powers, above about 10^6 W/cm^2
- "Dark-line defects," regions in the active layer which do not emit light and which gradually grow within a laser
- Degradation of optical facets, caused by oxidation and/or light-induced damage to the facet surface near the active region
- Failure of the metallization and/or bonding solder which makes electrical contacts to the chip

The user can generally prevent the first type of damage by being careful not to overload the laser. The other effects depend mainly on manufacturing processes, although operation at excessive powers or under particularly hostile conditions could speed degradation. Failures due to dark-line defects generally are weeded out during the "burn-in" of new devices by laser makers.

Possible Repairs Once a laser diode has failed, the user can only replace it and correct any operational problems which might have caused its failure.

Commercial Devices

Standard Configurations Diode lasers are sold in several configurations, the most important of which include

- Transistor or semiconductor rectifier cases which include a window through which the beam emerges
- Semiconductor device cases with the window replaced by an optical fiber "pigtail" which collects light from the laser diode, generally for delivery to an external fiber-optic system
- Fiber-optic connector housings, so the laser can be mated directly with a connectorized fiber-optic cable
- Complete fiber-optic transmitter, often including active cooling and perhaps even output power stabilization, with the emitted light either delivered to a connector interface or a fiber pigtail
- In a "light pen" unit that incorporates optics which convert the rapidly diverging diode laser beam into a collimated beam
- In a special-purpose optical head that focuses the beam to a small spot on an optical disk.

Some users have recently expressed their dissatisfaction with the usual ways of packaging commercial diode lasers. Most of the complaints center on lasers intended for applications other than fiber optics. Users say that some output windows are of such poor quality that they degrade beam quality. These problems should be alleviated as demand increases for diode lasers in applications other than fiber optics.

Options The major options available with commercial diode lasers are different types of packaging, as described above.

Pricing Prices of laser diodes currently span three orders of magnitude. Modest-quality pulsed single-heterostructure lasers can be bought for a few dollars each in quantity. Quantity prices are similar for stripe-geometry lasers manufactured for audiodisk playback, but high-performance devices—particularly monolithic arrays and InGaAsP lasers—can cost thousands of dollars, up to several thousand for 1.55-μm lasers. Some prices could come down as production increases, but arrays and long-wavelength lasers are likely to remain much more costly than their low-power GaAlAs audiodisk counterparts.

Suppliers About a couple of dozen companies in the United States, Japan, and western Europe manufacture diode lasers. Almost all of them make continuous-wave GaAlAs lasers, and about half produce InGaAsP lasers. About half a dozen make pulsed single-heterostructure GaAlAs lasers, and as of mid-1985 only a single company was offering monolithic arrays. Some of the companies tend to specialize in areas such as InGaAsP lasers or monolithic arrays, but many of them offer a broad range of products, and most areas of the market are competitive.

The market is being affected strongly by the mass production of low-power continuous-wave GaAlAs lasers emitting at 780 or 800 nm for use in digital compact disk audio players. The players, and the lasers used in them, are manufactured in Japan and Europe. The economies of scale for production of hundreds of thousands of the lasers have reduced prices to a few dollars each, in large quantities. Custom-packaged versions of those lasers should be inexpensive enough to be attractive for many applications.

Higher-power and longer-wavelength diode lasers probably will remain in more limited supply and be concentrated in applications such as fiber-optic communication and optical data storage which require a particular wavelength or output power above that of audiodisk lasers.

Applications

The range of applications for diode lasers has broadened rapidly beyond fiber-optic communications to include many other fields of "information handling." The single biggest in number of lasers used is the digital audiodisk player, a market which accounted for well over a million lasers in 1985. That application has a symbolic importance in being the first one that literally puts a laser in the home, as well as a practical importance in raising production quantities enough to sharply reduce laser prices.

Interest in fiber-optic communications was a major driving force behind the growth of diode laser technology during the 1970s and early 1980s. The first fiber-optic systems operated at the 800- to 900-nm wavelengths of GaAlAs lasers, but long-haul systems have shifted to longer wavelengths where attenuation is lower. The 1300-nm wavelength is widely used in current commercial systems because present optical fibers have wide bandwidth and low loss at that wavelength. The 1550-nm wavelength is attractive because fibers have even lower attenuation there, but the required lasers have not been perfected, and fibers with broad bandwidth at that wavelength have only just reached the market. Fiber-optic communications remains the dominant use of InGaAsP lasers, and it is likely to remain so. However, it now accounts for only a tiny fraction of the GaAlAs lasers.

The bulk of GaAlAs lasers go for other applications that fall under the broad heading of information handling, including reading prerecorded information and writing for printing or data storage. The shorter-wavelength beam is desirable for reading or writing because it can be focused to a smaller spot than a longer-wavelength beam. A further advantage for writing is that most light-sensitive materials respond more strongly at shorter wavelengths. (System manufacturers would like even shorter wavelengths than available from GaAlAs lasers, but they can live with 780-nm output.) Final practical advantages are low cost and small size—the latter important in designing compact optical heads for optical disk systems.

Diode lasers now dominate applications in data storage, audiodisk, and videodisk systems. Early videodisk players were designed around gas lasers, but in recent years manufacturers have shifted over largely to diode lasers. Audiodisk players were designed around the diode laser, as was the CD-ROM (compact disk–read only memory), which reads prerecorded computer data from reflective disks made using audiodisk technology. Diode lasers also are used in write-once optical data-storage systems developed for computers (Hecht, 1985). However, that application requires higher power than is available from audiodisk lasers. Likewise, laser printers, which write on electrostatic drums for transfer to plain paper using photocopierlike technology, use diode lasers (Hecht, 1984).

Semiconductor diode lasers have begun replacing helium-neon lasers in some of their traditional applications. One example is in automated construction-alignment systems which do not require visible red beams. Another example is in certain systems for reading bar-coded symbols, where the difference in wavelength between the two types is not critical. Diode lasers also are used in a handful of atmospheric communication systems.

Military applications, including range finding and battle simulation, account for most pulsed single-heterojunction lasers and arrays. Coded pulses from diode lasers can be detected to keep score in military war games; the pulse coding indicates the type of weapon being simulated, so a simulated rifle shot cannot "kill" a tank. Air-to-air missiles include diode-laser range finders, which detect the range to targets, and trigger detonation of the warhead when the missile approaches close enough to destroy the target. High-power monolithic arrays of diode lasers are being studied for possible use in intersatellite communications.

New civilian and military applications continue to be developed. The emergence of monolithic arrays has stimulated interest in using them to pump neodymium-YAG lasers, an approach that appears more efficient than flashlamp pumping (Cross et al., 1985). Diode lasers also are candidates for some types of medical therapy. Further applications are likely to emerge as prices decline and performance continues to improve.

BIBLIOGRAPHY

Dan Botez: "Single-mode lasers for optical communications," *IEEE Proceedings 129:*237–251, December 1982. Review of research work.

Peter S. Cross, Ralph R. Jacobs, and Don R. Scifres: "High power diode laser arrays: lifetests and applications," *Lasers & Applications 4*(4):89–91, April 1985.

P. Günter et al.: "Second harmonic generation with GaAlAs lasers and $KNbO_3$ crystals," *Applied Physics Letters 35:*461, 1979.

Robert N. Hall: "Injection lasers," *IEEE Transactions on Electron Devices ED-23:*700–704, July 1976. Historical account.

G. L. Harnagel et al.: "Lifetime of diode laser arrays operating at 200 mW CW," postdeadline paper ThZl at Conference on Lasers and Electro-Optics, Baltimore, May 21–24, 1985.

Jeff Hecht: "Printing with a laser beam," *PC 3*(9):252–256, May 15, 1984.

Jeff Hecht: "Optical storage for personal computers," *Lasers & Applications 4:*8, August 1985, pp. 71–76.

H. Kobayashi, Y. Horikoshi, and C. Uemura: "Room temperature operation of the InGaAsSb/AlGaAsSb DH laser at 1.8 μm wavelength," *Japan Journal of Applied Physics 19:*L30–L32, January 1980.

H. Kressel et al.: "Laser diodes and LEDs for fiber optical communication," in H. Kressel (ed.): *Semiconductor Devices for Optical Communication*, 2d ed., Springer-Verlag, Berlin and New York, 1982, pp. 9–62.

"The Laser Marketplace—1986," *Lasers & Applications 5:*1, January 1986, pp. 45–56.

Kam Y. Lau and Amnon Yariv: "High-frequency current modulation of semiconductor injection lasers," in W. T. Tsang (ed.): *Semiconductors and Semimetals*, vol. 22, *Lightwave Communications Technology, Part B, Semiconductor Injection Lasers, I*, Academic Press, Orlando, 1985, pp. 70–152.

Y. Nakano et al.: "Reliability of semiconductor lasers and detectors for undersea transmission systems," *IEEE Journal on Selected Areas in Communications SAC-2* (6):985–991, November 1984.

G. Osbourn et al.: "A GaAsP/GaP strained-layer superlattice," *Applied Physics Letters 41*:172–174, July 15, 1982.

D. R. Scifres et al.: "Phase-locked (GaAl)As laser emitting 1.5 W cw per mirror," *Applied Physics Letters 42*:645–647, April 15, 1983.

David Sliney and Myron Wolbarsht: *Safety with Lasers and Other Optical Sources*, Plenum, New York, 1980. Comprehensive handbook on laser safety.

C. B. Su et al.: "12.5-GHz direct modulation bandwidth of vapor-phase regrown 1.3-μm InGaAsP buried heterostructure lasers," *Applied Physics Letters 46*(4):344–346, February 15, 1985.

Yasuharu Suematsu et al.: "Dynamic single-mode semiconductor lasers with a distributed reflector," in W. T. Tsang (ed.) *Semiconductors and Semimetals*, vol. 22, *Lightwave Communications Technology, Part B, Semiconductor Injection Lasers I*, Academic Press, Orlando, 1985, pp. 205–255.

W. T. Tsang: "The cleaved-couple-cavity (C^3) laser," in W. T. Tsang (ed.): *Semiconductors and Semimetals*, vol. 22, *Lightwave Communications Technology, Part B, Semiconductor Injection Lasers I*, Academic Press, Orlando, 1985, pp. 258–373.

E. W. Williams and R. Hall: *Luminescence and the Light Emitting Diode*, Pergamon, Oxford, 1978. Technical monograph.

CHAPTER

19

lead salt and other long-wavelength infrared semiconductor diode lasers

The family of near-infrared semiconductor lasers made of compounds from columns III and V of the periodic table described in the previous chapter are by far the most common diode lasers. However, there is also another family of semiconductor diode lasers emitting at wavelengths between about 2.7 and 30 micrometers (μm). These mid- or far-infrared diode lasers are available commercially and have virtually all of their applications in spectroscopy.

These lasers are often called *lead salt* types because many (but not all) of the semiconductor compounds contain lead. The first lead salt diode laser, made of lead telluride, was made in 1963 by Jack F. Butler and colleagues at MIT's Lincoln Laboratory (J. F. Butler et al., 1964). That demonstration came only a year after the first report of the operation of any kind of diode laser, but for many years it stimulated little interest. Fiber-optic and military system requirements for compact sources of near-infrared light drove research in III–V diode lasers, but there was no such demand for lead salt lasers at longer wavelengths.

Interest in high-resolution infrared spectroscopy, which developed during the 1970s, has been the major impetus behind lead salt laser development. Technological development has focused almost exclusively on spectroscopic applications, although research on mid-infrared optical fibers (Klocek, 1982) has led to some consideration of potential applications in fiber-optic communications. Serious technological limits remain, however, including the need for cryogenic cooling and limited output powers from narrow-line lasers. Performance is being improved by adapting technology originally developed for the more widely used shorter-wavelength semiconductor lasers described in the previous chapter.

Internal Workings

Like III–V semiconductor lasers, lead salt lasers produce light when carriers recombine at the junction of *p*- and *n*-type semiconductor materials. Recombination occurs only when current flows through the forward-biased diode, so light emission depends directly on the drive current. The internal physics of the two types of semiconductor diode lasers are similar in many ways, but there are also some important differences.

The most obvious one is the materials which make up the diode. Lead salt lasers are fabricated from elements drawn from columns IIB, IVB, and VI of the periodic table, including lead, tin, sulfur, selenium, and tellurium. Near-infrared diode lasers contain elements from columns III and V, including gallium, arsenic, aluminum, indium, and phosphorus. Most lead salt lasers are made of ternary (three-element) compounds such as lead-tin telluride ($Pb_xSn_{1-x}Te$) where alteration of the fraction x changes the proportions of two of the elements and hence the wavelength emitted. The combination of valence states 2 and 6 in lead salt lasers leads to bonds that are somewhat more ionic than those in III–V compounds, and the crystalline structure of lead salt lasers is of the rock salt type.

The differences in composition between the two types of diode lasers cause major differences in laser characteristics. The most obvious difference, in wavelength, is due to differences in energy-level structure arising from composition. In lead salt lasers, the gap between conduction and valence bands is small, less than half an electronvolt. Lasers in the gallium arsenide family generally have bandgaps of around 1 eV. Thus while III–V diode lasers typically have wavelengths of 0.7 to 1.8 μm, the wavelengths of lead salt lasers are 2.7 to about 30 μm.

Like other laser diodes, lead salt types exhibit a threshold for laser action. At lower drive currents the diode emits low levels of incoherent light, but above the threshold current it produces stimulated emission. Typically lead salt lasers have threshold currents of 100 to 500 mA and voltage drop of a fraction of a volt. Lead salt lasers can be directly modulated by changing drive current, pulsed with 10- or 100-nanosecond (ns) pulses, or driven with a steady bias current to produce continuous output.

Internal Structure

Conceptually a diode laser requires only two semiconductor layers with opposite dopings and a junction between them. Such a laser is not practical, however, and commercial lead salt diode lasers contain additional layers to improve light confinement within the active layer, thus improving efficiency and reducing threshold current.

Like III–V lasers, lead salt diode lasers typically are small, 400 μm or less long and smaller in width and height. Their internal structures have been less complex than those used for III–V lasers, but the sophisticated structures developed for III–V lasers are finding increasing use. Lead salt lasers with output below a milliwatt still use the broad-area double-heterostructure design, in which the active region is optically confined on the top and bottom but not on the sides. However, mesa-stripe designs such as the type shown in Fig. 19-1 also are used, in which the active layer is a narrow stripe rather than a broad plane. This approach allows laser output powers to reach 10 mW and also improves output quality. Even higher powers are possible when several mesa-stripe lasers (typically four in commercial versions) are arranged in an array.

Other recent design innovations also are intended to improve quality and tuning. Lasers with cavities only 120 or 250 μm long, rather than the usual 400 μm, permit higher operating temperatures, operation in a single longitudinal mode, and increased tuning range in a single mode (Linden, 1985). In short-cavity as well as traditional types, the cavity is defined by cleaved facets at the end of the semiconductor chip, which reflect most of the light back into the active layer to stimulate additional emission. Two other laser designs originally developed to produce narrow-line output from III–V lasers, the cleaved-coupled-cavity laser and the distributed-feedback laser, also have been used for lead salt lasers.

External resonator optics are possible in theory, but rarely used in practice. The reason is that the short cavity length—typically 100 wavelengths or less,

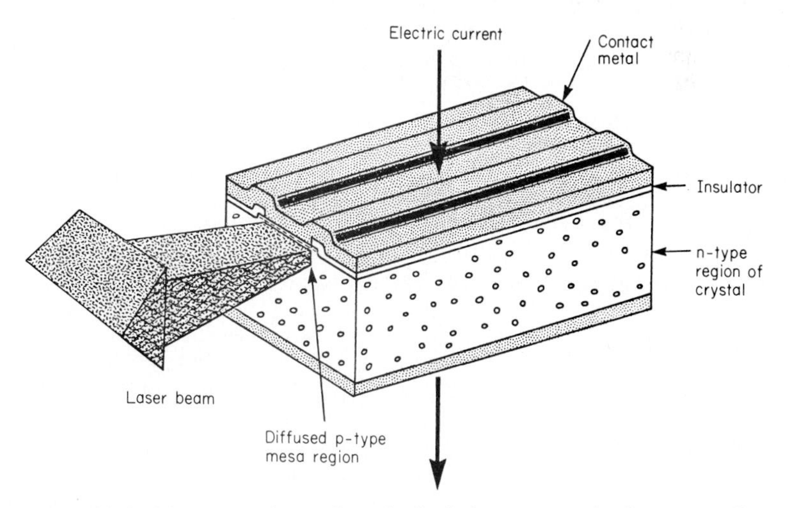

Figure 19-1 Mesa-stripe design for a lead salt laser uses technology originally developed for shorter-wavelength III–V diode lasers. *(Courtesy of Laser Analytics division of Spectra-Physics Inc.)*

extremely short by laser standards—leads to narrow resonance lines, which would require very careful adjustment of an external cavity. Linewidths as narrow as 22 kilohertz (kHz) have been measured in the laboratory (Freed et al., 1983). Note that the cavity length measured in wavelengths of light is much greater for III–V lasers because of their much shorter wavelength, so this constraint is not as severe for them.

Inherent Trade-offs

The lead salt family of diode lasers suffers from some serious limitations on performance which are largely inherent in the technology.

- As wavelength increases, cooling requirements become increasingly stringent. This occurs because much of the wavelength range of lead salt lasers covers wavelengths where blackbody emission is strong at room temperature. Internal losses go up drastically with temperature, and cryogenic cooling is needed to keep the laser at temperatures where it can operate properly. Liquid nitrogen cooling is generally adequate at short wavelengths, but at the long-wavelength end of the lead salt diode range, liquid helium may be necessary for proper operation.
- Although the family of lasers can span an order-of-magnitude range in wavelength, the tuning range of individual lasers is limited to several hundred wave numbers for an individual laser and to around 0.5 cm^{-1} for a single longitudinal mode from a long-cavity laser. However, short-cavity lasers can remain in a single longitudinal mode over a broader tuning range, 2 to 3 cm^{-1}. A laser's nominal center wavelength depends on its composition; that wavelength can be tuned over a limited range by means such as changing temperature, as described in more detail below.
- Output powers tend to be higher in the middle of the lead salt laser range (1200 to 2300 cm^{-1} or about 4 to 8 μm) than at shorter or longer ones. Those powers are modest—10 mW for the highest-power single devices—but they are concentrated in a narrow range of wavelengths, so spectral brightness is high. Arrays of mesa-stripe lasers give the most power, but their spectral bandwidth is much broader because of the difficulty of tuning all of the stripes to emit at the same wavelength.

Variations and Types Covered

The basic variable in the lead salt family of semiconductor lasers is chemical composition, which affects emission wavelength. The basic semiconductor material used in lead salt lasers is a compound made of elements from two groups: the metals lead, tin, and cadmium, and the nonmetals (often called chalcogenides) sulfur, selenium, and tellurium. Recently, the rare earth metals europium and ytterbium have been incorporated into lead salt lasers in the laboratory (Partin, 1985). Typically two elements from one group are combined with one element from the other; changing the relative proportions changes the center wavelength. The most commercially important lead salt lasers are lead sulfide, selenide, and telluride (PbS, PbSe, and PbTe, respectively), plus alloys of these compounds with each other and with tin selenide (SnSe), tin telluride (SnTe), and cadmium sulfide (CdS). Other types are in development, but this chapter will concentrate primarily on versions in or near commercial use.

Beam Characteristics

Wavelength and Output Power Center wavelengths of lead salt lasers cover an order-of-magnitude range from 2.7 to over 30 μm. Spectroscopists, the main users of lead salt lasers, typically measure this range in wave number or inverse centimeters, and in those terms the range is 4000 to 340 cm^{-1}. Individual lead salt lasers can be tuned over 20 to 200 cm^{-1}, or even to 400 cm^{-1} at the cost of lower linearity and pushing temperature limits. Typical operating ranges for the major compositions offered commercially are shown in Fig. 19-2. Each laser is tunable over only a small part of the entire range for a compound, but the whole range can be covered by many lasers of different compositions.

Laser wavelength can be tuned by adjusting hydrostatic pressure, magnetic field, or temperature, all of which affect the laser's effective cavity length. Temperature tuning is the most straightforward and is the most common approach. Coarse tuning is by adjusting diode temperature directly in the range 15 to 100 K. Fine tuning relies on making small changes in drive current, which indirectly changes device temperature because an increase in current increases heat dissipated within the crystal.

In practice, tuning is complicated by the laser's hopping between different longitudinal modes, as shown in Fig. 19-3. Although the gain curve normally allows oscillation on more than one longitudinal mode, one mode generally dominates. Changes in temperature cause two shifts, one in the peak of the gain curve, the other in the wavelengths of oscillating modes of the Fabry-Perot laser cavity. These accumulate to a point where the laser shifts from one dominant longitudinal mode to another. For broad-area lead salt lasers,

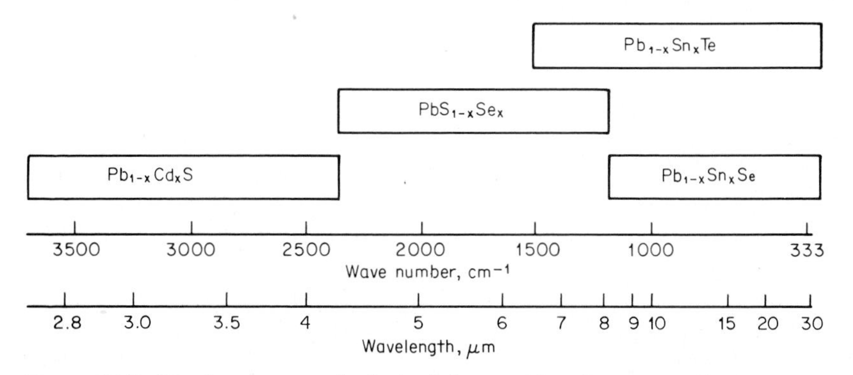

Figure 19-2 Wavelength ranges for lead salt lasers with various compositions.

the continuous tuning range of an individual longitudinal mode is only about 0.5 cm^{-1}. The range is much larger, 2 to 3 cm^{-1}, in commercial short-cavity lasers. Even broader continuous tuning ranges have been demonstrated in the laboratory, along with single-mode emission at wavelengths of 5, 10, and 30 μm (Linden, 1985).

Oscillation of a lead salt laser can be restricted to a single longitudinal mode with a cleaved-coupled-cavity (C^3) or distributed-feedback laser design, approaches originally developed for high-speed optical communications with III–V diode lasers. In both types, research indicates that tuning rates are slower than for multimode lasers, which can be an advantage because it allows more precise control of wavelength. Reported single-mode tuning ranges are comparable to those for short-cavity lasers (Linden and Reeder, 1984; Kapon and Katzir, 1985).

Output power in a multimode beam ranges from under a milliwatt to 100 mW, depending on type of laser, drive current, and part of the spectrum where the laser is operating. Powers tend to be highest in the center part of the lead salt range, 4.3 to 8 μm. Typical powers range from a fraction of a milliwatt for broad-area lasers to 1 or 10 mW for mesa-stripe devices. Powers to 100 mW have been obtained from arrays (Linden, 1984).

Efficiency The drive current which powers commercial lead salt diode lasers is typically converted to optical power with an efficiency well under 1 percent. The efficiency of the laser itself is a relatively minor consideration in practice,

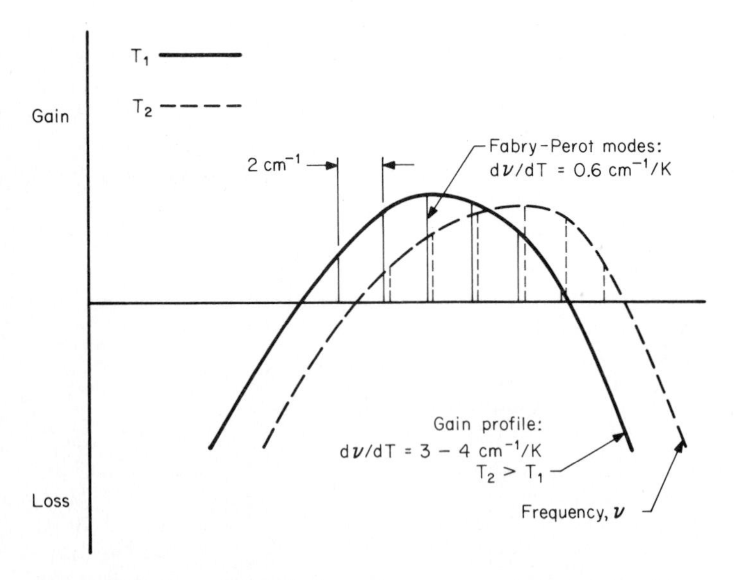

Figure 19-3 In a multimode laser, the gain curve includes several longitudinal modes of the Fabry-Perot resonator. Changes in temperature shift both the gain curve and the wavelength of the individual modes. *(Courtesy of Laser Analytics division of Spectra-Physics Inc.)*

because the laser is used as part of a system that actively cools the semiconductor. While the active cooling removes any excess heat up to reasonable power levels, it also consumes more power than the laser itself. The main interest in increasing efficiency of lead salt lasers is not to reduce power consumption per se, but rather to make lasers capable of operating at higher temperatures.

Temporal Characteristics Like other diode lasers, lead salt types rapidly respond to changes in electrical drive current and can produce pulsed or continuous output. In practice, the requirements of spectroscopic applications lead to operating conditions rather different than those for III–V diode lasers. In spectroscopy, a relatively slow modulation of the drive current can be superimposed on a steady dc bias current to sweep the laser output over a range of wavelengths. Commercial instruments offer such modulation as a standard feature at rates of 50 to 1000 Hz and with peak-to-peak amplitudes as large as 200 mA. Intensity modulation is incidental to the desired frequency modulation, and spectroscopic instruments have to accommodate for changes in laser output power. Modulation of lead salt laser intensity for spectroscopic uses normally is done externally with a chopper, to avoid drive-current induced changes in wavelength.

Spectral Bandwidth When broad-area lead salt lasers are operated significantly above threshold, they emit light in several narrow longitudinal modes, as shown in Figs. 19-3 and 19-4. Typical values for mode width and spacing are 3×10^{-4} cm^{-1} and 2 cm^{-1}, respectively, but both figures can vary with laser configuration and operating wavelength. Mode separation for lasers with short 120-μm cavities is larger, about 6 cm^{-1}, large enough that operation typically is in a single longitudinal mode (Linden, 1985). Cleaved-coupled-cavity and distributed-feedback lasers emit in only a single longitudinal mode.

For high-resolution measurements with multimode lasers, a monochromator can separate longitudinal modes, discarding all but the desired wavelength, but this reduces the amount of power available.

Frequency Stability The seemingly smooth spectral tuning curve of a multimode lead salt laser hides more complex characteristics, as shown in Fig. 19-4. The tuning process changes peak emission wavelength faster than the wavelength from any individual mode. As current increases, emission from a given mode first increases, then reaches a peak, and finally decreases, all while shifting gradually in wavelength. This can cause mode-switching instabilities. Phenomena similar in nature but different in degree also occur in short-cavity lasers, even though their emission is in a single longitudinal mode at most times. In that case, switching effects may be more dramatic because fewer modes are involved. The active stabilization of cleaved-coupled-cavity and distributed-feedback lasers limits their tuning ranges, so such effects would not normally be observed.

Otherwise, frequency stability of lead salt lasers is excellent, with the main concern being stability of the temperature controller and drive-current reg-

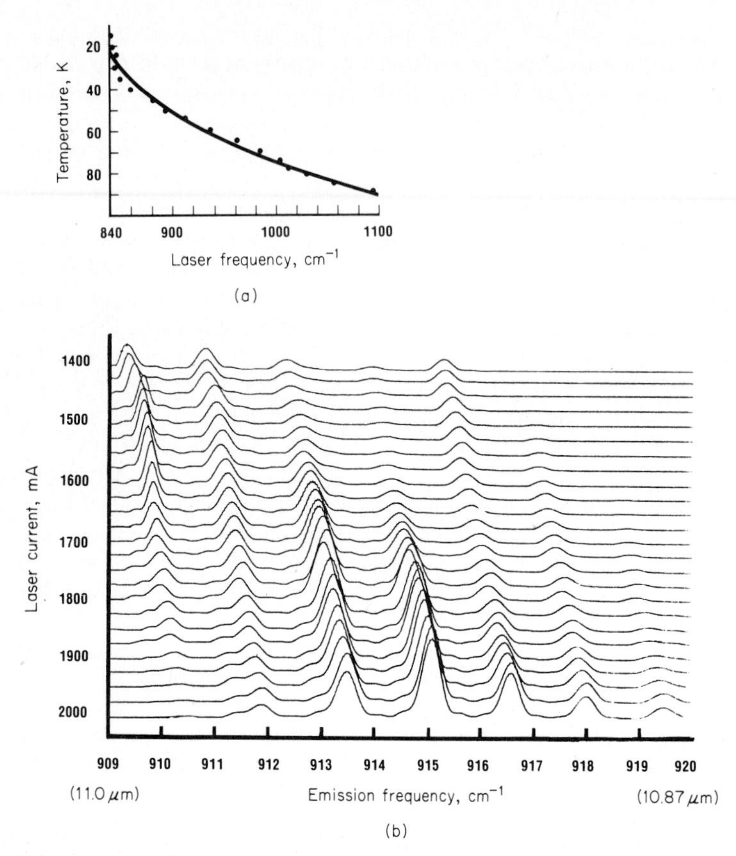

Figure 19-4 When viewed at low resolution (top), the tuning curve of a multimode lead salt laser looks smooth. However, more detailed analysis (bottom) shows that the distribution of energy among modes is changing, as are the wavelengths of the modes themselves. Mode shifting also occurs in short-cavity lasers, but at any one time the laser is likely to emit only a single mode. *(Courtesy of Laser Analytics division of Spectra-Physics Inc.)*

ulator. Frequency stability of a packaged laser source spectrometer is specified as 3×10^{-4} cm^{-1} for 1-s intervals, and 0.001 cm^{-1} over 30 minutes.

Amplitude Noise Because the main applications of lead salt lasers are in spectroscopy, amplitude stability is a lesser concern than frequency stability. However, the amplitude stability is good as long as the laser is not near mode-hopping wavelengths. Specified values for packaged instruments are ± 0.3 percent for 1 s and ± 2 percent for 30 minutes away from mode-hopping points.

Beam Quality and Polarization Lead salt diode lasers emit a broad, rapidly diverging beam that may have multiple lobes. Typically, a short-focal-length lens near the laser collects the output beam and collimates it before it can spread to a large diameter. Beam polarization typically is greater than 80 percent perpendicular to the emitting junction.

Beam Diameter and Divergence Like other diode lasers, lead salt lasers have small, rectangular light-emitting regions and short cavities, leading to large beam divergence. In practical instruments, the beam is collimated by a lens, and beam divergence may be specified as equal to that of an *f*/1 lens.

Stability of Beam Direction This quantity is not specified for lead salt lasers and is not crucial because of the nature of their applications and the focusing optics used with practical systems.

Suitability for Use with Laser Accessories The low-power continuous output from lead salt lasers makes them unsuitable for use with accessories such as harmonic generators which require high powers. It is at least conceptually possible to modelock lead salt lasers—something that has been done for shorter-wavelength III–V diode lasers—but there seems to be little interest in such possibilities.

Operating Requirements

Input Power Lead salt lasers typically have threshold currents of 100 to 500 mA and can be operated at currents as high as several amperes. Operating voltage is a fraction of a volt and drops at longer wavelengths. Thus the actual power consumption of the lead salt laser itself is under a watt. However, it is critical that the operating current be stable because fluctuations can change the laser's operating temperature and hence its wavelength.

In practice, the power consumption of the laser itself is insignificant compared with that of the equipment needed for it to operate. A commercial "laser source assembly" includes a compressor drawing up to 2 kW at 220 V, and a temperature stabilizer and laser control module which together draw 1 A at 110 V.

Cooling Requirements Lead salt lasers require cooling to cryogenic temperature for a combination of reasons: temperature control to stabilize operating wavelength, heat dissipation within the laser, and control of blackbody emission in the laser wavelength band. Dewars filled with cryogenic liquid helium were used in early experiments, but modern spectroscopic systems rely on closed-cycle mechanical refrigerators. These coolers can operate at temperatures between 12 K and room temperature, and can be stabilized to 0.0003 K. In practice, laser operation normally is at temperatures below 100 K.

Consumables There are no consumables needed for operation of lead salt lasers unless open-cycle cooling is used.

Required Accessories Lead salt lasers are usable only when incorporated into a system or subsystem because of the many accessories they require. These include

- Cooling equipment
- Temperature controller
- Driver with current stabilization
- Shutter or chopper for modulation
- Collimating optics
- Current sweep generator
- Reference and sample cells
- Detection and measurement equipment

Although all this equipment is not needed for every application, some of it is needed in all cases. Cooling and temperature-stabilization equipment are musts in all cases.

Operating Conditions and Temperature Lead salt laser systems are designed for normal laboratory conditions, although the lasers themselves must be maintained at cryogenic temperatures during operation. Some instruments have been operated under field conditions or in balloons for atmospheric measurements. This low-temperature requirement leads to a cool-down time before the system can be used. The lasers can be stored at room temperature without significant degradation.

Mechanical Considerations Unpackaged lead salt lasers are semiconductor chips less than a millimeter in all dimensions, but packaged versions are much larger. The standard commercial laser module is 0.75 by 0.312 by 0.25 in (19 by 8 by 6 mm). However, that module is usable only with a refrigeration system that weighs about 150 lb (70 kg). A full spectroscopic laser system is a benchtop laboratory instrument that weighs even more.

Safety From the standpoint of eye safety, lead salt lasers are completely innocuous. Their output is low in power and falls in wavelength range that does not penetrate into the eye. No high voltages are needed, and the cold temperatures of the compact laser assembly are sealed safely inside the system.

Reliability and Maintenance

Lifetime Cryogenic temperatures are a benign environment for diode lasers, which should have a long life if operated under the recommended conditions. However, operation at excessive currents could cause damage. In practice, the laser itself probably will have a much longer life than other elements of a lead salt laser system. One of the more vulnerable components is the refrigeration subsystem, typically specified as capable of 10,000 hours (a year) of maintenance-free operation.

Maintenance and Adjustments Needed Lead salt lasers themselves are maintenance-free, but the cooling system and other required accessories require periodic attention.

Mechanical Durability Although the laser itself is durable, the instruments housing it are designed for normal laboratory operation. However, the systems can be ruggedized for field use such as aircraft- or balloon-borne systems.

Failure Modes and Causes, Possible Repairs Failure of a lead salt laser is usually caused by progressive changes in device resistance, leading to irreparable degradation of the semiconductor device. However, the laser itself is far from the most expensive component of the system, and the modular design lets it be replaced easily.

The refrigeration system relies on compression and expansion of helium gas, which can leak through faulty connections. With proper sealing, the helium can be recharged, typically at intervals of one to several years as in any cryopumping refrigerator.

Commercial Devices

Standard Configurations Lead salt lasers are normally offered in three ways:

- As laser modules, packaged semiconductor lasers which require cooling equipment and other accessories
- As "laser source assemblies" which contain the laser itself, cooling equipment, electronic driver, and stabilization circuits for temperature control and drive electronics
- As part of complete systems such as laser spectrometers, which include the laser source assembly as above, plus sample cells, optics, chopper, monochromator, and detector assembly.

Designs are modular, and the laser modules are designed to plug into the laser source assemblies, which in turn plug into the spectrometers. Because the wavelength coverage of individual lasers is limited, several different lasers may be used with one laser source assembly or spectrometer. Generally the systems can accommodate two or four lasers, which can be switched as the operating wavelength is changed. Other lasers can be kept in reserve in case broader wavelength coverage is needed.

Options What is optional for a lead salt laser depends on how the laser is purchased. For a laser source assembly, collimating optics and certain controls are options. For the spectrometer, particular optics, measurement hardware, and signal-processing capabilities may be optional. The way the systems are offered, the choice of specific lasers turns out to be an important option. Wavelength coverage and time-response requirements are major factors influencing option choices.

Special Notes Lead salt lasers emitting near 16 μm are subject to export controls. They have such high resolution that they can be used to measure the proportions of different uranium isotopes in uranium hexafluoride by observing molecular absorption lines near 16 μm. This capability is important in uranium enrichment, making the technology sensitive from the standpoint of nuclear proliferation—and hence subject to stringent export controls.

Pricing Standard lead salt lasers currently have list prices of about $800 to $2500, with the highest prices for arrays. Individual mesa-ridge lasers able to produce 10 mW are about $1800. A laser source assembly, including refrigeration system and tuning equipment, is about $38,000. Detector assemblies also can be expensive, with prices in the $10,000 range. Complete spectroscopic systems, including a variety of equipment for general-purpose laboratory use, can run $100,000 or more.

Suppliers The market for lead salt lasers and systems built around them has long been dominated by a single company in the United States, although competition has begun to appear from overseas. The total market is too small to attract many competitors.

Applications

High-resolution infrared spectroscopy is the only commercially important application of lead salt diode lasers. As shown in Fig. 19-5, lead salt lasers offer resolution vastly superior to conventional infrared spectroscopy. However, this ultrahigh resolution is vital only for certain spectroscopic problems; in some cases it may supply far more detail than the user needs, wants, or can interpret. Most uses of high-resolution lead salt laser spectroscopy remain in the laboratory, but there are a few in industrial settings with demanding process-control or measurement requirements. One is the measurement of water content in integrated circuit packages. Others include monitoring plastics, halogen-lamp quality, automotive emissions, or hydrogen-fluoride concentration in UF_6.

In the long term, successful development of ultra-low-loss optical fibers transmitting in the mid-infrared could create a place for lead salt lasers in long-distance communications. However, both the laser and the fiber technology have a long way to go.

BIBLIOGRAPHY

J. F. Butler et al.: "PbTe diode laser," *Applied Physics Letters 5*:75, 1964.

J. F. Butler, R. E. Reeder, and Kurt J. Linden: "Mesa-stripe $Pb_{1-x}Sn_xSe$ tunable diode lasers," *IEEE Journal of Quantum Electronics QE-19*(10):1520–1524, October 1983.

D. L. Carter and R. T. Bate (eds.): *Semimetals and Narrow-Gap Semiconductors*, Pergamon Press, New York, 1971. Contains chapter on lead salt lasers.

R. S. Eng, J. F. Butler, and K. J. Linden: "Tunable diode laser spectroscopy: an invited review," *Optical Engineering 19*:945–960, November-December 1980.

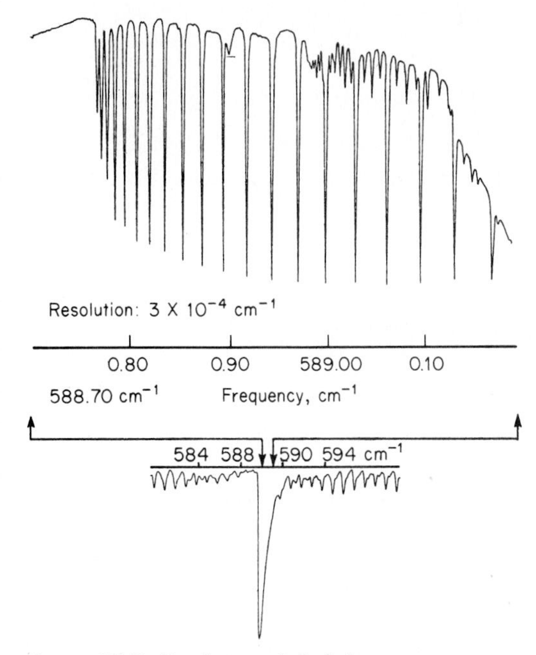

Figure 19-5 Resolution of diode-laser spectroscopy *(top)* compared with that of conventional infrared spectroscopy *(bottom)*. *(Courtesy of Laser Analytics division of Spectra-Physics Inc.)*

Charles Freed, Joseph W. Bielinsi, and Wayne Lo: "Fundamental linewidth in solitary, ultranarrow output $PbS_{1-x}Se_x$ diode lasers," *Applied Physics Letters 43*:629–631, October 1, 1983.

Eli Kapon and Abraham Katzir: "Distributed Bragg-reflector $Pb_{1-x}Sn_xTe/PbSe_yTe_{1-y}$ diode lasers," *IEEE Journal of Quantum Electronics QE-21*(12):1947–1957, December 1985.

Paul Klocek: "Infrared fiberoptics," *Lasers & Applications 1*(2):43–46, October 1982.

Laser Analytics Division, Spectra-Physics Inc., Bedford, Mass., literature package. Collection of literature describing products and applications, including a bibliography of over 380 research papers.

Kurt J. Linden: "Diode laser array with high power in the 4–5 μm infrared region," *Optical Engineering 23*(5):685–686, September/October, 1984.

Kurt J. Linden: "Single mode, short cavity Pb-salt diode lasers operating in the 5, 10, and 30 μm spectral regions," *IEEE Journal of Quantum Electronics QE-21*(4):391–394 April, 1985.

K. J. Linden, K. W. Nill, and J. F. Butler: "Single heterostructure lasers of $PbS_{1-x}Se_x$ and $Pb_{1-x}Sn_xSe$ with wide tunability," *IEEE Journal of Quantum Electronics QE-13*:720–725, 1977.

Kurt J. Linden and Robert E. Reeder: "Operation of cleaved-coupled-cavity Pb-salt diode lasers in the 4–5-μm spectral region," *Applied Physics Letters 44*(4):377–379, February 15, 1984.

D. L. Partin: "Heterojunction stripe geometry lead salt diode lasers grown by molecular beam epitaxy," *Optical Engineering 24*(2):367–370, March/April, 1985.

H. Preier: "Recent Advances in lead-chalcogenide diode lasers," *Applied Physics 20*:189, 1979.

CHAPTER

20

neodymium lasers

The neodymium laser is the commonest member of a family generally grouped together as *solid-state* lasers, a term which in the laser world does not encompass semiconductor devices. Qualitatively, neodymium lasers operate much like ruby (the first working laser) and other solid-state types. Atoms present in impurity-level concentrations—roughly 1 percent—in a crystalline or glass host material are excited optically by light from an external source, producing a population inversion in a rod of the laser material. The rod is mounted in an optical cavity which provides the optical feedback needed for laser action. The result is a simple and versatile laser which can emit a pulsed or continuous beam and which has become a standard tool for diverse applications.

The concept of optical pumping was conceived in the late 1950s. After protracted litigation, a patent on the concept was issued in 1977 to laser pioneer Gordon Gould. The validity of that patent, number 4,053,845, has been contested, and it remains the subject of litigation at this writing.

Neodymium lasers themselves are really more a family than a single type of device. The neodymium may be incorporated into various host materials, either synthetic crystals or glasses of different compositions. Accessories can shift the output wavelength from the near-infrared into the visible or ultraviolet.

Neodymium lasers can deliver continuous beams of less than a hundred milliwatts or short pulses of many megawatts, and find applications ranging from esoteric laboratory research to industrial materials working. No single device can do all these jobs, but all are possible with lasers of different commercially available designs.

Active Medium

Strictly speaking, the active medium in a neodymium laser is triply ionized neodymium, which is incorporated into a crystalline or glass structure. In a crystal, the neodymium is essentially an impurity, which takes the place of another element with roughly the same ion size (most often yttrium, another rare earth element). In glasses and the common crystalline hosts, the typical neodymium doping is around 1 percent by weight, giving a neodymium concentration of the order of 10^{20} atoms per cubic centimeter, which is generally the best for laser action. It is possible to grow laser crystals in which neodymium is an integral component of the crystalline structure rather than an impurity, such as neodymium pentaphosphate, NdP_5O_{14} (Chinn, 1982). However, the theoretical attractions of such materials so far have been offset by practical problems, and they remain in the laboratory.

By far the commonest host for neodymium lasers is yttrium aluminum garnet, a synthetic crystal with a garnetlike structure and the chemical formula $Y_3Al_5O_{12}$ that is known in the laser world by its acronym YAG. YAG is a hard and brittle material, and its growth is best characterized as a black art, but it has desirable optical, mechanical, and thermal properties. Although not an ideal laser medium, it is the best available for many practical neodymium lasers. Its most dramatic advantages are in thermal characteristics, which allow an Nd–YAG laser to produce a continuous beam of good quality—something nearly impossible at room temperature with most other solid-state laser materials.

Many other crystal hosts have been tested for neodymium, and a handful have even been offered commercially, but none has ever achieved anything remotely like the commercial acceptance of YAG. The two best-known alternatives to YAG at this writing are yttrium lithium fluoride, generally known as YLF, and yttrium aluminate, generally known as YALO because its chemical formula is $YAlO_3$.

Crystal growth problems limit the maximum length of YAG rods to on the order of 10 cm for most practical applications. YAG rods are generally 6 to 9 mm in diameter. Glass can be made in much larger blocks and is the choice for disks, slabs, or rods of laser material which must be larger than standard YAG rods to provide higher output power or energy. Glass rods as long as 18 in (46 cm) are used in some commercial materials-working lasers. Slab-geometry lasers, in which the beam follows a zig-zag path through a slab of neodymium-glass in the laser cavity, are under development (Eggleston et al., 1982). The Nova fusion laser at the Lawrence Livermore National Laboratory uses segmented elliptical disks 46 by 85 cm in its final amplifier stage; the segmentation is along the long axis and is intended to prevent premature depletion of the population inversion by amplified spontaneous emission (Martin et al., 1981). The larger size and surface area of some designs can offset some thermal problems, but repetition rates of large glass lasers still must be limited to lower levels than YAG to prevent harmful effects, often manifested in beam degradation because of thermal gradients in the laser material.

Energy Transfer

In optical pumping, absorption of light from an external source raises neodymium atoms to an excited energy level. In commercial neodymium lasers, the light source is a tungsten or arc lamp shining continuously, or a flashlamp producing pulses of light. The helical lamps used in early lasers have largely been supplanted by linear lamps. One or more such pump lamps are housed in a hollow reflective cavity designed to transfer the pump light to the laser rod. Several cavity configurations are possible, as shown in Fig. 20-1. One approach is to put the linear lamp next to the laser rod in a "close-coupling" configuration. Another is to place the lamp and laser rod at the two foci of an ellipse and let the geometrical properties of the reflective elliptical cavity carry the pump light from the lamp to the rod efficiently. Two lamps and a rod can be put into a dual elliptical cavity, which in cross section looks like two overlapping ellipses with the rod at the focus they share. High-power disk amplifiers used in fusion experiments may be surrounded by arrays of many flashlamps.

Continuous lamps are used to pump continuous-output lasers and many pulsed neodymium lasers with low or moderate output power. Where higher peak powers are required, pumping is with flashlamps, which can generate more intense light, but which require more complex pulsed electrical power supplies. Some inefficiencies are inherent in the energy storage and transfer hardware needed to generate high-voltage pulses for flashlamps.

Pump lamps emit a broad spectrum of light, but neodymium ions—whether in YAG, glass, or other hosts—tend to absorb most strongly at a limited range of pump wavelengths around 0.7 to 0.8 micrometer (μm), as shown in Fig. 20-2. Absorbing photons in this range raises the neodymium ions from the ground state to a higher energy level, from which they decay to a metastable level, producing a population inversion between the $^4F_{3/2}$ and $^4I_{11/2}$ states. The latter level decays by a fast nonradiative process to the ground state. Thus neodymium is a four-level laser system and is, as theory predicts, a more efficient laser material than the three-level ruby system. The strongest emission is at 1.06 μm, although energy-level splitting allows weaker emission on other lines. The actual energy-level kinetics are complex (see Svelto 1982, pp. 203–205, or Danielmeyer, 1976).

The interaction of natural decay time and emission and absorption cross sections limits the amount of energy that can be stored in a typical neodymium-YAG rod at any one instant to about 500 millijoules (mJ). That effectively limits energy in a *Q*-switched YAG pulse to under half a joule. The stored energy can be replenished on a time scale longer than the tens of nanoseconds of a *Q*-switched pulse but shorter than the millisecond length of a flashlamp

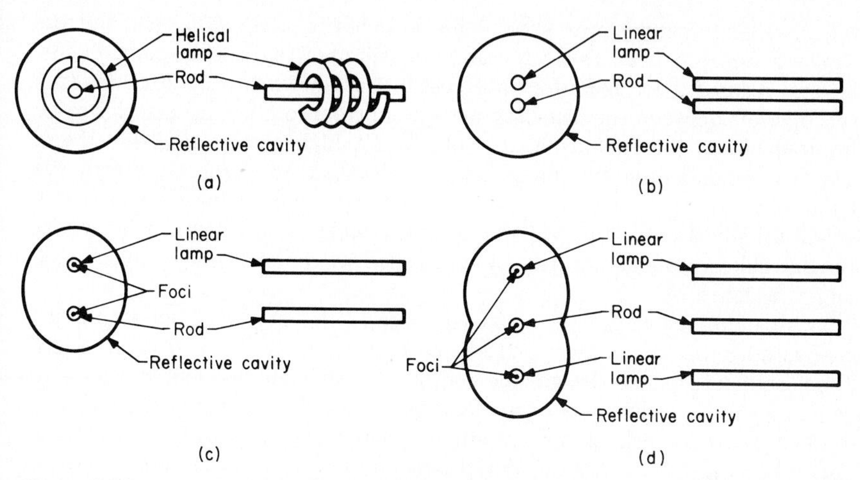

Figure 20-1 Representative pump cavity configurations. *(a)* Helical lamp, *(b)* closely coupled lamp, *(c)* elliptical cavity, *(d)* dual elliptical cavity.

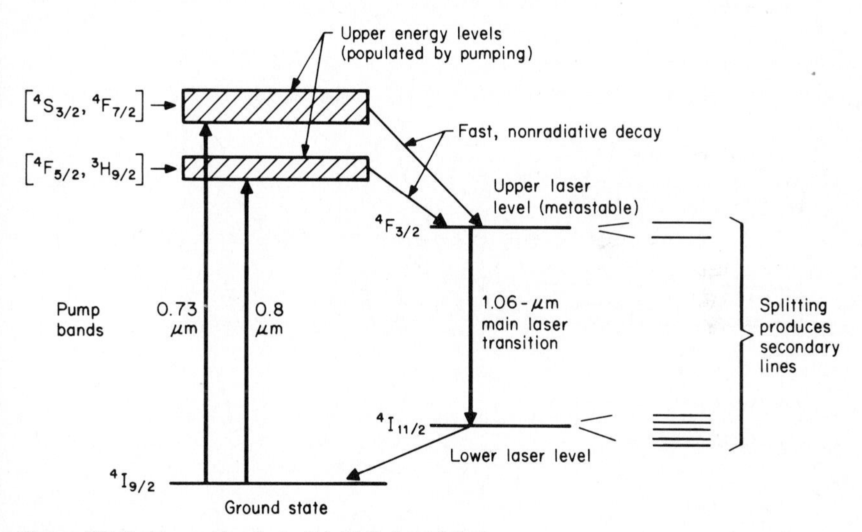

Figure 20-2 Energy levels in Nd–YAG (simplified).

pulse, so longer pulses can produce higher energies. Higher energies are also possible by using larger blocks of laser material, such as neodymium-glass disks, to store the energy.

Internal Structure

The internal structure of today's neodymium lasers is much more complex than that of Theodore Maiman's first ruby laser, in which a ruby rod with reflective end coatings was slipped inside a helical flashlamp. The mirrors which define the laser resonator are now discrete, adjustable components. Several resonator configurations are possible, as shown in Fig. 20-3. Often a space is left between the laser rod and the mirrors to allow accessories to be inserted within the laser cavity, to utilize the high powers available there. The cavity containing the pump lamps is designed to efficiently reflect the light from the pump lamps to the laser rod. YAG rods themselves are quite small, usually 6 or 9 mm in diameter and around 10 cm long, so the cavity can be quite small in portable systems such as military range finders and target designators.

Neodymium lasers can be arranged in series in an "oscillator-amplifier" configuration to obtain more pulsed power than is available from a single laser oscillator. In that design, the rod in the oscillator segment is equipped with a pair of resonator mirrors. A pulse from the oscillator then passes through a rod or disk in the amplifier, a separate cavity without resonator mirrors pumped by a separate pump lamp. When high powers are involved, optical isolators are inserted between amplifier stages to control the light and prevent any fraction of it from going backward toward the oscillator. When extremely high powers are needed, such as in laser fusion experiments, several

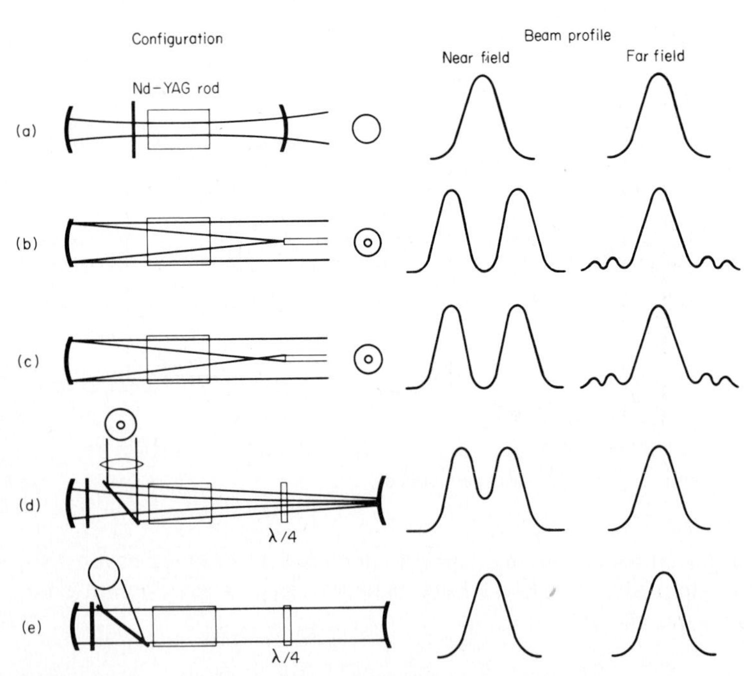

Figure 20-3 A sampling of resonator cavity designs with beam profiles they produce. *(a)* TEM_{00}; *(b)* diffraction-coupled positive unstable; *(c)* diffraction-coupled negative unstable; *(d)* polarization-coupled stable; *(e)* polarization-coupled unstable. λ/4 denotes a quarter-wave polarization rotator. *(Courtesy of Cooper LaserSonics.)*

stages of amplification may be used, each with a successively larger beam cross section, and the overall beam path may be long and complex, as shown in Fig. 20-4.

Inherent Trade-offs

The major trade-offs with neodymium lasers deal with output power. Peak power in an emitted pulse tends to decrease as repetition rate increases beyond

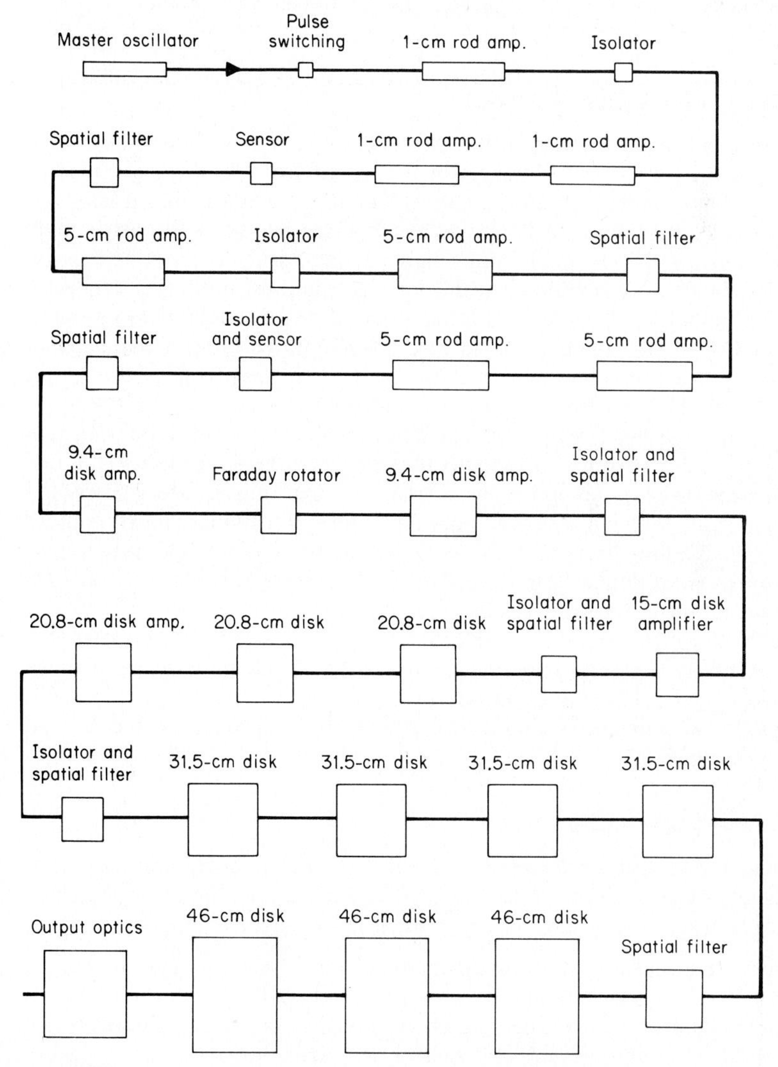

Figure 20-4 A simplified (!) version of the beam path in Livermore's Novette fusion laser. *(Adapted from George, 1982.)*

a certain point. In general, the higher the peak power, the longer the interval needed between pulses to dissipate excess heat that might otherwise distort the rod and degrade the quality of the output beam.

Saturation effects limit the amount of energy that can be stored in and extracted from neodymium laser rods. A resonator restricted to oscillate in only a single transverse mode extracts only a small fraction of the energy in a normal rod because it creates a narrow-diameter beam. More power can be extracted in a multimode beam, but the beam quality is poorer. In many cases, designers strike a balance by turning to an unstable resonator, which extracts energy from most of the rod and generates a beam of reasonable quality.

Variations and Types Covered

Neodymium lasers are versatile and can operate in many different configurations and host materials. Versions in use range from small single rods in hand-held range finders, to long amplifier chains in massive fusion lasers.

Pumping is generally with lamps, although there has been extensive study of other sources which might emit light more closely approximating the absorption bands of neodymium. At this writing, the most promising candidate for a new pump source is a monolithic array of semiconductor lasers, mentioned in Chap. 18, which operates near neodymium's 0.8-μm pump band. Much work remains to be done, however, and the diode-pumped laser is not yet on the market and hence will not be covered here.

As indicated earlier, many host materials have been studied for the neodymium ion, but YAG remains virtually the only crystalline host in commercial use. Two major glass compositions are on the market: silicate glasses and phosphate types, which offer advantages for some applications. The differences between neodymium lasers made of the two types of glass are relatively minor, except to the most demanding users. This chapter focuses exclusively on YAG and glass as the only commercially important neodymium laser materials.

Many of the most important variations among neodymium lasers are consequences of packaging. Adding harmonic generators to the laser can change the wavelength. Adding a *Q* switch or modelocker can change the pulse duration. These accessories, and the changes they can produce, are described in more detail below.

Beam Characteristics

Wavelength and Output Power The wavelength of neodymium lasers normally is quoted as 1.06 μm. This is a good working value, but there are three factors which can cause differences, ranging from slight to major.

- Interactions between the neodymium ion and the host material have a slight effect on energy levels and can shift wavelength by about 1 percent
- The neodymium ion can operate on weak transitions near 1.06 μm which are normally overwhelmed by emission on the main line

- Harmonic generation can multiply the frequency by a factor of 2, 3, or 4, thus dividing the wavelength by the same amount and shifting emission into the visible or ultraviolet

Output wavelengths of the major neodymium materials are:

- Nd–YAG: 1.064 μm
- Nd–YLF: 1.053 μm polarized, 1.047 μm unpolarized
- Nd-silicate glass: 1.062 μm
- Nd-phosphate glass: 1.054 μm

For most practical purposes there is no significant difference in these wavelengths, although care should be taken in matching the wavelengths of oscillators and amplifiers to obtain good performance.

Nd–YAG has significant laser gain at several wavelengths between 1.052 and 1.338 μm, but normally these are overwhelmed by 1.064-μm emission. Some companies offer 1.318-μm operation as an option on YAG lasers. One company produces a model in which the user can select eight wavelengths between 1.052 and 1.338 μm.

The second, third, and fourth harmonics of the neodymium laser's fundamental frequency, corresponding to wavelengths of 0.532, 0.335, and 0.266 μm, can be produced by passing the fundamental wavelength through nonlinear crystals. Some energy is lost in the process, which becomes increasingly efficient as the power level rises, and is most common with pulsed beams. Strictly speaking, harmonic-generation crystals are separate from the laser itself, but they sometimes are packaged together with the laser and offered as an integral device producing green or ultraviolet output. High-power pulses from neodymium lasers also can be used to drive other nonlinear processes which change light wavelength, including sum-and-difference frequency mixing, Raman shifting, and parametric conversion.

A wide range of output powers is available from YAG and glass lasers, depending on wavelength and configuration. The 1.064-μm line is by far the strongest in YAG; commercial lasers operating at 1.318 μm produce only around 20 percent as much power, while those operating at other neodymium wavelengths are even less powerful.

Output powers of continuous 1.064-μm YAG lasers range from a tenth of a watt to hundreds of watts. Similar average powers can be obtained from pulsed lasers. Peak power during a pulse can be much higher, from tens or hundreds of kilowatts in a "normal" pulse lasting on the order of a millisecond to over a hundred megawatts in a *Q*-switched pulse of some 10 to 20 nanoseconds (ns). Output also can be measured in pulse energy—tens of millijoules to perhaps a couple of joules for *Q*-switched pulses, and 0.1 to 100 joules (J) for pulses in the millisecond range. Obtaining more than a fraction of a joule in a *Q*-switched pulse requires an unusually large laser rod or an oscillator-amplifier configuration. Output power and energy are lower at harmonic wavelengths, with the losses being higher for the higher harmonics at the shorter wavelengths.

Because glass can be made in larger sizes and has lower gain, glass lasers can store more energy than YAG and produce pulses with more energy. Thus both peak power and energy of glass lasers can be higher than those of YAG. For example, one commercial oscillator-amplifier, with a 6-in (15-cm) oscillator and 12- and 16-in (30- and 40-cm) amplifiers can produce *Q*-switched pulses of 100 J with peak power of 5000 MW and duration of 20 ns. Even higher pulse energies can be produced with a larger laser; Livermore's Nova laser produces 10-kJ pulses from each of its 10 arms, which are frequency-doubled to generate about 70 kJ at the 0.532-μm second-harmonic wavelength. Commercial glass lasers also can be used with harmonic generators, although their lower power generally means lower efficiency. Glass's thermal problems limit it to average powers much lower than YAG and prevent continuous operation.

Efficiency Overall efficiency of a commercial neodymium laser—measured as laser power out divided by electrical power in—typically is 0.1 to 1 percent. The "slope efficiency," the fraction of added input power converted into laser energy, is generally somewhat higher, reflecting the need to surpass a pumping threshold to get the laser to operate at all. Efficiency is sometimes defined as laser power divided by power into the pump lamp, which is higher than overall or "wall-plug" efficiency because it neglects losses in the power supply.

A hierarchy of loss mechanisms limits the efficiency of neodymium lasers:

- Losses in the power supply which drives the pump lamp, which generally are more severe for pulsed supplies than for continuous-output types
- Incomplete conversion of the lamp-driving energy into light
- Losses in transferring the pump light to the laser rod
- Incomplete absorption of the pump light by neodymium atoms
- Losses inherent in the energy-level structure of neodymium because the excitation energy is greater than that of the emitted photon
- Losses caused by the failure of all neodymium atoms to emit the energy they absorbed in the form of light

Temporal Characteristics of Output Neodymium lasers can operate in several pulsed modes or, in the case of YAG, deliver a continuous output. Representative pulse lengths for various operating modes are shown in Table 20-1.

As can be seen from the table, pulse duration and repetition rate are strongly dependent on the type of pumping. In general, the repetition rate of the pumping flashlamp, 1 to 100 Hz, determines the repetition rate of *Q*-switched and cavity-dumped lasers. The situation is more complex with modelocking, which generates a train of short pulses. For flashlamp-pumped modelocked lasers, the repetition rate of the pulse train is the same as that of the flashlamp, while the interval between pulses in the train equals the cavity round-trip time (the speed of light divided by twice the length of the laser cavity). It is also possible to select a single modelocked pulse from a longer train of pulses. Both active and passive modelockers can be used with YAG lasers; passive

TABLE 20-1 Duration and Repetition Rate of Pulses from Neodymium-YAG Lasers

Excitation	Modulation	Typical repetition rate	Typical pulse duration
Continuous	None	Continuous	Continuous
Continuous	Q switch	0–100 kHz	100–700 ns
Continuous	Cavity dump	0.5–5 MHz	15–50 ns
Continuous	Modelocking	100–500 MHz	30–200 ps
Pulsed lamp	None	0–200 Hz, lamp-limited	0.1–10 ms
Pulsed lamp	Q switch	0–200 Hz, lamp-limited	3–30 ns
Pulsed lamp	Cavity dump	0–200 Hz, lamp-limited	1–3 ns
Pulsed lamp	Modelocking	Modelocked pulse trains, 0–200 Hz, lamp-limited	30–200 ps
Continuous	Q switch and modelocking	Depends on Q-switching rate; modelocked pulse trains	30–200 ps

modelocking gives shorter pulses, while active modelocking gives better stability.

Harmonic generation generally shortens pulse duration. This is due to the strong dependence of the process on optical power; the sharp drop in harmonic-generation efficiency at lower powers serves to cut off the less-intense portions of the light pulse.

Repetition rates of glass lasers are slower than those of YAG and typically run between one pulse per minute and a few pulses per second for commercial models. Much slower repetition rates, to one every several hours, are required for the ultra-high-power glass lasers used in fusion experiments.

Duration of glass-laser pulses is similar to that of YAG pulses except for modelocked operation. Because glass lasers have a much broader natural linewidth than YAG, modelocking them produces much shorter individual pulses, lasting as little as a few picoseconds. Thus glass becomes the clear choice if ultrashort pulses are required.

Spectral Bandwidth Most manufacturers do not list spectral bandwidth in their specifications for neodymium lasers, and for many purposes the lasers can be considered monochromatic. Linewidth of commercial Nd–YAG lasers is generally in the range of 1 to 5 inverse centimeters (cm^{-1}) (at neodymium's 1.06-μm wavelength, 1 cm^{-1} corresponds to roughly 1 part in 10^4). Scientific YAG lasers may be offered with line-narrowing accessories. An intracavity etalon can reduce linewidth below 0.2 cm^{-1}, and an additional electronic line-narrowing accessory can reduce linewidth to 0.02 cm^{-1}.

The natural linewidth of Nd-glass lasers is much broader, around 50 cm^{-1}.

Amplitude Noise Continuous and pulsed neodymium lasers suffer from different types of amplitude noise. In continuous lasers, a major cause is fluctuation of pump-lamp intensity, often at a harmonic of the 60-Hz line current which is rectified to power the lamp. In pulsed lasers, the prime variation is in peak power from pulse to pulse.

The magnitude of the fluctuations is comparable for the two types of lasers. Amplitude of continuous-output neodymium lasers typically fluctuates by a few percent to 10 percent. In commercial pulsed lasers, pulse-to-pulse variations are in the 1 to 10 percent range, with control depending on such techniques as "simmering" flashlamps—keeping a low, steady current flowing through them between pulses to avoid startup instabilities.

Beam Quality, Polarization, and Modes Neodymium lasers are available with output in a single (TEM_{00}) mode, in multiple transverse modes, or in modes determined by unstable resonators. Multimode output offers higher powers, while single-mode output offers better beam quality. Unstable resonators have become increasingly popular because they offer a combination of good beam quality with high power and efficient extraction of energy deposited in the laser rod, although they can only be used with pulsed lasers.

Commercial neodymium lasers are offered with polarized or unpolarized output. Polarized output generally is obtained by using polarizing components inside the laser cavity. Polarized output comes at a cost of reduced output power, but it may be required in certain cases, notably when the fundamental beam is to be used for harmonic generation or other nonlinear processes.

Beam quality can be degraded by operating at repetition rates or power levels too high for thermal stresses within the laser rod to be dissipated between pulses. Thermal gradients can cause beam breakup.

Coherence Length Coherence length of neodymium lasers is generally unspecified in data sheets. In practice, the output of Nd–YAG lasers is coherent enough for harmonic generation, sum-and-difference frequency mixing, and other applications which require coherent light. Based on their normal linewidth, Nd–YAG lasers can be estimated to have coherence length around 1 cm, while Nd-glass lasers have estimated coherence length of 0.2 mm. Line-narrowing options can greatly extend the coherence length; one company specifies a coherence length of 4 m for a scientific laser operating in a single longitudinal mode with spectral width of only 33 MHz.

Beam Diameter and Divergence Divergence of neodymium laser beams can run from a fraction of a milliradian to around 10 mrad. Beam diameter from YAG lasers generally is 1 to 10 mm, with larger diameters possible if beam expanders are used. Glass lasers, which can be made with larger cross-sectional area, can produce larger-diameter beams, particularly from high-power amplifiers at the end of a chain. Beams are usually circular, although some glass lasers have rectangular rods which produce rectangular beams.

Stability of Beam Direction Beam direction is generally stable to within around a milliradian, although this quantity is often unspecified.

Suitability for Use with Laser Accessories The neodymium laser is so well suited for use with laser accessories that many models have space within their optical cavities for them. Other models are sold in a package including such accessories as *Q* switches and harmonic generators.

Continuously pumped YAG lasers can be used with *Q* switches, modelockers, pulse selectors, or cavity dumpers to generate pulses of various durations and peak powers. In addition to these accessories, pulsed YAG lasers can be used with intra- or extracavity harmonic generators to produce shorter wavelengths. Because of their lower peak powers, continuous YAG lasers require intracavity harmonic generators. Typical harmonic-generation configurations are shown in Fig. 20-5; pulse durations for YAG lasers are listed in Table 20-1.

Glass lasers can be used with harmonic generators and modelockers, but their poor thermal qualities make them a poor choice for *Q* switching unless repetition rate is limited to a few pulses per minute. At higher repetition rates, thermal effects cause the output beam to become irregular in cross section. Because Nd-glass has much wider linewidth than Nd–YAG, it can produce much shorter modelocked pulses, with durations of only a few picoseconds compared with tens of picoseconds for YAG.

Operating Requirements

Input Power Small neodymium lasers can plug into an ordinary wall outlet and draw about 500 W from a 117-V supply. Large commercial models draw tens of kilowatts from three-phase, 220-V, four-wire power lines. Low-power military range finders based on YAG lasers are designed to operate from a 24- or 28-V battery pack, with the number of shots and degree of portability depending primarily on the size of the batteries.

Cooling Very low power YAG lasers in compact range finders, where portability is a paramount concern, may rely on purely conductive cooling, but

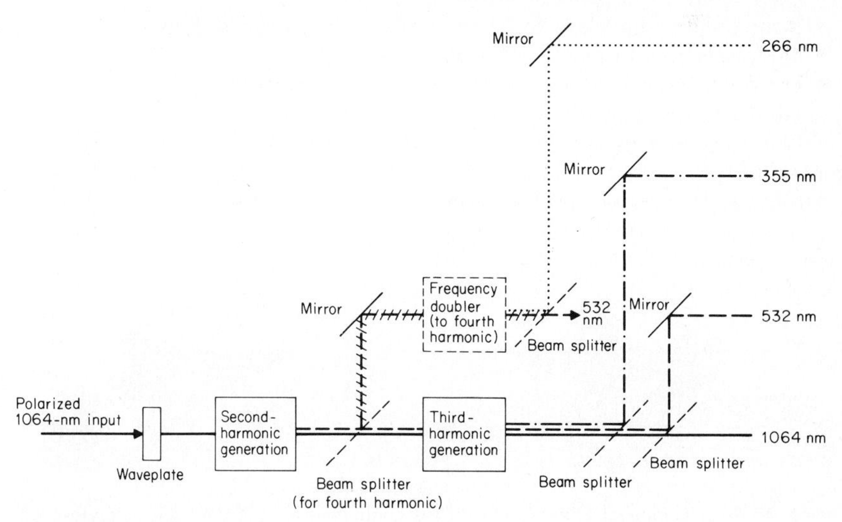

Figure 20-5 Typical arrangement for generation of second, third, and fourth harmonics.

this limits operation to low repetition rates. Most military range finders and target designators employ closed-cycle cooling systems. Laboratory and industrial lasers typically rely on open-cycle cooling with tap water. A typical low-power (0.1 to 1 W) YAG laser would require about 3 L (0.8 gal) of tap water per minute, while a massive industrial system delivering average power of 400 W would need as much as about 57 L (15 gal) of water per minute. Some small YAG lasers offer closed-loop water cooling as an option.

Consumables Solid-state neodymium lasers are designed to be self-contained units which do not require replenishment of supplies, other than a continuing supply of cooling water when needed. The only component which might be considered consumable is the pump lamp, which requires periodic replacement at intervals or a few hundred hours or several million shots.

Required Accessories Neodymium lasers generally are marketed as self-contained units which can operate directly from available wall current and are basically ready to begin operation. Some of these systems include components such as *Q* switches and harmonic generators, which could be considered accessories. In addition to the harmonic generators, *Q* switches, mode-lockers, and cavity dumpers mentioned earlier, the following accessories may be useful in working with neodymium lasers for certain applications:

- Shutters and beam isolators, to prevent undesired feedback in large amplifiers
- Optical benches, for stable mounting of accessories and cavity optics
- Temperature control systems to enhance laser stability
- A "beam dump" to absorb energy remaining at the laser's fundamental wavelength after harmonic generation
- Collimators for alignment of the laser optics
- Beam-delivery optics to focus the beam onto a remote object
- Mode selectors to pick a particular output mode
- Beam expanders to enlarge the output beam, particularly before passing a small-diameter beam through an amplifier
- Polarizers and polarization rotators to prepare the beam for harmonic generation
- Infrared viewers or viewing cards

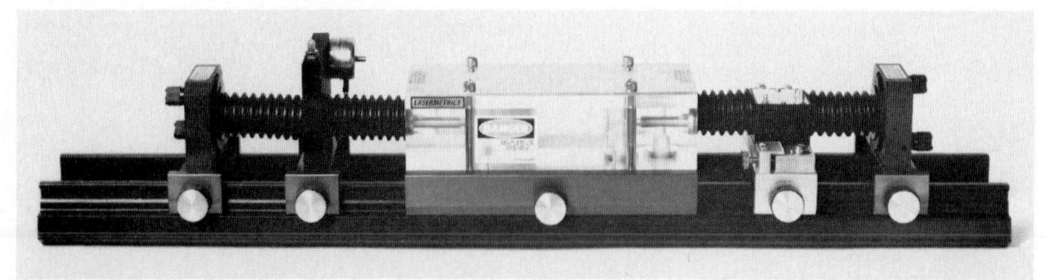

Figure 20-6 A commercial 50-W, *Q*-switched continuous-wave YAG laser. *(Courtesy of Lasermetrics Inc.)*

In some cases, some of these accessories may be supplied as standard equipment; in others they may be offered as options. Generally they are needed only for particular applications.

Operating Conditions and Temperature Most YAG and glass lasers are designed for the normal operating conditions of a laboratory or factory. Military range finders and target designators are designed to meet the much broader range of conditions encountered in military operations. There has been some work on developing space-qualified YAG lasers (Ross, 1978), with much of the attention concentrating on pump sources.

Mechanical Considerations The smallest YAG range finders are smaller than a pair of binoculars. Laboratory models are larger, but still compact for the amount of power they can deliver. One model weighs only 2.7 kg (6 lb) and measures 27 by 12 by 10 cm, but can deliver hundreds of milliwatts in multimode operation, although it does require 3 L/min (0.8 gal/min) of cooling water. Especially at high powers, the laser head, such as shown in Fig. 20-6, is separate from the power supply and controls. High-power models can require power supplies the height of a person, bulky cooling equipment, and laser "heads" that would cover a desktop. A special building was built to house large glass fusion lasers at the Lawrence Livermore National Laboratory.

Safety Like all lasers sold commercially in the United States, neodymium lasers have to meet a stringent safety standard established by the federal Center for Devices and Radiological Health (formerly the Bureau of Radiological Health) to minimize the dangers to users. Many industrial lasers designed for materials working are packaged in ways that carefully separate the laser beam from the operator. Such beam isolation is often impossible in the laboratory, and a few eye-damaging accidents have happened—with carelessness a major contributing factor.

Although the human eye cannot see the 1.06-μm emission of the neodymium laser, that wavelength can penetrate all the way into the eyeball and cause eye damage in the same way as visible light (Sliney and Wolbarsht, 1980). The danger of an operator being zapped in the eye by a beam that he or she cannot see is worsened by the fact that most neodymium lasers used in the laboratory produce short intense pulses. There has been more than one case in which a single neodymium laser pulse, reflected unexpectedly from an optical component, has caused an eye injury. The fact that the user cannot see where the invisible beam is going increases the likelihood of such reflections. The most effective protection against such accidents is to wear safety glasses or goggles which transmit most visible light while blocking the near-infrared output of neodymium lasers.

Harmonics of neodymium lasers also come in short, intense pulses, and they, too, present serious eye hazards. The second harmonic at 532 nm is in the visible green portion of the spectrum and can penetrate the eye in the same way as the fundamental wavelength. Although the visibility of the beam lets the user see where it is going, it also requires the use of safety goggles which block part of the visible spectrum. Hazards from the 355-nm third harmonic and the 256-nm fourth harmonic are different from those at longer wavelengths and from each other. The shorter of the two ultraviolet wavelengths can cause a sunburnlike effect on the skin as well as endangering the eye (Sliney and Wolbarsht, 1980, pp. 108–116). Ultraviolet-blocking safety goggles can provide eye protection against the third and fourth harmonics.

Users should be aware that the harmonic-generation process does not automatically get rid of all longer wavelengths. Some light at the fundamental wavelength may remain in a second-harmonic beam unless the optical system is explicitly designed to get rid of the fundamental. Some commercial lasers come with optional "beam dumps" for just this purpose. Multiwavelength laser output can present a severe hazard to the eye because of the difficulty in designing safety goggles to block light at two or more wavelengths.

Flashlamps used to pump neodymium lasers pose two distinct hazards: high voltages in the power supply and explosions of overloaded lamps. Lasers are designed to minimize high-voltage hazards, but caution is still required. Flashlamps can explode if pulsed at excessive power levels, but the shrapnel should stay inside the reflective cavity designed to direct the pump light to the laser rod.

Special Considerations Neodymium-YAG, neodymium-glass, and ruby lasers are similar enough that some companies offer a single basic laser system which can be adapted for use with rods of any of the three materials.

Because the lifetimes of pump lamps are limited, neodymium lasers generally should not be left operating for long periods if they are not being used. This warning does not apply to leaving flashlamps in a "simmer" mode, in which they are not actually firing pulses. Some warm-up time may be necessary, depending on the model of laser and the application; one manufacturer lists warm-up time of a minute, while another recommends a 15-minute warm-up for stable output. Some time may be needed to set up and align sensitive accessories such as line-narrowing and resonator optics, harmonic generators, *Q* switches, and modelockers; nonlinear optical elements used in *Q* switches and harmonic generators may require angle or temperature tuning.

Reliability and Maintenance

Lifetime Complete neodymium laser systems can last for a decade or more, in some cases outliving their manufacturers by several years. However, some components must be replaced during that period.

The shortest-lived component in a neodymium laser is the pump lamp. Continuous-output lamps have rated lifetimes of only a few hundred hours of operation. Flashlamps are typically rated for a few million to tens of millions of shots. To put that lifetime into perspective, a flashlamp firing a dozen shots per second—not uncommon for YAG—will fire a million times in 24 hours of operation.

The lifetimes of pulsed high-voltage components also may be comparatively short, depending on their nature and the laser design. The lifetime of the laser rod and optics generally are limited only by abuse—mechanical damage or operation at damage-producing power levels.

Maintenance and Adjustments Needed The most frequent maintenance needed for neodymium lasers is pump-lamp replacement, and most commercial lasers are designed to make such replacement simple. All exposed optical surfaces need to be cleaned periodically to prevent accumulation of dirt, which can cause excess light absorption leading to optical damage. High-voltage electronics may require their own periodic maintenance. Cooling system filters require periodic cleaning at intervals that may run in the range of a thousand hours.

Periodic alignment and adjustment of the cavity optics and optical accessories are also necessary to make sure the laser is producing the desired output. Major adjustments may require a low-power helium-neon alignment laser, but the visible laser is not normally needed for routine periodic alignment.

Mechanical Durability Neodymium lasers can be packaged well enough to meet stringent military specifications for portable field equipment. Laboratory lasers are not packaged that way, and even a bumped mirror can knock the laser cavity out of alignment. Users should also remember that pump lamps are made of fragile glass, and that the laser rods themselves are brittle.

Failure Modes and Possible Repairs A number of things can go wrong with neodymium lasers, some easy to fix and others requiring more extensive repairs:

- The pump lamp can burn out, requiring replacement, which generally is easy because most neodymium lasers are designed with lamp replacement in mind.
- The optical cavity can become misaligned, reducing output power and/or degrading beam quality, a problem that can be solved by realigning the optics.
- The laser rod or other optical components can suffer optical damage, either because the laser was operated at excessive powers or because accumulated dirt absorbed enough light energy to trigger damage. In many cases damage to the rod, or long-term degradation, can be repaired by repolishing and recoating the surface.
- Faults in high-voltage electronics can reduce power delivered to the pump lamp, thereby reducing laser output.
- Flashlamps can explode at high power levels, potentially damaging the laser rod and other components in the pump cavity. During the first 3 years of operating its Shiva fusion laser, the Lawrence Livermore National Laboratory reported that 40 of the 2000 flashlamps exploded, and that repairs accounted for about 3 percent of Shiva operating costs (Martin et al., 1981). Although the Shiva experience is by no means typical because of the extremely high powers involved, users should realize that flashlamp explosions are possible at high power levels.

Generally neodymium and other solid-state lasers are assembled from separate components and subassemblies rather than built as an integral unit (such as a gas laser tube). This makes refurbishment of an old laser reasonable, and a few companies offer refurbishment services for old neodymium lasers made by other manufacturers. Most active solid-state laser manufacturers maintain an inventory of spare parts for their products.

Commercial Devices

Standard Configurations Neodymium lasers are generally offered as complete packages, including laser rod, pump lamp, pumping cavity, cavity optics, power supply, cooling system, and controls. Some lasers are packaged as part of laser machining systems, which may also include beam-delivery systems, parts-handling equipment, beam enclosures, and computer controls. The difference between the two types of systems is usually obvious from the sales literature: descriptions of machining systems stress machining capabilities, while literature on general-purpose lasers specifies optical characteristics. Both types of systems may incorporate components such as *Q* switches which could be considered as accessories.

Materials-working neodymium lasers generally emit at 1.06 μm, delivering a continuous beam, long pulses in the millisecond range, or *Q*-switched pulses. General-purpose and scientific models typically operate continuously, *Q*-switched, cavity-dumped, or modelocked; some models are designed for operation at harmonic wavelengths. Scientific lasers often include room within the laser cavity for accessories such as harmonic generators and *Q* switches, which may be standard or optional in particular product lines. In general, models designed for scientific or general-purpose use have a broader range of options.

Options The usual range of options for neodymium lasers includes harmonic generators, *Q* switches, and modelockers. Options on some scientific lasers include optics to narrow the spectral linewidth below the natural level and output at near-infrared wavelengths other than the usual 1.06 μm. Some manufacturers allow the choice of single- or multimode output beams by changing the cavity optics. Makers of general-purpose and scientific neodymium lasers generally provide ways to accommodate accessories inside or outside the laser resonator cavity.

External amplifiers can be used to boost pulsed output power to high levels.

Special Notes The versatility of neodymium lasers, particularly YAG, has led to a wide range of commercial devices. Design tends to be modularized, so small companies can offer a wide variety of "standard" products that can be assembled from a small stock of standardized components to offer different capabilities. This increases flexibility and the range of user choice.

YAG is a hard material to grow, and there was a time around 1980 when it was in short supply because of limited production capacity and heavy military demand for range finders and designators. Both production capacity and the number of crystal growers has increased since then, and YAG supply has not been a problem recently. (There is virtually no overlap between growers of YAG crystals and makers of lasers; the laser manufacturers buy the material from outside vendors.)

Pricing Low-power laboratory YAG lasers can currently (1985) be bought (singly) for as little as $3000, but prices rise with output power, number of accessories, and performance requirements. Typical prices are in the five-figure category, and a highly sophisticated scientific laser or high-power materials-working system can run in the $200,000 range.

Suppliers The past few years have seen an influx of companies into the neodymium laser business. Industry directories list a few dozen firms which say they supply YAG lasers, and perhaps half that number as glass laser suppliers. A few of these companies are not actually active in the field, but there are plenty of active suppliers aggressively marketing well-designed products. There is enough competition to make comparison shopping well worthwhile.

Only a handful of companies supply YAG lasers for military range finders and target designators, and there is little overlap between that market and the market for civilian neodymium lasers.

Applications

As a steady and reliable source of laser energy, the YAG laser has found a wide range of applications in science, civilian industry, medicine, and military equipment. Glass lasers have found a somewhat narrower range of applications, because of their lower repetition rate and thermal problems. In many cases, the two are interchangeable. The range of applications is far too extensive to cover completely here; the following touches only on major and/or particularly interesting ones.

Many scientific applications of neodymium lasers involve generating other wavelengths of light, either by pumping tunable dye lasers or by nonlinear processes such as harmonic generation, sum-and-difference frequency mixing, parametric oscillation, or Raman shifting. YAG and glass lasers are also standard laboratory tools for generating high peak powers for a variety of types of research, including plasma physics, materials interactions, and nonlinear optics. Because they can generate higher peak powers, glass lasers are favored for the highest-power applications, such as laser fusion experiments.

YAG lasers are finding increasing uses in medical treatment and research as a reliable source of moderate to high powers in the near-infrared, visible, and ultraviolet. Although the carbon dioxide laser remains the preferred type for surgery, the near-infrared output of neodymium lasers has significant advantages for some types of treatment. For example, ordinary glass optical fibers can be used to direct the 1.06-μm wavelength of neodymium lasers but not the longer CO_2 wavelength. Thus flexible fiber-optic endoscopes can be used with YAG lasers to treat gastrointestinal bleeding and to deliver high laser powers to regions of the body which could not be reached with the bulky articulated arms now used with CO_2 lasers. YAG beams also can penetrate the lens of the eye to perform intraocular procedures without opening the eye.

The main industrial uses of YAG and glass lasers have been in materials working. Engineers often view neodymium and CO_2 lasers as complementary because of their different wavelengths and output characteristics. The high peak powers and short wavelength of neodymium lasers make them a better choice for drilling, particularly the many metals which are strongly reflective at CO_2's 10-μm wavelength. For similar reasons, neodymium lasers are often a better choice for spot welding. On the other hand, the higher continuous output of CO_2 lasers and the strong absorption of some materials (notably titanium and nonmetals) at 10 μm make CO_2 the better choice for many types of continuous cutting. The detailed trade-offs are complex and beyond the scope of this book (Duley, 1983).

The single largest application of YAG lasers is as military range finders and target designators. Comparatively little of this technology has been transferred to the civilian sector, although military agencies have sponsored much fundamental research. Neodymium lasers are not considered ideal for range finding and target designation because they have problems penetrating smoke and fogs, and because their pulsed output presents a serious eye hazard to friendly troops in training exercises. Several alternatives to YAG are under study, but YAG remains the standard in the range finders and designators which the Pentagon is buying in quantity.

BIBLIOGRAPHY

S. R. Chinn: "Stoichiometric lasers," in M. J. Weber (ed.), *CRC Handbook of Laser Science and Technology*, vol. 1, CRC Press, Boca Raton, Fla., 1982, pp. 147–169.

S. R. Chinn and W. K. Zwicker: "Flash-lamp excited NdP_5O_{14} laser," *Applied Physics Letters 31*:178, 1977.

H. G. Danielmeyer: "Progress in Nd–YAG lasers," in Albert K. Levine and Anthony J. DeMaria (eds.), *Lasers*, vol. 4, Marcel Dekker, New York, 1976, pp. 1–72.

W. W. Duley: *Laser Processing and Analysis of Materials*, Plenum Press, New York, 1983.

J. M. Eggleston et al,: "Slab-geometry Nd:glass laser performance studies," *Optics Letters 7*(9):405–407, September 1982.

E. Victor George, scientific ed.: *1981 Laser Program Annual Report*, Lawrence Livermore National Laboratory, Livermore, Calif., 1982. Review of the laser fusion program at Livermore, giving many details on glass laser development for fusion.

Walter Koechner: *Solid State Laser Engineering*, Springer-Verlag, Berlin and New York, 1976.

W. E. Martin et al.: "Solid-state disk amplifiers for laser fusion systems," *IEEE Journal of Quantum Electronics QE-17*:1744–1755, September 1981.

Peter F. Moulton: "Paramagnetic ion lasers," in M. J. Weber (ed.), *CRC Handbook of Laser Science and Technology*, vol. 1, CRC Press, Boca Raton, Fla., 1982, pp. 21–146.

Monte Ross et al.: "Space optical communications with the Nd–YAG laser," *Proceedings of the IEEE 66*:319–344, March 1978.

David Sliney and Myron Wolbarsht: *Safety with Lasers and Other Optical Sources*, Plenum, New York, 1980.

S. E. Stokowski: "Glass Lasers," in M. J. Weber (ed.), *CRC Handbook of Laser Science and Technology*, vol. 1, CRC Press, Boca Raton, Fla., 1982, pp. 215–264.

Orazio Svelto: *Principles of Lasers, 2d ed.*, Plenum Press, New York, 1982, pp. 203–205.

CHAPTER

21

ruby lasers

The ruby laser has proved surprisingly long-lived. A quarter-century ago, it was the first laser demonstrated, in pioneering experiments by Theodore H. Maiman (Maiman, 1960)—much to the surprise of some prominent physicists who had said a ruby laser would not work. Today, ruby lasers remain viable commercial products, while most of the other types envisioned by laser pioneers are at best little more than laboratory curiosities.

Ruby emits at 694.3 nanometers (nm) in the deep red when a rod of the material is pumped optically by an incoherent lamp. Commercial models operate in pulsed mode with flashlamp pumping, both because of the high laser threshold and the material's thermal characteristics. Continuous-wave operation has been demonstrated in the laboratory, but there is essentially no practical demand for continuous ruby lasers.

Ruby was the first laser to be used in materials working, but except for certain special applications has largely been replaced by other types. However, ruby remains a valuable source of high-power pulses of coherent visible light and is widely used in pulsed holography, interferometry, nondestructive testing, and plasma measurements.

Qualitatively, ruby lasers are similar to the neodymium lasers described in Chap. 20, although the maximum repetition rates are much lower than those of Nd–YAG. The similarities are close enough that in some cases ruby, Nd–YAG, and Nd-glass rods can all be operated in the same laser, switching only the optics for the ruby wavelength, although performance of the laser is not as good as when the laser is optimized for a particular material and wavelength.

Internal Workings

Active Medium The active medium in a ruby laser is a rod of synthetic ruby, a sapphire crystal to which a small quantity of chromium has been added during growth from an aluminum oxide (Al_2O_3) melt. Chromium, with the same +3 valence as aluminum, takes the place of some of the aluminum in the crystal. The chromium concentration is typically in the range of 0.01 to 0.5 percent by weight, corresponding to on the order of 10^{19} atoms per cubic centimeter, enough to color the rod pink or red. Ruby rods are available in diameters ranging from 3 to 25 mm (⅛ to 1 in) and lengths to about 20 cm (8 in).

Ruby is resistant to optical damage at normal power levels if the surface is kept clean. Heat conductivity is better than that of glass and even YAG, but because ruby is a three-level laser, its laser characteristics degrade much more rapidly with increasing temperature than four-level neodymium lasers and thus limit ruby's repetition rate.

Light at the laser wavelength is strongly self-absorbed in the three-level ruby system. This effect occurs in unpumped parts of the ruby rod, making it important to illuminate as much of the rod as possible with pump light.

Energy Transfer Visible-wavelength photons from the pump lamp raise chromium ions to one of the two excited levels shown in Fig. 21-1. These decay in about 100 nanoseconds (ns) to a pair of metastable levels, each with room-temperature lifetime about 3 milliseconds (ms). Both metastable levels can serve as the upper level of a laser transition, but if emission is allowed on both lines, the lower-energy 694.3-nm transition from the $\overline{E}$ state will dominate. This occurs because the higher-energy 2A metastable state is linked to the $\overline{E}$ state by a nonradiative decay with 1-ns lifetime.

High optical pump energies are needed to invert the population of chromium ions because the ground state is the lower laser level in the three-level system. This leads to a high laser threshold and limited efficiency because the number of chromium ions in the metastable state must exceed the number in the ground state for laser action to occur. Gain and efficiency are lower in ruby than in the four-level neodymium laser, but energy storage in ruby is higher, allowing production of higher-energy *Q*-switched pulses.

The excess pump energy remains in the rod as heat, which must be removed because laser gain drops with temperature, and because the heat can cause lensing in the rod that degrades beam quality. This limits repetition rate to a few hertz (except for very small rods), and to lower levels for large rods.

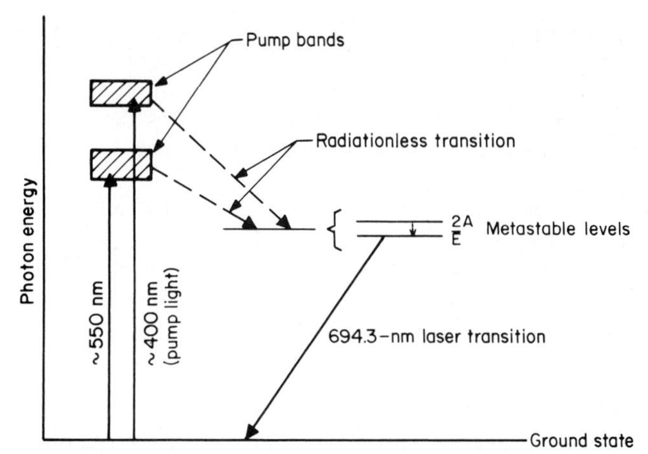

Figure 21-1 Energy levels of the chromium ion in ruby, showing the pumping and laser transitions. This is called a three-level system because only three energy levels are involved in laser action: the ground state (which is both the initial state and the lower laser level), an excited state produced by the pump light, and the metastable upper laser level.

Internal Structure

Optics In Maiman's first ruby laser the ends of the rod were silvered, one for total reflection, the other for partial reflection and output coupling. In modern ruby lasers, separate mirrors are usually used, with the ends of the rod polished flat. One mirror is totally reflecting; the other partly reflective and partly transmissive for output coupling. In most cases they are flat, but in some cases the rear mirror may be concave with a long focal length, around 10 m, to compensate for lensing in the laser medium.

Cavity Length and Structure As with other solid-state lasers, the ruby laser has two kinds of cavities: the pump cavity which transfers pump light from a lamp to the rod, and the resonant cavity in which laser oscillation occurs.

The pump cavities used for ruby are the same types used for neodymium lasers, as shown in Fig. 20-1. Because of the high pump power requirements, efficient coupling and reflection are more crucial for ruby than for neodymium lasers. Thus some cavities use silver coatings for the cavity, because they offer high reflectivity despite their need for periodic refinishing. Other designs may use glazed ceramic surfaces with high diffuse reflectivity, which are less sensitive to surface quality than coated cavities and provide more diffuse illumination. Because of the crucial importance of cooling, in many cases cooling water may flow *through* the coupling cavity, either filling the cavity, or flowing annularly around the rod and linear flashlamps. Helical flashlamps are used in some models.

The two mirrors at the ends of the laser rod usually form a stable resonator. Cavity lengths can range from slightly over the length of the rod to somewhat longer, if intracavity accessories such as *Q* switches are used. Many commercial ruby lasers are assembled on optical rails or similar structures that let the user adjust positions of cavity optics; these structures typically are 1 to 2.5 m long.

Ruby operates well in an oscillator-amplifier configuration, in which a pulse from a ruby oscillator passes through a second ruby rod that amplifies it. Commercial versions can include one to four amplification stages. This oscillator-amplifier configuration generally is the best approach to producing pulses with high energy and good beam quality.

Variations and Types Covered

All commercial ruby lasers are pulsed, with output characteristics varying over a considerable range. The technology is well developed, and many models are built primarily for specific applications, such as double-pulse holography, plasma diagnostics, or hole drilling. Some plans have been published for homemade ruby lasers (Iannini, 1983; McAleese, 1979), but those devices are not covered here. Neither does this chapter cover the continuous-wave, low-power ruby laser that has been demonstrated experimentally.

Beam Characteristics

Wavelength and Output Power The ruby laser wavelength is 694.3 nm. Oscillators can produce millisecond fixed-*Q* pulses with energies to 50 or 100 joules (J), while oscillator-amplifier configurations can generate well over 100 J. Those high energies are for output in multiple transverse modes; if emission is *Q*-switched or limited to TEM_{00} mode, pulse energy is much lower. Average powers are limited by the low repetition rates—no more than a few hertz except for the very smallest rods, and often just a few shots per minute. Maximum average power level is around 100 W for a single-rod oscillator, and that requires careful cooling with fast-flowing water.

Efficiency Wall-plug efficiency of ruby lasers typically is in the 0.1 to 1 percent range. This is rarely a major concern because ruby lasers are not normally used in high-volume production.

Temporal Characteristics Ruby lasers normally operate in one of three pulse modes: long-pulse (often called "normal" or "conventional"), *Q*-switched, or modelocked.

If no pulse-control accessories are added to the laser, pulse duration is set by the length of the light pulse from the pump lamp. The laser pulses in this operating mode are from 0.3 to a few milliseconds long. Ruby lasers produce the highest-energy pulses in this mode, although average power during the pulse is no more than tens of kilowatts. A variety of internal effects lead to generation of spikes during the pulse, as shown in Fig. 21-2, so instantaneous pulse energy can vary considerably. This spiking is not purely a nuisance, and in fact is considered desirable by some engineers for drilling holes.

Q-switching limits pulse energy from a single oscillator to a few joules, but by compressing pulse duration to 10 to 35 ns can raise peak power to the 100-MW range. With an oscillator-amplifier configuration, peak power in a pulse of similar length can reach 1 GW. Passive (saturable dye) *Q* switches can be used if only a single *Q*-switched pulse is needed. For multipulse operation, an active *Q* switch must be used to generate pulses spaced by 1 to 1000 microseconds (μs) within the duration of the pumping flashlamp pulse. Ruby can generate higher-energy *Q*-switched pulses than YAG because of its better energy-storage capacity.

When modelocked, ruby lasers can produce a train of 20 to 30 pulses, each as short as 3 to 4 picoseconds (ps), lasting a total of a few hundred nanoseconds. Each pulse has energy around 1 mJ, but peak power during its brief duration

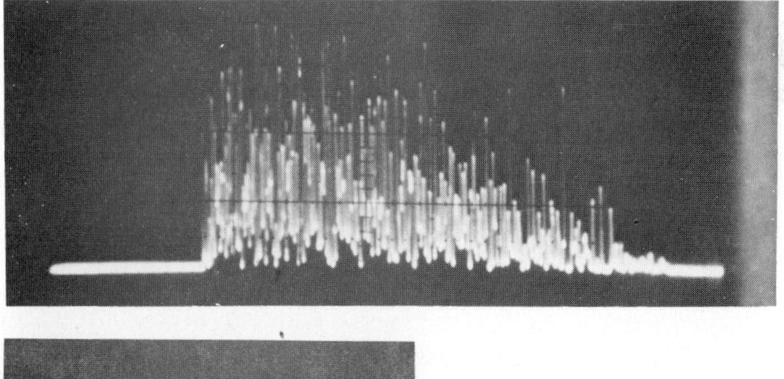

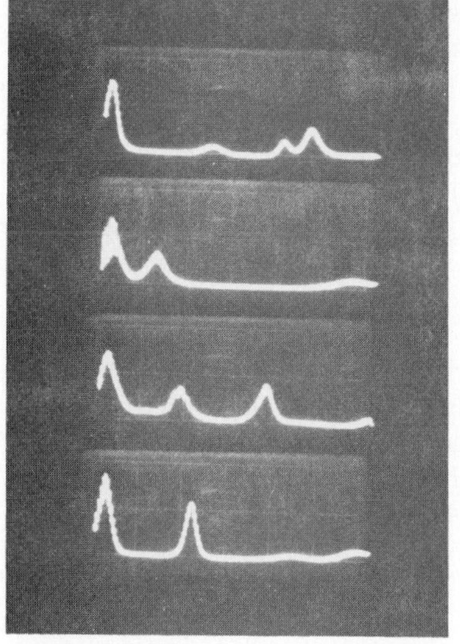

Figure 21-2 Power spiking in long-pulse output of a ruby laser showing entire 600-μs pulse *(top)* and detail for 6-μs intervals *(bottom)* indicating that output is zero at times. *(Courtesy of Lumonics Inc.)*

can reach hundreds of megawatts or even higher with amplification, but such operation is not common because of poor pulse-to-pulse stability.

Spectral Bandwidth Interaction of chromium ions with lattice phonons homogeneously broadens spectral bandwidth of the ruby laser transition to 330 GHz, or 0.53 nm, about one part in a thousand of the transition frequency. Use of transmission and reflection etalons in the laser cavity can reduce bandwidth to on the order of 30 MHz in lasers used for holography.

Amplitude Noise Pronounced power spiking occurs during long-pulse operation of ruby lasers, as shown in Fig. 21-2. Most manufacturers do not specify pulse-to-pulse variations, although one claims that with cooling *Q*-switched pulse energy is repeatable within ± 5 percent.

Beam Quality, Polarization, and Modes Ruby lasers have beams ranging in quality from near diffraction-limited TEM_{00} beams for holography to much broader multiple-transverse-mode beams for applications which require brute-force energy. Operation at high repetition rates or without proper cooling can cause lensing effects in the rod which degrade beam quality. Lasers designed for holography may emit in one or sometimes two longitudinal modes, but those built to generate high-energy pulses operate in multiple transverse modes and many longitudinal modes. The ruby crystal tends to impose linear polarization on the beam, but strongly pumped ruby will also emit in the orthogonal plane, so some polarization-sensitive element may be needed to assure linear polarization.

Coherence Length Holographic ruby lasers can have coherence lengths of 10 cm to 10 m. High-energy models designed for other applications have shorter coherence lengths.

Beam Diameter and Divergence Commercial ruby lasers have beam diameters ranging from 1 to 25 mm, and divergences ranging from 0.25 to 7 milliradians (mrad). Holographic lasers typically have beams close to the diffraction limit, but some multimode lasers have both large divergence and large-diameter beams.

Suitability for Use with Laser Accessories As mentioned earlier, *Q*-switching of ruby lasers is common. Modelocking is done less often. The high peak powers of *Q*-switched ruby lasers can readily be frequency-doubled in RDA (rubidium dihydrogen arsenate, RbH_2AsO_4), with conversion efficiency typically around 30 percent, to produce 347-nm ultraviolet pulses.

Operating Requirements

Input Power Ruby lasers typically operate from single-phase 110-V ac lines, but some work from 220-V ac supplies. Peak line currents of 10 to 20 A may be drawn to start charging the capacitor bank that discharges 5 to 10 kV through the flashlamp(s). Once the capacitor bank is charged, much lower line currents can maintain the charge.

If a separate refrigeration unit is used to cool the laser, its power requirements are comparable with those of a moderate-power ruby laser—around a dozen amperes at 110 V.

Cooling Some form of active cooling is necessary for ruby lasers. Excess heat will not damage the ruby crystal, but it can increase laser threshold and decrease output power. Typical designs circulate deionized water through the laser head and through a heat exchanger that transfers the excess heat to air or flowing tap water. Deionized water is needed if the coolant passes over electrical connections as it removes heat directly from the laser rod and flashlamp; otherwise distilled water will suffice. Because of the short distances involved, clean water will absorb little of the pump light, and because the refractive index of water is between that of air and the rod, it can help reduce surface reflections.

Consumables Flashlamps must be replaced periodically, but because the repetition rate of ruby lasers is low, the period is long. If a flashlamp lasts a typical rated life of 1 million shots, it could generate one pulse per minute for 24 hours a day for two full years. Closed-cycle lasers also consume water filters and deionizer cartridges.

The focusing lenses of drilling lasers are inevitably exposed directly to debris from the hole-drilling process. Some dust eventually will be deposited on the lens, and in practice that lens may be considered a consumable item because its surface and coating become pitted.

Required Accessories *Q*-switches are needed for short-pulse operation. Line-narrowing etalons are needed to achieve the coherence lengths required for holography.

Operating Conditions Ruby lasers are built to function in laboratory or industrial environments. Stability is important in the operation of holographic lasers, because of the good beam quality needed to produce high resolution holograms. Cleanliness is important in applications where beams are focused to high power densities on optical surfaces, because energy absorption by dirt can lead to surface damage. Good temperature control of the laser rod is important because of the drop in ruby gain with increasing temperature.

Mechanical Considerations Typical ruby laser heads are 1 to 2 m long and much smaller in cross section and include an integral optical rail or similar stabilizing structure. In high-power lasers, the head and optics usually are enclosed in a sealed housing; in lower-power systems components outside of the laser cavity itself often are exposed for user adjustment. In many models, particularly high-power ones, optical elements are connected by hollow tubes of plastic or aluminum that prevent dust from settling on surfaces and contain scattered and reflected light. Heads of holographic lasers may be tripod-mounted; general-purpose ruby heads (e.g., Fig. 21-3) usually rest on an optical table. Supporting structures account for most of the weight of laser heads, typically in the 20- to 50-kg range for oscillators.

The real bulk and weight come in the power supply and controls, typically housed in a console 4 or 5 ft (1.3 to 1.6 m) high that rests on the floor. Separate coolers normally are boxes a couple of feet high. Total shipping weights can range from a few hundred pounds for a small oscillator to nearly a ton for a large oscillator-amplifier unit.

Safety Ruby lasers are a serious eye hazard because their high-power pulses can penetrate the eye and cause permanent damage to the retina. As with neodymium-YAG lasers, a single pulse can permanently impair vision. Ruby pulses also can carry enough energy to cause skin burns.

The 5- to 10-kV potentials applied across the flashlamps are accompanied by large enough currents to electrocute a person. Because the laser operates in pulsed mode, with the power supply charging up between pulses, many components may retain large charges after the laser is turned off. Care should be taken to discharge capacitors and other components likely to retain high voltages before investigating the power supply. (Most current commercial lasers automatically dump capacitor charge when power is off or the case is opened.)

Flashlamps can explode if driven beyond their rated input powers, but a sealed pump cavity should contain any shrapnel.

Special Considerations Pulse-timing jitter is an important consideration in double- or triple-pulse holography, because of the need to know the interval between recorded images. Typical ratings for jitter of actively Q-switched pulses are ± 10 ns.

Figure 21-3 Q-switched ruby laser assembled on an optical rail. The laser rod is in the box just left of center with the DANGER label; to its right is the Q switch. The tubes running between optical elements keep them free of dust and control reflections. *(Courtesy of Lasermetrics Inc.)*

Reliability and Maintenance

Ruby lasers should be reliable when given proper care. Some early models have lasted many years longer than the companies that made them. For high-power operation, cleanliness is critical because absorption of energy by dust or dirt can damage optical surfaces. Such damage is a leading cause of service calls. It is not enough just to dust off the optics once in a while. The laser head and beam path should be closed or sealed, and some manufacturers suggest passing a slow flow of clean nitrogen gas through the head to maintain positive pressure.

Silver coatings are used in some pump cavities because their high reflectivity increases pump energy delivered to the ruby rod. However, silver tarnishes, and the coatings must be repolished or reapplied periodically to maintain high reflectivity. Flashlamp replacement is infrequent because of the low repetition rates.

The resonant cavity may require periodic realignment to maintain maximum output. Many ruby laser makes offer optional helium-neon lasers and autocollimators for that purpose.

Commercial Devices

Standard Configurations Ruby lasers may be designed either for general-purpose use or for a specific application. The commonest specialized versions are built for holography, where factors such as coherence length, beam quality, and pulse timing are crucial. For materials working, the goals are delivering high pulse power and energy to the workpiece. The packaging of the laser head depends on factors such as the power densities at optical surfaces and the user's need for access to internal components. Generally in all variations the laser head is separate from the power supply and control console, with any cooling or refrigeration unit also packaged separately.

Makers of solid-state lasers often take a modular design approach. Ruby oscillators may be packaged separately, or sold together with one or two amplifier stages. Similar cavities, components, controls, and optics may be used with ruby and neodymium lasers. Indeed, ruby and neodymium lasers often can operate in the same cavity if the resonator optics are changed, although neither gives its best performance in such a configuration.

Options Common options offered for ruby lasers include

- He–Ne alignment laser
- Pulse energy monitors
- *Q* switches
- Autocollimators (for alignment)
- Pulse slicers
- Pulse counters
- Double- or triple-pulse output
- Selection of cooling system
- Single-transverse-mode operation

- Etalons for single-longitudinal-mode operation
- Remote control of laser operation
- Integration with positioning equipment of holographic cameras for end-user applications

Pricing Ruby laser prices in 1985 range from $14,000 to about $200,000, with the most costly models specially packaged oscillator-amplifier designs.

Suppliers About a dozen companies produce ruby lasers. All major ruby suppliers also make neodymium lasers, but many makers of neodymium lasers do not bother with ruby because the market is much smaller. Recent years have seen refinements in holographic ruby lasers, and in design of power supplies, pump cavities, and controls, but the basic technology of ruby lasers is well established.

Applications

The biggest single application of ruby lasers is in holography. A *Q* switch can produce one, two, three, or more highly coherent pulses during a single flashlamp pump pulse. When these 10- to 30-ns pulses illuminate a moving object, they can record holograms that freeze its motion. Double pulses can record a pair of holograms on the same plate, and reconstruction of the hologram generates an image with a superimposed interference pattern that shows motion or changes in shape, as shown in Fig. 21-4. The concentrations of dark lines show regions of rapid displacement; where there is little motion, there are few or no dark lines. Similar effects can be used in triple-pulse holography. These holographic techniques have been developed widely for use in nondestructive testing (Abramson, 1981).

Figure 21-4 A holographic interferogram such as this one records displacement between a pair of pulses from a laser, in this case *Q*-switched pulses from a ruby laser. *(Courtesy of Apollo Lasers Inc.)*

The density and temperature of electrons in high-energy plasmas can be recorded by observing Thomson scattering of light from high-peak-power pulses from ruby lasers. These plasma diagnostics are used in fusion research, but recent improvements in detector sensitivity have led to increasing interest in taking advantage of the much higher repetition rates possible with neodymium lasers to speed data collection.

Ruby lasers also can be used in atmospheric ranging, scattering, and lidar measurements, where the high peak power and red wavelength are important.

The low repetition rate severely limits the use of ruby lasers for materials working, but they do find some special applications. In some cases, such as trimming resistors or integrated-circuit masks, the short wavelength is important because it allows the beam to be focused to a finer spot than the 1.06-μm fundamental wavelength of neodymium lasers. The spiking of output power in long-pulse mode is considered desirable by some manufacturing engineers for drilling high-quality holes. The ability to generate high energies in *Q*-switched pulses can be important in certain applications, such as dynamic balancing by removing small chunks of material from rotating objects by pulses from a laser.

Ruby lasers have been used in range finders and target designators which remain in military arsenals. However, virtually all range finders and designators now being produced use 1.06-μm Nd–YAG lasers because of their higher efficiency, faster repetition rate, and invisible wavelength.

Ruby lasers also are used in general-research applications, including plasma production and fluorescence spectroscopy. High-power ultraviolet pulses can be produced by frequency-doubling *Q*-switched pulses. Although the repetition rate and average power are much lower than an excimer laser, peak powers are comparable, operation is simpler, and the capital investment of as little as $20,000 is significantly less.

BIBLIOGRAPHY

Nils Abramson: *The Making & Evaluation of Holograms,* Academic Press, New York, 1981. See especially chap. 7.

L. Allen: *Essentials of Lasers,* Pergamon Press, Oxford, 1969.

Robert E. Iannini: *How to Build Your Own Laser, Phaser, Ion Ray Gun and Other Working Space-Age Projects,* TAB Books, Blue Ridge Summit, Pa., 1983. Hobbyist project book, includes plans for a ruby laser.

Walter Koechner: *Solid-State Laser Engineering,* Springer-Verlag, New York and Berlin, 1976.

Theodore H. Maiman: "Stimulated optical radiation in ruby," *Nature 187:*493, 1960. First report of a laser.

Frank G. McAleese: *The Laser Experimenter's Handbook,* TAB Books, Blue Ridge Summit, Pa., 1979. Hobbyist project book, with ruby laser plans.

Joseph T. Verdeyen: *Laser Electronics,* Prentice-Hall, Englewood Cliffs, N.J., 1981. Description of ruby laser physics on pp. 255–261.

CHAPTER

22

vibronic solid-state lasers

One of the newer additions to the laser world is a family of solid-state lasers which emit on "vibronic" transitions, in which the active species (an atom in a solid-state host) changes both electronic and vibrational states. Dye lasers also emit on vibronic transitions, and like dye lasers, solid-state vibronic lasers can emit light at a range of wavelengths. Solid-state vibronic lasers also promise high output power and operation simpler than that of dye lasers, a combination attractive enough to have stimulated much development work. However, the technology has yet to prove itself by establishing a place in the world of commercial laser applications.

The first vibronic laser to reach the market was alexandrite, chromium-doped $BeAl_2O_4$, which is tunable between 701 and 826 nanometers (nm), although not across the entire range under the same conditions. Other types are on the way, including chromium-doped crystals of garnetlike structure, synthetic emerald, titanium-doped sapphire, and cobalt-doped magnesium fluoride. Developers have envisioned a range of applications from materials working to spectroscopy, and there have even been suggestions that some might be useful as drivers for laser inertial-confinement fusion. In the near term, the most likely applications are in research and development, although such work could lead to more extensive use in industry, medicine, or other fields. So far, however, even alexandrite, the best known and most promoted type, has yet to move beyond the laboratory into any commercial-scale applications.

Introduction and Description

Vibronic lasers are superficially similar to other solid-state lasers in which the active species is an impurity in a crystalline matrix of another material. Chromium, the active species in nontunable ruby lasers, is also the active species in several vibronic lasers, including alexandrite. Indeed, alexandrite was first studied as a possible replacement for ruby as a host crystal for chromium, but laser performance on the 680.4-nm fixed-wavelength transition was poor. Laser action on the longer-wavelength ${}^4T_2 \rightarrow {}^4A_2$ vibronic transition was only discovered later (Walling, 1982).

The tunability of vibronic lasers comes from their energy-level structure. As shown in Fig. 22-1, the vibronic laser transition is between two electronic states which are split into vibrational substates, creating a continuum over which emission can be tuned. In contrast, the laser levels in conventional solid-state lasers are discrete, not part of a continuum. The range of energies covered by the continuum is large enough that some vibronic lasers can be tuned by ±20 percent from a central wavelength. Thus both the tunability and energy-level structures of vibronic solid-state lasers are similar to those of liquid dye lasers.

On the other hand, vibronic solid-state lasers are similar to fixed-wavelength solid-state types in their operation. Both types require optical pumping, typically with a flashlamp for the more practical types, although some vibronic lasers

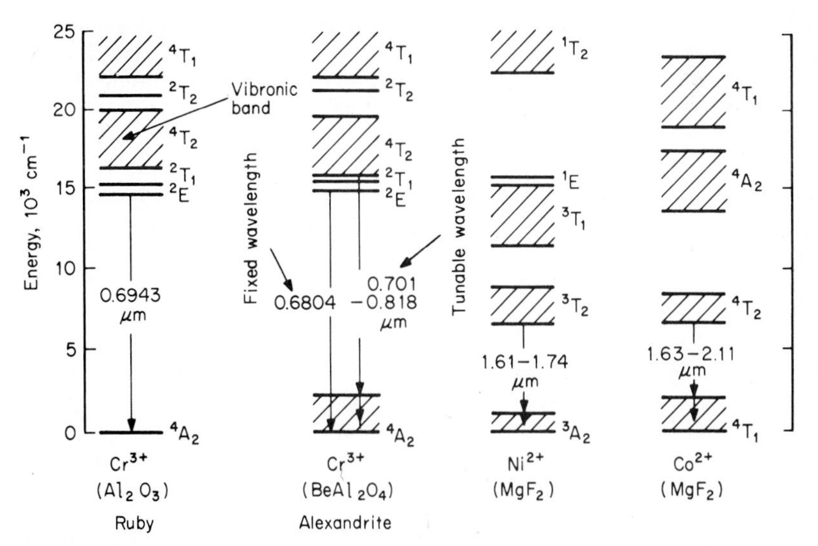

Figure 22-1 Energy levels and laser transitions in ruby, alexandrite, and Ni- and Co-doped MgF_2 lasers, showing the difference between discrete energy-level lasers (ruby) and vibronic lasers (others).

require laser pumping. It is even possible to operate an alexandrite laser and a neodymium-YAG laser in the same cavity, although this does not produce optimal results. Because of major differences in the internal kinetics, energy storage, and gain cross section for the two materials, best performance is obtained from a cavity specifically designed for one material. Developers are still fine-tuning the optical and pumping characteristics of laser cavities to match vibronic laser materials, further improving the performance of vibronic lasers. The differences in design requirements are a major reason why two decades after the demonstration of the first vibronic solid-state laser (Johnson et al., 1963), only a handful were in the hands of laser users.

Internal Workings

Some of the more important of the many vibronic lasers being developed are listed in Table 22-1. In general, their internal workings are similar, although details differ. This chapter will use the best known of the vibronic lasers, alexandrite, as a starting point, then describe major considerations for other vibronic laser materials.

The alexandrite used in laser rods contains about 0.01 to 0.4 percent chromium. Vibration of the chromium atoms in the crystalline matrix creates a set of vibrational energy levels that are superimposed upon electronic states, so transitions between electronic states may be accompanied by vibrational transitions—a so-called vibronic transition (from a contraction of vibrational-electronic). The energy of a vibrational transition is sometimes called a phonon. In vibronic lasers the range of possible vibrational states is so large that they form a continuum of energy levels, as shown in Fig. 22-1.

Alexandrite is pumped by an external source emitting in an absorption band at 380 to 630 nm that is similar to that of ruby. The optical pumping and subsequent fast relaxation processes populate a mixture of electronically excited levels, primarily a short-lived $^4T_2$ level and a longer-lived ^2E level with

TABLE 22-1 Types of Vibronic Solid-State Lasers

Type	Pump	Operation	Wavelengths	Maximum power
Alexandrite	Flashlamp	Pulsed	701–826 nm	100 W average
Alexandrite	Mercury arc	CW	—	50 W
Ce–LaF_3	KrF excimer laser	Pulsed	286 nm	—
Ce–YLF	KrF excimer laser	Pulsed	309–325 nm	—
Co–MgF_2	1.32-μm Nd-YAG laser	50 Hz	1.5–2.3 μm	0.15 J pulses, 7.5 W average
(80 K)	1.32-μm Nd-YAG laser	CW	1.5–2.3 μm	4.3 W CW
Cr–GSGG	Ar or Kr laser	CW	742–842 nm	250 mW
Emerald	Kr laser	CW	728.8–809 nm	320 mW
Ti-sapphire	Ar laser	CW	660–986 nm	—
Ti-sapphire	532-nm doubled YAG laser	Pulsed	—	—
Ti-sapphire	Enhanced flashlamp	Pulsed	—	—
Cr–$KZnF_3$	Kr laser	CW	780–850 nm	50 mW

Note: CW = continuous wave.

slightly lower energy. Together these levels act as an upper laser level with lifetime and emission rate depending on temperature. This combination is unique among vibronic lasers and leads to an increase in laser gain with an increase in temperature, rather than the decrease that is typical of most solid-state lasers. In other vibronic lasers the upper laser level is a single level or vibronic band.

In all vibronic lasers, the laser transition terminates in a vibronic band. In alexandrite, the lower laser level is the lower vibronic band of the ground electronic state, ${}^4A_2$, as shown in Fig. 22-1. The laser transition goes to a point in the middle or upper part of the vibronic band, with vibrational relaxation depopulating those states to maintain a population inversion. The tunability of vibronic lasers comes from the fact that the transition can terminate at many different energy levels within the lower vibronic band.

Because of the nature of the lower laser levels, the emission wavelengths of all vibronic lasers depend somewhat on temperature. As temperature rises, so does population of the lower vibrational levels, making it harder to produce a population inversion between them and the upper laser level. These lower sublevels correspond to shorter-wavelength, more-energetic laser transitions, so a temperature increase shifts emission toward longer wavelengths. Near room temperature, the shift is typically in the range of 0.2 to 0.3 nm/°C.

The emission cross section of alexandrite is only 6×10^{-21} cm^2 at room temperature, an order of magnitude lower than that of neodymium-glass. This leads to much lower gain than in neodymium lasers and makes energy-extraction efficiency a much more important design consideration. However, the lower emission cross section also allows alexandrite to store much more energy per unit volume than neodymium lasers. The picture changes at higher temperatures; at 250°C the peak cross section of alexandrite reaches about 3×10^{-20} cm^2, comparable to that of Nd-glass.

Values of emission cross sections, vibronic bandwidths, and upper-laser-level lifetimes differ among vibronic lasers, but the three factors are interrelated, and all vibronic lasers suffer from the same trade-offs among them. For example, titanium-doped sapphire, which has broader vibronic bandwidth and higher emission cross section than alexandrite, also has a much shorter excited-state lifetime, and thus poorer energy-storage capability. Although Ti-sapphire has been flashlamp-pumped, in general such short lifetimes can make flashlamp pumping inefficient or impossible. Narrow absorption bandwidths also reduce efficiency of flashlamp pumping. Because of such limitations, some vibronic lasers have so far been pumped only with external lasers, often continuous-wave types which can sustain a population inversion even for a short excited-state lifetime. Considerations of cross section and excited-state lifetime enter into design of vibronic lasers and can cause significant differences among them.

Material growth is an important concern for all solid-state lasers. Some vibronic lasers, such as titanium-doped sapphire, are based on crystals which are easy and inexpensive to grow. Others, such as the garnet family of crystals, can be grown by adapting techniques already developed for related materials. In some cases, particularly chromium-doped emerald, large uncertainties remain about the feasibility of crystal growth. Concerns also have arisen about the cost and availability of some materials, notably the scandium used in chromium-doped GSGG (gadolinium scandium gallium garnet, $Gd_3Sc_2Ga_3O_{12}$).

Internal Structure

In alexandrite lasers, as in neodymium and ruby types, a laser rod is mounted in the same reflective optical cavity with a pump lamp, which transfers light to the laser rod. Typical alexandrite rods are 0.3 to 0.7 cm in diameter and 7.6 to 10 cm long; linear or helical flashlamps may be used. It is possible to operate a neodymium laser rod and an alexandrite rod in the same cavity, but the best performance is obtained if the cavity is designed specifically to match the low-gain, high-storage characteristics of alexandrite. A multipass amplifier configuration is particularly attractive for extracting energy from alexandrite.

Alexandrite is unusual in that its laser performance improves as temperature rises beyond room temperature, as shown in Fig. 22-2. The performance of most lasers degrades with increasing temperature. The improvements in alexandrite are due to its internal energy-transfer kinetics and to the material's good thermal properties, which make it less subject to thermal lensing than many other solid-state laser materials. This also lets alexandrite operate continuously with a suitable pump lamp.

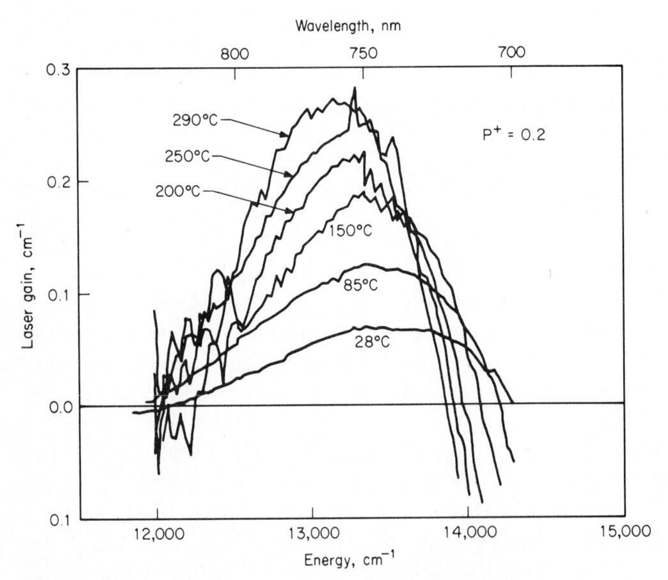

Figure 22-2 Gain of alexandrite as a function of wavelength for various temperatures, showing how value of gain increases and wavelength of peak gain increases slightly with temperature. *(From Shand and Jenssen, 1983, © 1983 IEEE. Reproduced with permission.)*

Optimum laser cavity design differs for other vibronic lasers, depending on energy-transfer kinetics. As mentioned earlier, some require pumping with an external continuous-wave laser. A few, notably cobalt- and nickel-doped magnesium fluoride, must be cooled well below room temperature to create favorable energy-transfer kinetics because excited-state lifetime drops dramatically as temperature increases.

The cavities used with vibronic lasers are typically somewhat longer than the laser rods, but because the lasers are so new, design rules have yet to be firmly established. Low-gain materials such as alexandrite are less suitable for traditional unstable-resonator designs than high-gain materials such as neodymium. As with other solid-state lasers, thermal lensing effects must be considered to obtain best performance from vibronic lasers.

Inherent Trade-offs

As in other tunable lasers, the gain of vibronic lasers is a function of wavelength, with a peak near the center of the laser band. Gain, and hence output power, drop off at both longer and shorter wavelengths, as shown in Fig. 22-2.

As mentioned earlier, certain trade-offs affect the choice of vibronic laser materials. Three fundamental characteristics are interrelated: excited-state lifetime (which determines energy storage and other characteristics), tuning bandwidth, and stimulated emission cross section. For example, Ti-sapphire offers a gain bandwidth 2.5 times that of alexandrite, and an emission cross section about 10 times higher. However, the lifetime of the upper laser level is 3 microseconds (μs), much lower than that of alexandrite, making Ti-sapphire less suitable for conventional flashlamp pumping. Although different materials offer different sets of characteristics, none can offer the most desirable values for all three characteristics.

Variations and Types Covered

The past few years have seen a rapid proliferation of solid-state vibronic lasers. A comprehensive listing of all types under investigation is beyond the scope of this book, although at least one review of the field has been prepared (Moulton, in press). However, it is possible to list some of the most promising types, whose laser characteristics are summarized in Table 22-1.

- Alexandrite, Cr-doped $BeAl_2O_4$: The archetype of the family, developed by the Allied Corp., which has pushed strongly to make it a commercially viable laser. The tunable output can reach high powers when pumped with a suitable pulsed or continuous lamp. Crystal growth requires care but appears tractable.
- Emerald, Cr-doped $Be_3Al_2(SiO_3)_6$: Offers attractive laser characteristics, including the potential of higher gain than alexandrite, but suffers from very serious problems in growing the crystals (Shand and Lai, 1984).
- Cr-doped GSGG ($Gd_3Sc_2Ga_3O_{12}$): Output properties are attractive (Struve and Huber, 1984), but so far only two groups have been able to pump with

a flashlamp. Others have used a 647-nm krypton ion laser for continuous-wave pumping. Growth of the garnet-type crystal appears tractable, and this material has become the second vibronic laser to be offered commercially. However, the need for scandium could be a concern because of its high cost and present limited availability.

- Cr-doped Perovskite ($KZnF_3$): A four-level system continuously tunable between 780 and 850 nm with peak output power of 50 mW. Pump source is the 647-nm krypton ion laser line.
- Ti-doped sapphire ($Ti^{+3}:Al_2O_3$): One of the newest and most promising materials (Moulton, 1984, and in press). Its tuning range and emission cross section are both much larger than those of alexandrite, but its short upper level lifetime prevents conventional long-pulse flashlamp pumping, although it can be pumped by flashlamps augmented with fluorescence converters. Continuous-wave pumping is with an argon ion laser; laser output can also be produced when pumping with frequency-doubled pulses from a neodymium-YAG laser. Crystal growth is considered relatively easy, and this is considered a particularly promising type.
- Co-doped MgF_2: Optical pumping with a 1.3-μm laser produces output tunable between 1.5 and 2.3 μm. The output characteristics are attractive, but this type can operate only at temperatures well below room temperature. The highest reported operating temperature is 225 K, compatible with thermoelectric cooling, but laboratory experiments normally are done at the 77 K temperature of liquid nitrogen for convenience. Some related lasers have been demonstrated, including nickel in MgF_2 and in a complex garnetlike crystal (Moulton, 1982a).
- Cerium-doped materials (lathanium fluoride, LaF_3, and yttrium lithium fluoride, YLF): Pumping with an excimer laser has produced tunable pulses in the ultraviolet (Ehrlich et al., 1978 and 1980), but the material has received little further attention.

Because alexandrite is the best characterized member of the vibronic laser family, the rest of this chapter will use it as an example of vibronic laser characteristics, noting differences from other types when that information is available. However, many vibronic lasers are only incompletely understood, and new research results could lead to important changes in their performance. New types also are likely to emerge.

Beam Characteristics

Wavelength and Output Power Alexandrite has emitted at wavelengths between 700 and 826 nm in the laboratory, although the entire range may not be possible under the same conditions. Commercial alexandrite lasers so far have had much more limited tuning, between about 720 and 780 nm, or operate untuned at the peak emission wavelength of 755 nm. These commercial versions are rated to produce 0.2- to 0.8-joule (J) pulses at repetition rates to 10 Hz, corresponding to average power as high as several watts. Higher repetition rates and pulse energies have been achieved in the laboratory, and average power in repetitive long pulses has approached 100 W (Heller and Walling, 1984). Continuous-wave output, not available commercially as of mid-1984, has passed 50 W with mercury lamp pumping in the laboratory (Samelson and Harter, 1984).

Other vibronic lasers began appearing on the market in packaged form in 1984. The Cr-doped GSGG laser was followed by Cr-doped Perovskite. Crystal suppliers also have begun producing vibronic laser rods. New types are likely to continue appearing.

Table 22-1 lists output wavelengths and powers that have been obtained from selected other vibronic lasers. Because these figures are based on laboratory demonstrations, they could well change as researchers gain a better understanding of the laser materials.

Efficiency Overall electrical-to-optical efficiencies in the 1 percent range have been reported for alexandrite. Slope efficiency of alexandrite—the ratio of the increase in output power to the increase in input electrical power above laser threshold—has passed 5 percent. Optical slope efficiencies of 34 percent for continuous-wave emerald pumped by a krypton laser and of 50 percent for continuous-wave alexandrite have been reported (Shand and Lai, 1984), but this refers only to conversion of optical pump energy, not overall efficiency. Quantum efficiency of nearly 100 percent has been reported for laser-pumped Co–MgF_2 (Moulton, 1982a), meaning that almost all the photons from the pump laser led to emission of photons from the Co–MgF_2. However, overall electrical-to-optical efficiency in both the latter two cases was much lower, probably lower than that of alexandrite, given the low efficiencies of the pump lasers.

Temporal Characteristics of Output Alexandrite can operate continuously in long pulsed mode with 100- to 400-μs pulses, or in *Q*-switched mode with output in 30- to 200-ns pulses. Repetition rates of commercial *Q*-switched alexandrite lasers range between one pulse every 15 s to 10 pulses per second; repetition rates a few times higher have been demonstrated in the laboratory.

Spectral Bandwidth Spectral linewidths of commercial alexandrite lasers are 0.2 to 0.5 nm for tunable models and 3 nm for untuned models. Linewidth of about 0.001 nm has been achieved in an alexandrite laser custom-built for narrow linewidth (Walling, 1982). Functionally the spectral bandwidth of tunable vibronic lasers depends on the cavity optics, which vary widely among laboratory-built demonstration lasers.

Suitability for Use with Laser Accessories Alexandrite lasers can be *Q*-switched readily, producing the high peak power pulses required for nonlinear optics, harmonic generation, and Raman frequency shifting. They also are suitable for oscillator-amplifier operation. Alexandrite has been modelocked actively and passively, and modelocking of two other vibronic lasers, Co–MgF_2 and the related nickel-doped MgF_2, has been demonstrated (B. C. Johnson et al., 1984), but little work has been done with other vibronic lasers.

In general vibronic lasers with long excited-state lifetimes and good energy storage should be amenable to *Q* switching. They should be able to produce pulses with high peak power useful for harmonic generation, Raman frequency shifting, and nonlinear processes. Harmonic generation and Raman shifting have been demonstrated with alexandrite, but most current emphasis is going to development of the lasers, with comparatively little attention devoted to their use with laser accessories.

Operating Requirements

Operating requirements for vibronic lasers have not been well-quantified because most of them remain in development. Present commercial models come with data sheets which do not list such quantities as input power and cooling requirements. Such characteristics as lifetimes, maintenance needs, and durability also remain to be established.

Input Power Vibronic lasers require optical input intense enough to drive them above laser threshold. Flashlamps or continuous-output mercury lamps are typical incoherent pump sources. Lasers that have been used for pumping of vibronic lasers include continuous-wave argon, krypton, and 1.3-μm Nd–YAG; near-ultraviolet vibronic lasers have been pumped with excimer laser pulses. Current commercial alexandrite lasers use power supplies drawing 1 to 10 kW of electrical power.

Cooling Most vibronic lasers operate at room temperature and may require water cooling for operation at high average powers. A few, notably Co–MgF_2 and Ni–MgF_2, operate only well below room temperature and require cooling with liquid nitrogen or thermoelectric coolers; for convenience, liquid nitrogen is the usual choice in the laboratory.

Consumables Pump lamps are the only obvious items that could be considered consumables with vibronic lasers, other than perhaps coolants. Lamp lifetime is 10^6 to 10^8 shots, depending on operating conditions, which is similar to that with other solid-state lasers pumped by flashlamps.

Safety Users of vibronic lasers at wavelengths shorter than 1.4 μm should obey the usual laser-safety precautions for lasers with wavelengths able to penetrate to the retina of the eye. They should also be aware of the following special hazards:

- Short, powerful *Q*-switched pulses are especially dangerous because even a single pulse can cause permanent eye damage.
- Because vibronic lasers are not widely available, no standard safety goggles are designed for use with them. Standard safety goggles designed for use with other lasers may not block all the wavelengths emitted by vibronic lasers, so careful checking of specifications is crucial. This is particularly true if the

laser is to be tuned in wavelength, because such tuning could take the emission to a part of the spectrum not blocked by the goggles.

- Seemingly faint red beams may not really be low in power. As wavelength increases beyond about 650 nm, retinal sensitivity drops off logarithmically, as shown in Fig. 22-3. Although 700 nm is often given as the long-wavelength end of the visible spectrum, the eye can perceive longer wavelengths, but only very inefficiently. Thus what the eye sees as a faint red beam might instead be a much more intense beam at 780 or 800 nm. Care is essential because even near-infrared radiation which cannot be perceived by the eye could cause permanent damage to the retina (Sliney and Wolbarsht, 1980).

The eye is much less vulnerable to wavelengths longer than 1.4 μm because they cannot penetrate far enough into the eyeball to reach the retina. However, intense infrared radiation at such longer wavelengths could burn the outer eye just as it could burn the skin. Long-term effects of exposure to infrared radiation at high to moderate intensities are not well understood.

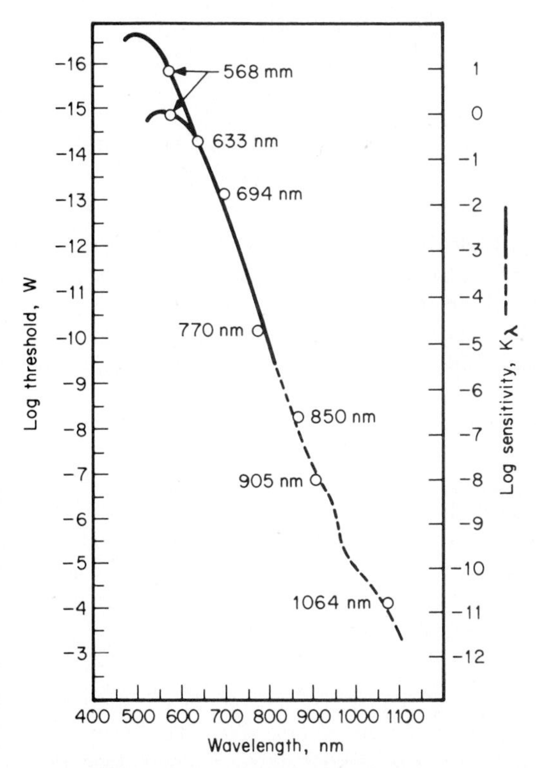

Figure 22-3 Spectral sensitivity of the eye as a function of wavelength in the near-infrared. Response drops off logarithmically as wavelength increases. *(From Sliney and Wolbarsht, 1980, with permission.)*

Commercial Devices

As of mid-1985, commercial vibronic lasers were few and far between. Laser crystals were coming on the market, but packaged lasers, ready for use by people not interested in developing new vibronic lasers, were offered by only two companies. Most new vibronic lasers remained in the development laboratory and seemed likely to stay there for a while. Some developers have said that Ti-sapphire may be an exception that reaches the market soon. However, market-dynamic forces, such as the association of particular research groups with commercial laser makers, probably will play an important role in those decisions as well. Indeed, it was such an association that brought alexandrite to the market first.

Prices for vibronic lasers are not well established because both sales and production have been small. Complete alexandrite laser systems carry prices above $50,000, although stripped-down heads have been advertised for as little as $10,000. Alexandrite is commercially unique in that a single company, the Allied Corp., holds patents that dominate the technology. The patent situation is less clear with other materials, but it seems unlikely that any type will be as completely controlled by a single company as alexandrite.

Applications

Because their technology is new, most present applications of vibronic lasers are in research and development. However, a number of other potential applications have been suggested for alexandrite:

- Materials working, particularly drilling holes in materials such as copper which strongly reflect the longer-wavelength output of neodymium lasers
- Annealing of semiconductor devices
- Spectroscopy, using alexandrite as a source of tunable, narrow-band light. Harmonic generation and Raman shifting can shift alexandrite's output wavelengths through the visible and into the near-ultraviolet
- Laser radar and atmospheric measurements from space, air, or the ground
- Pumping of dye lasers
- Medical applications requiring light at near-infrared wavelengths not obtainable at comparable power levels from other lasers
- Military applications such as wavelength-agile laser range finders and target designators. By shifting wavelength, they could foil laser countermeasures which rely on a precise knowledge of laser wavelength to detect laser illumination and possibly direct fire against the laser
- Photochemistry and isotope enrichment requiring specific wavelengths emitted by vibronic lasers

The potential for applications of other vibronic lasers, other than in research and development, remains to be established. Possibilities could include materials working, spectroscopy, and medicine. Vibronic lasers emitting at wavelengths longer than 1.4 or 1.5 μm could be attractive to the military for use as eye-safe laser range finders and target designators—a topic of increasing interest because of problems encountered in training soldiers to use lasers that require eye protection.

BIBLIOGRAPHY

D. J. Ehrlich, P. F. Moulton, and R. M. Osgood Jr.: "Optically pumped Ce:LaF_3 Laser at 286 nm," *Optics Letters 5*:339, 1980.

D. J. Ehrlich, P. F. Moulton, and R. M. Osgood Jr.: "Ultraviolet solid-state Ce:YLF laser at 325 nm," *Optics Letters 3*:184, 1978.

Donald F. Heller and John C. Walling: "High-power performance of alexandrite lasers," paper WI4 at Conference on Lasers & Electro-Optics, Anaheim, Calif., June 19–22, 1984.

B. C. Johnson et al.: "High average power mode-locked operation of Co:MgF_2 and Ni:MgF_2 lasers," paper WC2 at Conference on Lasers & Electro-Optics, Anaheim, Calif., June 19–22, 1984.

L. F. Johnson, R. E. Dietz, and H. J. Guggenheim: "Optical maser oscillation from Ni^{+2} in MgF_2 involving simultaneous emission of phonons," *Physical Review Letters 11*:318, 1963.

Peter F. Moulton: "Pulse-pumped operation of divalent transition-metal lasers," *IEEE Journal of Quantum Electronics QE-18*(8):1185–1188, August 1982a.

Peter F. Moulton: "Paramagnetic ion lasers," in Marvin J. Weber (ed.), *Handbook of Laser Science & Technology,* vol. 1, *Lasers & Masers,* CRC Press, Boca Raton, Fla., 1982b, pp. 21–146.

Peter F. Moulton: "New developments in solid-state lasers," *Laser Focus 19*(5):83–88, March 1983.

Peter F. Moulton: "Recent advances in solid-state lasers," paper WA2 at Conference on Lasers & Electro-Optics, Anaheim, Calif., June 19–22, 1984.

Peter F. Moulton: "Tunable paramagnetic-ion lasers," in Michael Bass and Malcolm Stitch (eds.): *Laser Handbook,* vol. 4, North Holland, Amsterdam, in press.

Harold Samelson and D. J. Harter: "High-pressure mercury-arc-lamp-excited continuouswave alexandrite lasers," paper WI5 at Conference on Lasers & Electro-Optics, Anaheim, Calif., June 19–22, 1984.

Michael L. Shand and H. P. Jenssen: "Temperature dependence of excited-state absorption of alexandrite," *IEEE Journal of Quantum Electronics QE-19*:480–484, March 1983.

Michael L. Shand and S. T. Lai: "CW laser pumped emerald laser," *IEEE Journal of Quantum Electronics QE-20*(2):105–108, February 1984.

David L. Sliney and Myron Wolbarsht: *Safety with Lasers and Other Optical Sources,* Plenum Press, New York, 1980.

B. Struve and G. Huber: "Tunability of the Cr^{3+}:GSGG laser," paper WI2 at Conference on Lasers & Electro-Optics, Anaheim, Calif., June 19–22, 1984.

Topical Meeting on Tunable Solid State Lasers, Arlington, Va., May 16–17, 1985, *Technical Digest,* Optical Society of America, Washington, D.C.

John C. Walling: "Alexandrite lasers: physics & performance," *Laser Focus 18*(2):45–50, February 1982.

John C. Walling, D. G. Peterson, and R. C. Morris: "Tunable CW alexandrite laser," *IEEE Journal of Quantum Electronics QE-16*:1302, 1980.

CHAPTER

23

color-center lasers

Color-center lasers are types in which optical pumping of a crystal generates wavelength-tunable output in the near infrared. They sometimes are called *F*-center lasers from the German word for color, *Farbe*. Some of their characteristics resemble those of continuous-wave dye lasers, but there also are important differences. Color-center lasers have an intrinsically narrow spectral linewidth that makes them attractive for spectroscopy. Experimental versions have operated near 0.4 micrometer (μm) and at 0.8 to 4.0 μm, but commercial models operate in a more restricted range, near 1.5 μm and beyond 2 μm.

Continuous-wave color-center lasers differ markedly from the family of flashlamp-pumped crystalline or glass solid-state lasers described in Chaps. 20 to 22 and 24. The active medium is a thin crystal, into which point defects have been introduced deliberately. The resulting microscopic defects in the crystals tend to absorb light, coloring the normally colorless crystals and earning the name color centers. Color centers are found in many types of normally colorless crystals. Laser action in color centers, first seen in 1965 (Fritz and Menke, 1965) is mostly in alkali halide crystals.

Like continuous-wave dye lasers, color-center lasers require pumping with another laser. The usual choices are ion, dye, or neodymium-YAG lasers.

Normally operation is continuous, but color-center lasers also can be synchronously pumped by modelocked lasers. Pulsed operation of color-center lasers has been demonstrated experimentally with both laser and flashlamp pumps. However, only continuous-wave and modelocked versions are marketed.

Internal Workings

Active Medium The active media in color-center lasers are crystals doped with selected impurities to generate the desired color-center (or *F*-center) flaws. The crystals then are processed in ways that cause holes in the lattice (which are filled by free electrons instead of by negative ions) to come together with positively charged impurities, which generally are smaller than the positive ions in the lattice. In some defects, a single electron occupies each hole; in others two adjacent holes share a single electron, giving the *F*-center a positive charge. These color centers tend to absorb light in the visible or near-infrared, an effect used to pump color-center lasers.

The type of color center depends on the atomic configuration of the defect, and that, in turn, depends on the type of crystal and the nature of the impurity. Figure 23-1 shows the normal configuration of one type of color center, the F_A(II) center, and its configuration after it has absorbed light. This type is formed only when lithium is added to crystals of potassium chloride or rubidium chloride. Figure 23-2 shows a second, similar type, the F_B(II) center, formed in KCl or RbCl crystals doped with sodium. These two types can operate as lasers, with emission wavelengths between 2.2 and 3.65 μm, although each center and crystal operates over a more limited range. Related Tl^0(I) color centers rely on doping alkali halides with thallium, producing Tl^+ defects in the crystals. These defects consist of a Tl^+ impurity adjacent to an anion vacancy, with an electron bound to the complex. Lasers based on this defect have shorter emission wavelengths, tunable between about 1.4 and 1.7 μm (Gellermann et al., 1982).

Another type of color center, the F_2^+ center, can produce laser emission at visible and near-infrared wavelengths. The F_2^+ center forms when two adjacent holes share a single electron; it is positively charged because the region contains one electron less than it should to balance the charge of the positive ions.

Color centers can form only under restricted conditions, and laser operation is possible only in a more limited range. Commercial models operate at the 77 K temperature of liquid nitrogen, with the low temperature essential because the fluorescence quantum efficiency decreases as the temperature increases. The defects in the crystals are stable at temperatures as high as about −10°C. The type II color centers gradually dissociate if the crystals are stored at room temperature, but no permanent damage is done and the color centers can be regenerated. Type II centers can be regenerated even if the crystal is exposed to light at room temperature. F_2^+ centers are harder to produce and stabilize than F_A(II) and F_B(II) centers (Gellermann et al., 1982).

The first commercial color-center lasers relied on the stabler F_A(II) and F_B(II) centers. Tl^0(I) lasers have been introduced more recently.

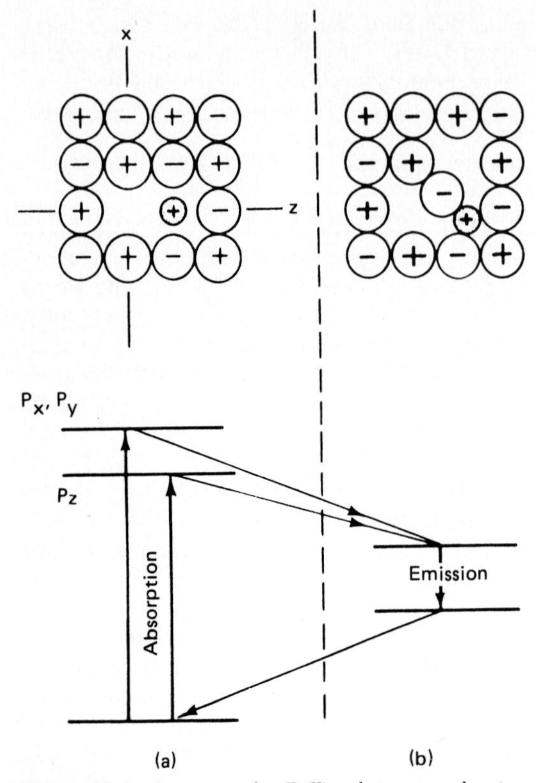

Figure 23-1 Structure of a F_A(II) color center showing *(a)* normal (vacancy) configuration and *(b)* configuration in upper laser level. The energy-level structure is shown at bottom. *(From Mollenauer, 1982, with permission.)*

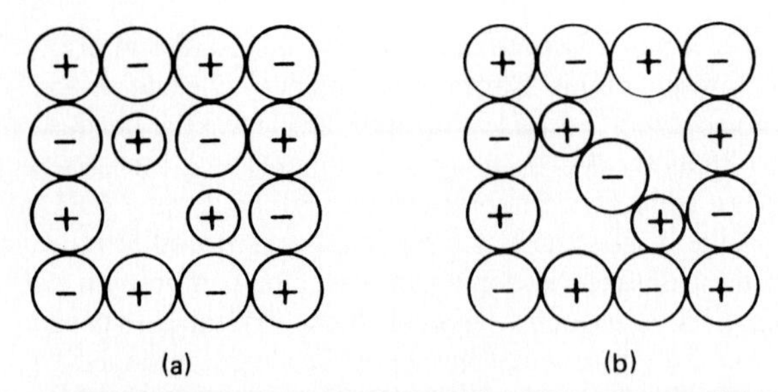

Figure 23-2 Structure of a F_B(II) color center, showing *(a)* normal configuration and *(b)* excited configuration. *(From Mollenauer, 1982, with permission.)*

Energy Transfer Color-center lasers are pumped by an external laser with output wavelength falling into the color center's absorption band. Figure 23-3 shows the pump bands of F_A(II) and F_B(II) center crystals in commercial lasers. The pump band of the Tl^0(I) center is at slightly longer wavelengths and includes the 1.06-μm output wavelength of the Nd–YAG laser.

Absorption of the pump photon excites the atoms at the defect site, causing them to rearrange themselves as shown in Figs. 23-1 and 23-2. The rearrangement, accompanied by a radiationless relaxation to a slightly lower overall energy level, puts the defect site in the upper laser level. The atoms stay in this rearranged state long enough to be stimulated to emit light on a transition to a lower energy level of the rearranged state. From that lower laser energy level, they then relax to the original configuration. As indicated at the bottom of Fig. 23-1, this corresponds to the standard four-level laser scheme, and it can operate quite efficiently.

The multitude of possible vibrational states in the crystal leads to homogeneous broadening of the laser transition. The broadening is large enough to permit tuning of the emission wavelength over a comparatively broad range, with total range about 15 to 20 percent of the central wavelength.

Internal Structure

Commercial color-center lasers are pumped longitudinally with a continuous beam (or modelocked pulse train) using the type of optical cavity shown in Fig. 23-4. The design bears a general resemblance to that used for many continuous-wave dye lasers. The beamsplitter reflects the pump wavelength through the laser crystal but transmits the color-center emission to the tuning cavity. Part of the pump beam also is transmitted by the beamsplitter and ultimately emerges through the output window along with the color-center laser beam.

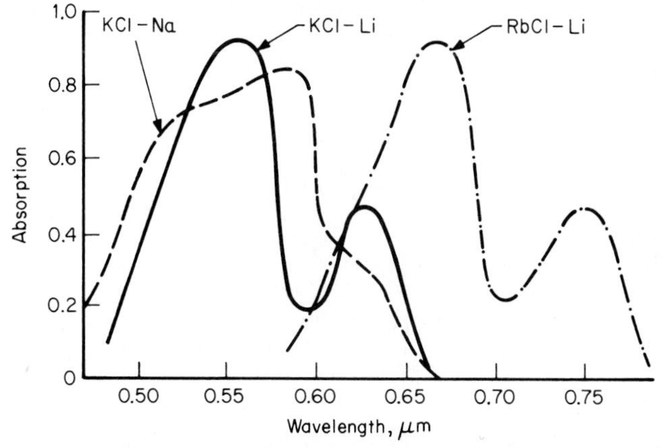

Figure 23-3 Absorption bands of three color-center crystals used in commercial laser. *(Courtesy of Burleigh Instruments Inc.)*

The color-center crystal is housed in a chamber that cools it to 77 K; the same chamber houses a pair of curved mirrors that direct the pump beam through the crystal. Wavelength is tuned by adjusting the angle of a grating in the tuning arm. The first-order reflected beam from the grating is fed back into the laser cavity; the zeroth-order beam is coupled out the output window as the laser beam. The grating design keeps direction of the output beam constant to within 0.25 milliradian (mrad) as wavelength is tuned by tilting the grating.

At low pump power, a color-center laser tends to emit a single longitudinal mode even without an intracavity etalon. However, as pump power increases, spatial hole burning within the crystal can produce a second longitudinal mode. Insertion of an etalon within the cavity can eliminate the second mode and assure reliable single-frequency operation.

Both crystal chamber and tuning arm are sealed, and when the laser is in operation they are evacuated to low pressures. In addition to helping avoid condensation on the crystal and other components at the low operating temperatures, this helps avoid atmospheric absorption. Water vapor in the atmosphere absorbs strongly at 2.5 to 3.0 μm, and other strong atmospheric bands lie at longer and shorter wavelengths; excessive absorption could reduce laser output or prevent lasing altogether.

The beam passes through only a small part of the crystal, a thickness of about 2 mm, and the crystal itself is oriented at Brewster's angle. The result is linear polarization of the pump and color-center beams that helps separate them at the beamsplitter that couples in the pump beam. The entire color-center laser is fairly compact, and the total length of the folded cavity is about half a meter.

Variations and Types Covered

Commercial color-center lasers are only a small subset of the color-center lasers that have been demonstrated experimentally. Laser action has been demonstrated in several types of color centers in a variety of alkali halides.

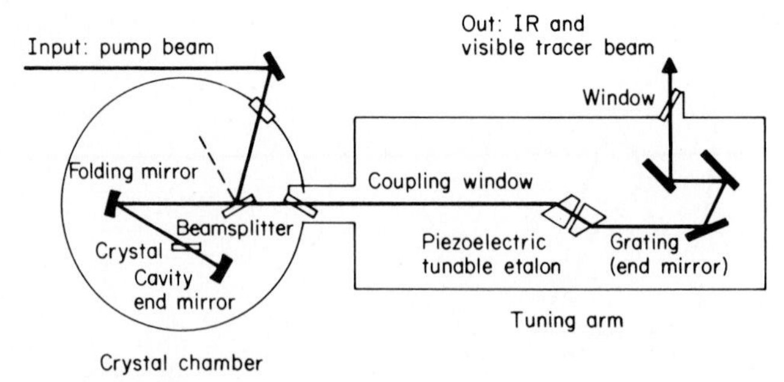

Figure 23-4 Internal structure of commercial color-center laser, showing configuration with an intracavity etalon to limit oscillation to a single frequency. *(Courtesy of Burleigh Instruments Inc.)*

Operation of F_2^+ center lasers has been demonstrated in LiF, NaF, KF, NaCl, KCl, KBr, and KI, as well as in sodium-doped KCl and RbCl and Li-doped KCl (Mollenauer, 1982). F^+ color centers in calcium-oxide crystals have produced laser output at 357 to 420 nm when pumped by pulsed or continuous-wave ultraviolet lasers (Henderson, 1981), but like many other results, that has been obtained only in experimental demonstrations.

Most color centers which can lase tend to photodegrade under illumination by a pump laser. Only the type II centers and the Tl°(I) center are stable.

Developers of commercial models concentrate on only a few color-center lasers that are stable enough to meet customer needs. This chapter concentrates on the commercial part of the color-center laser world.

Beam Characteristics

Wavelength and Output Power Figure 23-5 shows the tuning ranges and output powers available at wavelengths longer than 2 μm from a commercial color-center laser without an etalon. Three crystals can be used in the laser, with each providing tuning over part of the range. Insertion of an etalon into the laser cavity to restrict oscillation to a single frequency reduces the output power to 70 percent at the peak wavelength. If a KCl crystal doped to produce Tl°(I) centers is pumped with a 2-W YAG laser, the color-center laser can be tuned between 1.43 and 1.58 μm, with peak output over 100 mW at 1.51 μm in either a single mode or two modes.

Efficiency Power-conversion efficiencies of color centers range from under 0.1 percent off the peak emission wavelength for inefficient materials to 5

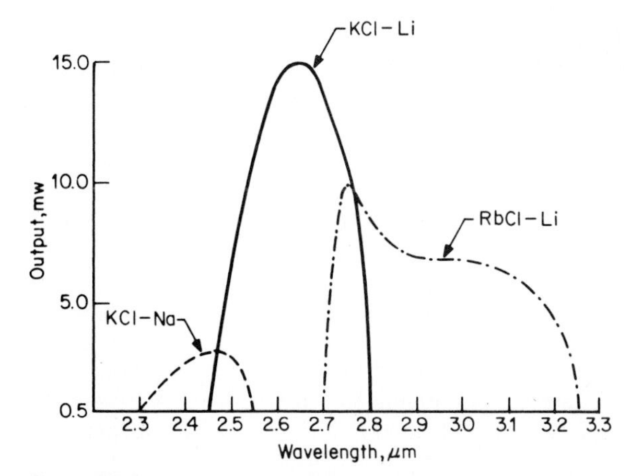

Figure 23-5 Tuning curves for the three color-center crystals used to produce wavelengths longer than 2 μm in a commercial laser, along with output power as a function of wavelength. Pump powers were 3 W at 488 and 514 nm for KCl–Na; 1 W at 647 nm for KCl–Li; and 1 W at 647 and 676 nm for RbCl–Li. *(Courtesy of Burleigh Instruments Inc.)*

percent at the peak of the most efficient types. Efficiency varies considerably between materials and is highest for F_2^+ center materials.

Temporal Characteristics Commercial color-center lasers normally operate continuously, as do their pump lasers. Modelocked pulses can be generated by synchronous pumping with a modelocked laser; pulses of 20 to 30 picosecond (ps) duration are produced at repetition rates of 82 to 100 megahertz (MHz), depending on the pump laser.

Spectral Bandwidth Individual longitudinal modes of color-center lasers have linewidths of about 1 MHz. If the laser operates in two longitudinal modes, each has 1-MHz linewidth, but the two are separated by about 1.5 GHz. In modelocked operation, the laser operates over a band of equally spaced longitudinal modes, with total linewidth on the order of 20 GHz. Active stabilization of laser frequency is possible and can reduce linewidth to 20 kHz.

Beam Quality, Polarization, and Modes Output beams are in a single transverse mode, 1.5 to 2 mm in diameter. Divergence is 1.5 to 2.5 mrad. Both figures vary with the square root of wavelength. The laser output is linearly polarized.

Operating Requirements

Input Power A pump laser with continuous output of one to a few watts is needed, with output wavelength in the visible or near-infrared matched to a color-center absorption band. The standard types used are argon ion, krypton ion, dye, and neodymium-YAG. Equivalent powers are needed for synchronously pumped modelocked operation.

Electrical input to 10 A from a 110- or 220-V source is needed to drive accessories including a vacuum pump, but not counting the pump laser.

Cooling The color-center laser crystal and its pumping chamber are cooled by a dewar which holds a two-day supply of liquid nitrogen, to maintain operation at 77 K. When the laser is not operating, the crystal must be stored in a refrigerator.

Required Accessories A vacuum pump with a cryogenic trap is needed to evacuate both the laser crystal cavity and the tuning arm to low pressure for laser operation. An intracavity etalon is needed for single-frequency operation. Various controls and measurement instruments may also be needed, depending on the application, including wavelength meters and spectrum analyzers.

Operating Conditions and Temperature Cryogenic operation is needed to avoid damage to color-center crystals. Some brief initial preparation may be required to regenerate color centers after the crystal has been stored. No degeneration occurs at cryogenic temperatures. Precautions also must be taken in shutting down the laser to avoid exposing the crystal to conditions that might cause damage. Details are given in manufacturers' instruction manuals (Burleigh, 1984.)

Mechanical Considerations The color-center laser itself measures 72 cm long, 23 cm wide, and 47 cm high. The height is largely due to the liquid-nitrogen dewar; the rest of the body of the laser is only about 15 cm high. The entire assembly weighs about 50 kg. (Those figures do not include the vacuum pump or pump laser needed for operation of a color-center laser.)

Safety Output wavelengths of commercial color-center lasers are in the "eye-safe" region beyond 1.4 μm where absorption within the eye prevents significant fractions of input light from reaching the retina. However, the output beam from a color-center laser includes a small fraction of the pump beam, which is at a shorter wavelength that is not in the eye-safe region. It would be wise to take the same precautions that would be used if the pump laser were in the laboratory by itself, not pumping the color-center laser.

Because of the cryogenic cooling, parts of the color-center laser are cold enough to freeze flesh. The liquid-nitrogen coolant poses similar hazards. However, in commercial models the very cold components normally are inside a narrow-necked dewar.

Special Considerations The output beam from a color-center laser includes a small part of the pump beam. If the pump beam is visible, it makes the entire output beam visible, a feature that can be useful in alignment. However, to assure spectral purity it may be desirable to filter out the pump wavelength at some point outside the laser cavity. A piece of silicon could block all pump laser wavelengths but transmit all *F*-center wavelengths.

Reliability and Maintenance

Color-center lasers are built as precision laboratory instruments and should be durable if operated in the manner recommended by the manufacturer. However, the laser crystal is particularly vulnerable to improper operation and handling. Instruction manuals give detailed directions on how to prepare the laser for operation, how to use it, and how to shut it down. It is essential to pay heed to those directions, particularly because the procedures differ markedly from those for other lasers.

Commercial Devices

Standard Configurations All commercial color-center lasers are variations on a single basic design, with the crystal chamber housed in a cylindrical dewar. That assembly and the tuning arm are mounted on a rigid metal base about three-quarters of a meter long. The pump laser and vacuum pump must be mounted separately.

Options Color-center lasers are offered separately or as part of integrated systems that include vacuum pump and other accessories. Specific pump lasers are recommended by the color-center laser manufacturer, although others with similar characteristics can be used. A sampling of other major options includes

- Synchronously pumped modelocked operation
- An intracavity etalon for single-frequency operation
- Accessories for wavelength tuning
- Measurement accessories

Pricing Current (1985) prices of commercial color-center lasers start in the $30,000 range and go upward, depending upon the types of options and accessories chosen.

Suppliers Only a single company actively produces and markets color-center lasers, an indication both of the limited market and of the technical sophistication needed to produce a viable product. Experimenters studying the physics of color-center lasers have built their own, but that is not an attractive prospect for the user not deeply involved in color-center physics.

Applications

Virtually all applications of color-center lasers are in spectroscopy or closely related fields which require modest powers and extreme spectral purity. Major applications have been in high-resolution molecular spectroscopy, involving both linear absorption measurements and some nonlinear studies. Intracavity measurements are possible (Schrepp et al., 1984) and are attractive where enhanced sensitivity is needed. The narrow linewidth of color-center lasers permits selective stimulation of excited states for chemical kinetics studies. The infrared wavelengths are useful in solid-state spectroscopy, particularly of semiconductors.

The $Tl^0(I)$ center laser emitting in the 1.5-μm region is useful for studies of optical fibers in this region, where the loss of conventional silica glass fibers is lowest. Modelocked models can be used to generate short pulses for measurements of pulse dispersion in fibers, an important issue in the 1.5-μm region.

Some research has been done on the prospect of using highly stabilized color-center lasers as frequency standards. Stabilized bandwidths below 20 kHz have been reported in the laboratory, and even lower bandwidths may be possible.

Some commercial applications for color-center lasers have been investigated, including evaluation of infrared detectors and of the optical characteristics of infrared systems.

BIBLIOGRAPHY

B. Fritz and E. Menke: "Laser effect in CCl with F_A(Li) centers," *Solid State Communications* 3:61, 1965.

Werner Gellerman, Klaus P. Koch, and Fritz Luty: "Recent progress on color-center lasers," *Laser Focus 18*(4):71–75, April 1982.
B. Henderson: "Tunable visible lasers using F_+ centers in oxides," *Optics Letters 6*(9):437–439, September 1981.
Instruction Manual Tech Memo for F-Center Lasers, Burleigh Instruments, Fishers, N.Y., 1984. Detailed description of the characteristics and operation of commercial color-center lasers. Burleigh also distributes a bibliography of about 100 articles on color-center lasers.
Linn F. Mollenauer: "Color center lasers," in Marvin J. Weber (ed.), *CRC Handbook of Laser Science & Technology,* vol. 1, *Lasers and Masers,* CRC Press, Boca Raton, Fla., 1982, pp. 171–178.
Linn F. Mollenauer and D. H. Olson: "Broadly tunable lasers using color centers," *Journal of Applied Physics 46*:3109, 1975.
I. Schneider: "Continuous tuning of a color center laser between 2 and 4 μm," *Optics Letters 7*:271–273, June 1982.
W. Schrepp, H. Figger, and H. Walther: "Intracavity spectroscopy with a color-center laser," *Lasers & Applications 3* (7):77–79, July 1984.

CHAPTER

24

other solid-state lasers

Far more solid-state lasers have been demonstrated experimentally than have ever received any kind of commercial development, let alone reached the market status of those described in Chaps. 20 to 23. Technical reviews (e.g., Weber, 1982) list hundreds of solid-state laser lines produced from dopants in crystalline and glass hosts. Laser emission has been produced from crystals in which the light-emitting species is an integral part of the crystalline compound, a "stoichiometric laser" (Chinn, 1982). The most extensive work has gone into developing hosts for the neodymium ion, but many other active species have been studied in a variety of hosts.

Most of these lasers have never gone much further than experimental demonstration and initial evaluation. Many suffer from serious inherent weaknesses, including low output power, high laser threshold, cryogenic cooling requirements, problems with crystal growth or quality, or poor heat dissipation. Others simply offer no compelling advantages over types already developed. Although military agencies have invested much effort in studies of prospective new solid-state lasers, no alternative materials have received anything remotely approximating the investment in the neodymium laser, particularly Nd–YAG.

Nonetheless, a few alternative types have received enough development to approach the threshold of commercial availability, although none could be considered a standard product yet. They are based on rare earth ions similar to neodymium: holmium, emitting at 2.06 micrometers (μm); crystalline erbium lasers at 0.85, 1.22, and 1.73 μm; and erbium-glass at 1.54 μm.

A major factor pushing development has been military interest in "eye-safe" laser range finders. The 1.06-μm line of Nd–YAG lasers is focused onto the retina by the eye, and that focusing combined with the high peak power in a range-finder pulse can readily cause eye damage. This presents a problem not in aiming at enemy soldiers, but in battle simulations and training exercises involving friendly troops. To avoid those problems, military planners would like to use range finders emitting at wavelengths longer than 1.4 μm, which are not transmitted well by the eye, and hence not focused to high intensity on the retina. At these eye-safe wavelengths, the maximum permissible exposure of the eye to laser power is much higher.

Holmium Lasers

The 2.06-μm holmium laser may rely on one of two crystalline hosts, YLF (yttrium lithium fluoride, $YLiF_4$) or YAG. Holmium is added to the raw material from which the crystal is grown along with two other impurity dopants, erbium and thulium. The latter two rare earths serve as "sensitizers," by absorbing the pump light from incoherent lamps and transferring the energy to the Ho^{3+} ions in the crystalline lattice. Without the sensitizers, the holmium itself cannot absorb enough pump energy to produce laser emission.

Holmium lasers operate in two distinct modes. At cryogenic temperatures, such as the 77 K of liquid nitrogen, they are four-level systems. However, at room temperature the lower laser level is thermally populated, making the laser a three-level system. As would be expected, the holmium laser is far more efficient as a four-level cryogenic laser than as a three-level room-temperature laser.

At liquid-nitrogen temperature, the holmium laser performs very well. Average powers of 10 W have been produced in a TEM_{00} beam. Pumping with a continuous linear quartz-iodine-tungsten lamp gave overall wall-plug efficiency of 3 percent. The laser produced *Q*-switched pulses of 100 millijoules (mJ), with duration under 50 nanoseconds (ns), at a 100-hertz (Hz) repetition rate (Chicklis, 1985). Alternatively, the Ho^{3+} laser can be operated continuously at the temperature of liquid nitrogen.

At room temperature, continuous-wave operation becomes impossible, and the high pump powers of a flashlamp become necessary to produce laser output. With flashlamp pumping, repetition rate can reach about 20 Hz. Energy storage is larger at the higher temperature, and the energy in *Q*-switched pulses can reach 500 mJ.

Although the overall efficiency of a 77 K holmium laser is excellent for solid-state lasers, the need for cryogenic operation is a very serious practical drawback. Despite its lower efficiency, the room-temperature version might be more likely to find applications.

The 2.06-μm wavelength of holmium does seem attractive for some medical applications, and the high peak and average power in certain cases might lead to applications in welding and drilling. It may also have some military applications outside of range finding—military planners seem to have dropped the idea of using holmium lasers for eye-safe range finding—but those other uses have yet to be identified in public.

Crystalline Erbium Lasers

The three major transitions in crystalline erbium lasers, 0.85, 1.23, and 1.73 μm, all are four-level lasers at room temperature and below. However, gain does decrease with increasing temperature, so higher pump powers are needed at room temperature. A variety of crystalline hosts have been tested, with YLF apparently the best. Er–YLF is pumped with a flashlamp, and efficiency can be increased by adding a sensitizing ion to absorb the flashlamp energy and transfer it to the Er ions.

The 1.73-μm transition did receive serious consideration for eye-safe range finding, although it eventually lost out to 1.54-μm sources, which pose even less danger to the eye and which may be easier to detect. Pulse energies of 10 to 20 mJ can be produced at that wavelength with overall efficiency in the 0.5 percent range. *Q*-switched 150-ns pulses can be produced at a 10-Hz repetition rate.

The 0.85-μm laser transition is about as efficient as alexandrite (see Chap. 22) and can generate 50-mJ pulses. However, in the face of the tunability of the alexandrite laser and the greater marketing push behind it, the fixed-wavelength erbium laser has attracted little attention.

Pulse energies are lower at the 1.23-μm Er line than at 0.85 μm, but the 1.23-μm wavelength may have some special advantages in certain medical applications.

Erbium-Glass

When erbium is doped into certain glasses, it can produce laser output on two closely spaced lines at 1.54 μm. This system has become a prime beneficiary of the military's interest in eye-safe laser range finders. Quirks of optical absorption in the eye raise the eye-hazard threshold at 1.54 μm about a factor of 100 higher than at other nominally eye-safe wavelengths. For that reason, military planners recently have focused specifically on that wavelength. Er-glass is one of two contenders; the other is Raman shifting of Nd–YAG output to 1.54 μm.

Erbium-glass was first tested in the late 1960s but attracted little interest then because no good 1.5-μm detectors were available and because the glasses then in use limited laser performance. Renewed interest was stimulated by the development of 1.5-μm detectors for fiber optics and the development of pentaphosphate glasses for laser fusion research. The pentaphosphate glasses turned out to be excellent hosts for erbium. Ytterbium is added to the glass as a sensitizer, which absorbs pump light from the flashlamp and transfers it to the erbium ions in the glass.

Figure 24-1 Handheld laser range finder in which an Er-glass rod is pumped by a flashlamp. Similar in appearance to Nd–YAG laser range finders, this 1.5-kg model poses much less threat to the eye because its 1.54-μm wavelength does not reach the retina. *(Courtesy of KEI.)*

Thermal characteristics of the glass limit repetition rate of Er-glass lasers to about 5 Hz, but this is adequate for range-finding applications. Proprietary techniques under development have made it possible to produce peak powers of 1 to 1.5 MW in 30-mJ *Q*-switched pulses lasting 20 ns (Johnson, 1985). That is adequate for the military requirement of range finding over distances of 4 to 6 km. The range-finder package shown in Fig. 24-1 is similar to that used for Nd–YAG range finders.

BIBLIOGRAPHY

Evan Chicklis (Sanders Associates): Private communication, 1985.

S. R. Chinn: "Stoichiometric lasers," in Marvin J. Weber (ed.): *CRC Handbook of Laser Science & Technology*, vol. 1, *Lasers & Masers*, CRC Press, Boca Raton, Fla., 1982.

V. P. Gapontsev et al.: "Erbium-glass lasers," in Wilhelm Waidelich (ed.): *Optoelectronics in Engineering*, Springer-Verlag, Berlin and New York, 1982, pp. 7–11.

Anthony M. Johnson, KEI: private communication, 1985.

Marvin J. Weber (ed.): *CRC Handbook of Laser Science & Technology*, vol. 1, *Lasers & Masers*, CRC Press, Boca Raton, Fla., 1982.

APPENDIX

A

glossary

Aberration Deviation from ideal behavior by a lens, optical system, or optical component. Aberrations exist in all optical systems, and designers have to make trade-offs among them.

Absorptance The fraction of the incident light absorbed.

Absorption spectroscopy The study of the wavelengths of light absorbed by materials and the relative intensities at which different wavelengths are absorbed. This technique can be used to identify materials and measure their abundances. Lasers are used because their narrow range of wavelengths permits high-resolution measurements.

Achromatic Optical elements which are designed to refract light of different wavelengths at the same angle. Typically achromatic lenses are made of two or more components of different refractive index and are designed for use at visible wavelengths only.

Acousto-optic An interaction between an acoustic wave and a light wave passing through the same material. Generally, the interaction deflects or otherwise modifies the light wave. Acousto-optic devices can perform beam deflection, modulation, signal processing, and *Q* switching.

Acousto-optic glass Glass with a composition designed to maximize acousto-optic interactions.

Active medium The material in a laser which produces the amplified stimulated emission. Generally, the name of the laser identifies the active medium; for example, a helium-neon laser has an active medium in which energy is transferred from helium atoms (which absorb the excitation energy efficiently) to neon atoms, which emit stimulated emission.

ADA Ammonium dihydrogen arsenate, a nonlinear crystal with chemical formula $(NH_4)H_2AsO_4$.

Adaptive optics Optical components which can be made to change the way in which they reflect or refract light. In practice, the term usually means mirrors with surface shapes that can be adjusted. Such "rubber mirrors" are used to adjust the output wavefronts of high-energy laser beams so they can be transmitted more readily through the atmosphere.

ADP Ammonium dihydrogen phosphate, a nonlinear crystal with chemical formula $(NH_4)H_2PO_4$.

Airy disk The central bright spot produced by a theoretically perfect circular lens or mirror. This spot is surrounded by a series of dark and light rings, produced by diffraction effects. The size of the airy disk is generally considered to be the focal spot size.

Alexandrite laser Alexandrite is a crystalline host which can be doped with chromium to serve as a solid-state laser material. Alexandrite lasers can produce pulsed or continuous output which is tunable between about 700 and 815 nanometers.

Amici prism Also known as a roof prism, a right-angle prism in which the hypotenuse has been replaced by a roof, where two flat faces meet at a 90° angle. Rather than simply reflecting an image, a roof prism performs "image erection" while deflecting the light by 90°. This is the same as rotating the image by 180°—reversing it left to right and at the same time inverting it top to bottom.

Amplifier (laser) A laser amplifier is a device which amplifies the light produced by an external laser but lacks the mirrors needed to sustain oscillation and independently produce a laser beam. Note that sometimes this term is used less rigorously and may denote a device capable of oscillation.

Amplitude noise Random fluctuations in the amount of power emitted by a laser, a reflection of the fact that the amount of power produced is not perfectly constant.

Analog A waveform is analog if it is continuous and can vary over an arbitrary range in a continuous fashion. For example, ordinary sound is an analog signal.

Angstrom A unit of length equal to 10^{-10} meter or 0.1 nanometer. Although in common use for wavelength measurements, the angstrom is not part of the SI system of units, and its use is sometimes frowned upon.

Anisotropic Exhibiting different properties when characteristics are measured along different directions or axes.

Antireflective coating A coating designed to suppress reflections from an optical surface. The simplest of many varieties is a layer of material with refractive index intermediate between that of air and the optical material.

Aperture A hole in a surface through which light is transmitted. Generally used to remove fringes of a laser beam; that is, to "clean up" and get rid of light scattered out of the beam and to produce an abrupt edge to the beam rather than the gradual falloff which a laser produces by itself. Apertures are sometimes called spatial filters, a more descriptive term.

Arc lamp A high-intensity lamp in which a direct-current electric discharge produces light. Output is continuous, as opposed to that of a flashlamp, which produces pulses. Arc lamps are used to pump some solid-state lasers.

Array A group of detecting or emitting elements usually arranged in a straight line (a linear array) or in a two-dimensional matrix (e.g., an imaging array).

Articulated arms (or waveguides) A beam-direction arrangement in which light passes through a series of jointed pipes containing optics. These pass light along to its eventual destination. Articulated arms have elbows, and look rather like the arms attached to dentists' drills. They are used for beam delivery in surgery and some types of materials working.

Aspheric For optical elements, surfaces are aspheric if they are not spherical or flat. Lenses with aspheric surfaces are sometimes called aspheres. The term aspheric is also used for mirrors.

Atmospheric communication Sending signals in the form of modulated light through the atmosphere, without the use of fiber optics to contain and direct the beam. This was the original form of optical communication, and remains valuable when desirable to avoid the need for a physical connection between two points. However, most optical communication is now via optical fibers.

Atmospheric monochromator A monochromator in which the optical path is through air. This is the standard type used for visible and infrared wavelengths transmitted by air.

Attenuator An optical device which reduces the intensity of a beam of light passing through it.

Autocollimator A telescopic sight including a light source and partially reflecting mirror, focused to infinity, for use in measuring small angular motion and checking alignment.

Avalanche photodiode A type of semiconductor detector operated at high voltages. When incident light generates one photoelectron from the material, the high voltage across the device accelerates the electron enough to cause an avalanche of other electrons, effectively amplifying the signal. An avalanche photodiode could be seen as similar to a solid-state photomultiplier.

Average power The average amount of power (usually measured in watts) emitted by a laser. This equals the energy per pulse (in joules) times the number of pulses per second, and is most meaningful for rapidly pulsed lasers.

Backscattering The scattering of light in a direction opposite to the original one in which it was traveling.

Beam diameter The diameter of a laser beam is usually considered to extend to a point where the intensity of light in the beam has dropped to $1/e^2$ (0.135) of the peak intensity. Note that according to this definition, some light in the beam falls outside of the beam diameter.

Beam expander An optical system which expands a narrow beam to a larger diameter, ideally without changing the divergence of the beam. Used with lasers because the diameter of the beam emitted by the laser is too small for some applications.

Beam integrator A device which integrates the energy in a beam to make it uniform across the beam cross section, instead of uneven as it typically is when produced by a laser.

Beamsplitter A device which divides a beam of light passing through it into two separate beams going in two different directions. Some types affect polarization of the beam, while others do not. Various splitting ratios are possible (e.g., 90–10, 70–30, 50–50, etc).

Bench (optical) An optical bench is a one-dimensional mounting surface which lets the user mount optical components along what is essentially a straight line. These range from lightweight units to massive ones carved from granite. Large, two-dimensional flat surfaces used for optics are generally called optical tables.

Biostimulation The use of lasers to stimulate the skin or nerves. One example is laser acupuncture, in which a low power laser beam stimulates acupuncture points on the skin. Other applications under study include stimulation of wound healing and alleviation of pain.

Birefringent element Has a refractive index which is different for light waves of different orthogonal polarizations. Because of this difference, light of the two orthogonal polarizations travels at different speeds and is refracted slightly differently. One result is rotation of the polarization direction, as light of one linear polarization gets farther ahead of that polarized in the other direction. Another can be the creation of double images within the birefringent material.

Bolometer A sensitive type of infrared detector whose operation is based on a change in temperature induced by infrared radiation. The bolometer is made up of two thin, blackened gratings of platinum, one of which is illuminated and the other kept in the dark. The absorption of heat changes the electrical resistance which is detected by comparing the resistances of the two gratings in an electrical circuit.

Bore The central hole in a laser tube or other type of tube (e.g., a capillary, waveguide, or hole in a microchannel plate). Laser action takes place in the bores of gas lasers.

Brewster angle window A window inserted into an optical path at Brewster's angle—the angle at which light polarized in one direction suffers no reflection losses. When this optical path is a laser cavity, insertion of a Brewster angle window causes laser emission to be polarized in that direction.

Calorimeter An instrument or detector which measures the amount of heat it absorbs. It can be used to measure incident radiation if the percentage of absorbed radiation is known.

Calorimetric detector A detector which operates by measuring the amount of heat absorbed—incident radiation must be absorbed as heat to be detected.

Candela The unit used to measure intensity of light visible to the human eye. It corresponds to the emission from 1/60th of a square centimeter of a black body operating at the solidification temperature of platinum, and emits one lumen per steradian.

CCD Charge-coupled device, which can be used as an optical detector, typically in linear or two-dimensional arrays.

CDA Cesium dihydrogen arsenate, a nonlinear crystal with chemical formula CsH_2AsO_4.

Chemical laser From a research standpoint, any laser in which the excited species is produced by a chemical reaction can be considered a chemical laser. Practically speaking, chemical lasers are those in which the active medium is hydrogen fluoride (HF) or deuterium fluoride (DF), with wavelengths of 2.5 to 3 micrometers (μm) and 3.5 to 4 μm, respectively.

Chromatic aberration The focusing of light rays of different wavelengths at different distances from the lens. This is not a significant effect with a single-wavelength laser source, but be wary if working at different or multiple wavelengths. Some lenses designed for use with lasers may not take chromatic aberration into account.

Circularly polarized light Light in which the polarization vector rotates periodically, but does not change in magnitude, describing a circle. Another way of looking at circularly polarized light is as the superposition of two plane-polarized (or linearly polarized) light waves, of equal magnitude, one 90° in phase behind the other.

Cleaver A device used to cut or break optical fibers in a precise way so the ends can be connected with low loss.

Coating (for optics such as lenses) A thin layer or layers applied to the surface of an optical component to enhance or suppress reflection of light, and/or to filter out certain wavelengths.

Coherence A property of electromagnetic waves that are all of the same wavelength and precisely in phase with each other.

Coherence length The distance over which light from a laser retains its coherence after it emerges from the laser.

Collimate To make parallel.

Collimator An optical system which focuses a beam of light so all the rays form a parallel beam. In other areas of optics, the term may indicate a small telescope that is used in aiming a larger telescope.

Color-center laser A solid-state optically pumped laser (see Chap. 23) in which the active medium is a crystal containing microscopic defects or "color centers" deliberately introduced into it. It is these color centers which emit laser light.

Color filter A filter containing a colored dye, which absorbs some of the incident light and transmits the rest. Such filters are typically used in photography; because they absorb unwanted light rather than reflect it, they are subject to damage by intense light.

Color separation The process of breaking up a full-color image into four component colors (black and three primaries) for use in printing. Separate negatives are produced for each of the four colors, and these negatives are used for four separate press runs, one in each color, to produce a printed full-color image. Lasers are often used to record the separate negatives for each color, and sometimes may be used to record intensity of the original image in various colors.

Coma A lens aberration in which light rays from an off-axis source which pass through the center of a lens arrive at the image plane at different distances from the axis than do rays from the same source which pass through the edges of the lens. It makes a point-source object appear like a comet-shaped blob in the image.

Complex lens A lens consisting of more than one element or simple lens.

Compound semiconductor A semiconductor material such as gallium arsenide which is made up of two or more elements, in contrast to simple single-element materials such as silicon and germanium.

Concave Curving inward. A concave surface is shaped like the inside of a bowl.

Continuous-wave laser A laser which continuously emits a beam. In practice, this means a laser which operates continuously, but strictly speaking the term "continuous" is used in reference to the physical process generating the laser beam. Thus a high-power chemical laser producing hundreds of kilowatts may be operating "continuous-wave" for a matter of seconds, because that is long compared with the scale of the physical processes which generate the light.

Convex Curving outward. A convex surface is shaped like the outside of a ball.

Corner-cube prism A prism in which three flat surfaces meet at right angles, as they would if they were the corner of the cube. Light incident through a planar face opposite the apex of the intersection can be totally reflected (via total internal reflection) back in precisely the same direction from which it came. Such a prism, in short, is a retroreflector. (See Retroreflector.)

Correlation Measurement of the degree of similarity of two images as a function of detail and relative position of the images. It is obtained by multiplying the Fourier transforms of the two images, then taking the Fourier transform of the product. (See Fourier transform.)

Crown glass An optical glass of alkali-lime-silica composition with index of refraction usually 1.5 to 1.6.

CW Continuous wave.

Cylindrical lens A lens which is cylindrical in cross section, so it is curved in one direction but not in the perpendicular direction. Such lenses bend light in one dimension but not in the perpendicular dimension.

DC Direct current. In many cases this abbreviation may be used to indicate continuous output even though the term "direct current" is not appropriate. For example, a light source with continuous output may be labeled "DC" even though it is not delivering current of any sort.

Decibel A logarithmic unit used in comparisons of power, which is defined as one-tenth of a bel—though nobody ever measures anything in bels. The difference of the power levels of two signals in decibels equals 10 times the base-10 logarithm of their ratio: $n = 10 \log_{10}(P_2/P_1)$. Abbreviated dB.

Delay line A transmission medium which delays a signal passing through it by a known amount of time; typically used in timing of events.

Densitometer An instrument which measures the density of an image on a photographic plate or film. This measurement of density indicates the intensity of the radiation which produced the image.

Depolarizers Optical components which scramble the polarization of light passing through them, effectively turning a polarized beam into an unpolarized beam.

Detector A device which detects light, generating an electrical signal which can be measured or otherwise processed.

Detector-amplifier A device in which an optical detector is packaged together with electronic amplification circuitry.

Dewars Insulated thermos-like containers for cryogenic liquids, which can be designed to house detectors or lasers requiring active cooling.

Diamond-turned mirror A mirror in which the surface has been formed by machining away material with a diamond tool. The technique is most often used for metal mirrors intended for infrared use, although in certain cases it can be used for other types of optics. The diamond machining process allows fabrication of complex surfaces which could not be made by ordinary polishing.

Dichroic filter A filter which selectively transmits some wavelengths of light and reflects others. Typically such filters are based on multilayer interference coatings.

Dielectric coating An optical coating made up of one or more layers of dielectric (i.e., nonconductive) materials. The layer structure determines what fractions of incident light at various wavelengths are transmitted and reflected. Typically the transmittance and reflectance of such coatings depend on wavelength and angle of incidence of the light. Dielectric coatings are used to suppress or enhance reflection of certain light wavelengths.

Difference frequency generation The subtraction of the operating frequency of one laser from that of another, to yield a third frequency that is lower than that of the first laser (and which may be lower than that of the second laser as well). From the standpoint of wavelength, the process yields a longer wavelength than that of the first laser.

Differential amplifier A device which compares two input signals and amplifies the difference between them. It is typically used in comparing two signals.

Diffraction-limited beam A beam with far-field spot size dependent only on the theoretical diffraction limit, which is a function of output wavelength divided by output aperture diameter.

Digital Encoded in the form of digital units, typically binary ones and zeros.

Diode laser A laser in which stimulated emission is produced at a *pn* junction in a semiconductor material. Sometimes called an injection or semiconductor laser. Only certain semiconductors are suitable for diode-laser operation, among them gallium arsenide, indium phosphide, and certain lead salts. LEDs are similar, but lack cavity mirrors and are generally operated at lower drive currents.

Diode laser array A device in which the output of several diode lasers is brought together in one beam. The lasers may be integrated on the same substrate, or discrete devices may be coupled optically and electronically. In some laboratory devices, individual diode lasers in an array can emit light independently of other lasers in the array.

Diopter A measurement of refractive power of a lens equal to the reciprocal of the focal length in meters. Thus a lens with 20-centimeter focal length has a power of 5 diopters, while one with a 2-meter focal length has a power of 0.5 diopter. Values for negative lenses generally are prefixed by minus signs.

Dispersing prism A prism designed to spread out the wavelengths of light to form a spectrum.

Divergence The spreading out of a laser beam with distance, measured as an angle.

Doped germanium A type of detector in which impurities are added to germanium to make the material respond to infrared radiation at wavelengths much longer than those detectable by pure germanium.

Doped silicon A type of detector in which impurities are added to silicon to make the material respond to infrared radiation at wavelengths much longer than those which pure silicon can detect.

Doppler shift A change in the wavelength of light caused by the motion of an object emitting (or reflecting) the light. Motion toward the observer causes a shift toward shorter wavelengths, while motion away causes a shift toward longer wavelengths.

Doublet (lens) A lens with two components of different refractive index, generally designed to be achromatic.

Dove prism A prism which looks like a symmetrical trapezoid in profile and is used to rotate images. Light is incident through one of the angled faces and emerges, after reflection by one of the flat surfaces, from the other angled face. As the prism is rotated, the image that has passed through it will rotate twice as fast.

Dry-processing film Photographic film which can be processed to develop an image without the use of wet chemical baths.

Dynamic balancing The process of balancing a rapidly rotating object, such as a gear, by removing small chunks of material from it. A short pulse from a high-power laser can be used to remove the material if the laser pulses are synchronized with the balancing system.

Edge filter An interference filter which abruptly shifts from transmitting to reflecting over a narrow range of wavelengths. Thus, for example, all wavelengths longer than a certain value may be strongly transmitted, while shorter wavelengths may be strongly reflected.

Efficiency For a laser, the percentage of the energy that is fed into the laser which emerges in the laser beam. The common-sense definition is "wall-plug" efficiency, the percentage of the power going into the power supply which emerges in the beam. However, some people measure efficiency based on the amount of energy emerging from the power supply, or actually absorbed by the laser medium—practices which lead to higher numbers because they ignore losses in the power supply and energy deposition processes.

Electronic transition A transition in which an electron in an atom or molecule moves from one energy level to another. Generally electronic transitions are more energetic (i.e., at a shorter wavelength) than vibrational transitions. (See Vibrational transitions.)

Electrical pumping Deposition of energy into a laser medium by passing an electrical current or discharge, or a beam of electrons into the material. The electrons transfer their energy to the laser medium to provide the excitation energy needed for laser action.

Electro-optic materials Materials in which the application of an electric field changes the refractive index of light. There are two types of electro-optic interactions: Pockels interactions in which the effect is linearly proportional to the electric field, and Kerr interactions, which are proportional to the square of the electric field. Such materials are used in modulators, *Q* switches, beam deflectors, and some other applications.

Ellipsometer An optical instrument which measures the ellipticity of polarized light. It is most often used in thin-film measurements.

Elliptically polarized light Light in which the polarization vector rotates periodically, changing in magnitude with a period of 360° so it describes an ellipse. Another way of viewing elliptically polarized light is as the superposition of two plane-polarized light waves, of unequal magnitude, one 90° out of phase with the other. The special case where the two light waves are of equal intensity is circular polarization.

Emissivity The ratio of the emission of electromagnetic radiation from an object to the emission from a theoretically perfect blackbody at the same temperature.

Etalon A type of Fabry-Perot interferometer in which the distance between the two highly reflecting mirrors is fixed. It can be used to separate light of different wavelengths when the wavelengths are closely spaced.

Excimer laser A laser in which the active medium is an "excimer" molecule—a diatomic molecule which can exist only in its excited state. The internal physics are conducive to high powers in short pulses, with wavelengths in the ultraviolet. Typical examples are argon fluoride, krypton fluoride, xenon chloride, and xenon bromide. Molecular fluorine lasers emitting in the ultraviolet are normally lumped under the same heading.

Fabry-Perot (interferometer) A pair of highly reflecting mirrors, which can be adjusted to select light of particular wavelengths. When used as a laser resonator, this type of cavity can narrow the range of wavelengths emitted by the laser.

Faraday rotation A rotation of the plane of polarization of light caused by the application of a magnetic field to the material transmitting the light.

Faraday rotator A device which relies on the Faraday effect to rotate the plane of polarization of a beam of light passing through it. Faraday rotator glass is a type of glass with composition designed to display the Faraday effect.

Far field Distant from the source of light. This qualification is often used in measuring beam quality, to indicate that the measurement is made far enough away from the laser that local aberrations in the vicinity of the laser have been averaged out.

Far-infrared laser Generically, this term could be taken to mean any laser emitting in the "far-infrared," a vaguely defined region of wavelengths from around 10 micrometers to 1 millimeter. In the laser industry, it is usually taken to mean any of a family of lasers which operate in this region and require optical pumping by an external laser (usually carbon dioxide). Typically a single laser device is sold for use with a variety of laser gases emitting in this region.

***F*-center laser** A solid-state laser in which optical pumping by light from a visible-wavelength laser produces tunable near-infrared emission from defects—called "color centers" or "F centers"—in certain crystals.

Fiber-optic gyroscope A device in which changes in the wavelength of light going in different directions through a long length of optical fiber wound many times around a ring is used to measure rotation speed. The physics is different from that involved in the laser gyroscope, and the fiber-optic gyroscope is less developed.

Fiber sensor A sensing device in which the active sensing element is an optical fiber or an element attached directly to an optical fiber. The quantity being measured changes the optical properties of the fiber in a way that can be detected and measured. For example, pressure changes induced in a fiber by acoustic waves can change the amount of light transmitted by a fiber.

Field curvature Formation of an image that lies on a curved surface rather than a flat plane. For single- and double-element lenses, curvature is always inward, but for other types the curvature can be in either direction.

FIR Far-infrared.

Flashlamp A gas-filled lamp which is excited by an electrical pulse passing through it to emit a short, bright flash of light. A broad range of wavelengths is produced, with their precise nature depending on the gas or gases used.

Flashlamp-pumped dye laser A dye laser in which the pump light comes from a flashlamp housed in the same reflective cavity as the dye cell. Some designs employ coaxial flashlamps, in which the light-producing flashlamp forms a ring around a central tube containing the dye. Others place linear flashlamps inside a reflectivity cavity with a separate dye cell.

Flint glass An optical glass which contains lead or other elements which raise its refractive index to 1.6 to 1.9, higher than other types of optical glass.

FLIR Forward-looking infrared, a type of military imaging system.

Fluorescence spectroscopy The study of materials by the light which they emit when irradiated by other light. For example, many materials emit visible light after they have been illuminated by ultraviolet light. The intensity and wavelengths of the emitted light can be used to identify the material and its concentration.

***F* number** The ratio of the focal length of a lens to its diameter, to a simple approximation.

Focal length For practical purposes, focal length can be defined as the distance from a lens at which rays initially parallel to the axis of the lens are brought to a point. In the case of a negative lens, which bends the rays outward instead of inward, the focal point to which the focal length is measured is considered to be the point from which the rays appear to diverge from behind the lens, and the focal length is assigned a negative value.

Focal ratio Also known as *f* ratio and *f* number, this is the ratio of focal length to lens diameter.

Fourier optics Optical components used in making Fourier transforms and other types of optical processing operations.

Fourier transform The conversion of information from the frequency domain to the spatial domain and vice versa. A positive lens can perform a Fourier transform on a two-dimensional image.

Free-electron laser A laser in which stimulated emission is produced by a beam of free electrons passing through a magnetic field which periodically (in space rather than time) alters its polarity. The magnetic field is produced by an array of magnets with the orientation of their poles alternating as the electron moves along its path. The free-electron laser is considered a promising new type for high-energy applications, but is not available commercially.

Frequency multiplication The generation of harmonics of the frequency of a light wave by nonlinear interactions of the light wave with certain materials. Frequency doubling is equivalent to dividing the wavelength in half. High power beams are needed for the nonlinear interaction to occur.

Frequency stability A measurement of how well the output frequency (or equivalently emitted wavelength) of a laser stays constant. In some types the emitted wavelength tends to drift because of factors such as changing temperature of the laser itself.

Fresnel lens A lens in which the surface is composed of a number of smaller lenses with the same focal length desired for the larger lens. Typically for high-quality optical applications, the smaller lenses are concentric circles. This technique is used to compress a short focal length optical component into a thin volume.

Fused silica The term usually applied to synthetic fused silica, formed by the chemical combination of silicon and oxygen to produce a high-purity silica. Optical glass is made in a different way, by the melting of high-purity sands, while fused quartz is made by crushing and melting natural quartz.

FWHM Full width at half maximum, a standard measurement of pulse width. For a laser pulse, for example, the full width at half maximum is the interval between the time power increases above 50 percent of the maximum value to the time that the power drops below 50 percent of the peak.

Gas-dynamic pumping The production of a population inversion by a gas-dynamic process, in which a hot, dense gas is expanded into a near-vacuum, causing the gas to cool rapidly. If the gas cools faster than energy can be redistributed, a population inversion is generated. (See Population inversion.)

Gaussian beam A laser beam in which the intensity has its peak at the center of the beam, then drops off gradually toward the edges. The intensity profile measured across the center of the beam is a classical gaussian curve.

Gimbal mount An optical mount which allows position of a component to be adjusted by rotating it independently around two orthogonal axes.

Golay cell An infrared detector in which the incident radiation is absorbed in a gas cell, thereby heating the gas. The temperature-induced expansion of the gas deflects a diaphragm, and a measurement of this deflection indicates the amount of incident radiation.

Graded-index fiber An optical fiber in which the refractive index changes gradually between the core and the cladding, in a way designed to refract light so it stays in the fiber core.

Graded refractive index lens A lens in which the refractive index of the glass is not uniform. Typically the index will differ with distance from the center of the lens.

Grazing-incidence monochromator A monochromator in which the spectrum is dispersed by grazing-incidence optics, allowing it to work in parts of the spectrum where conventional dispersing optics do not work well.

Half-wave plate A polarization retarder which causes light of one linear polarization to be retarded by a half wavelength (180°) relative to the phase of the orthogonal polarization.

Hard-clad silica fibers Silica optical fibers which are coated with a hard plastic material, not with the soft materials typically used in plastic-clad silica.

Harmonic generation The multiplication of the frequency of a light wave by nonlinear interactions of the light wave with certain materials. Generating the second harmonic is equivalent to dividing the wavelength in half.

Heat-absorbing filter A glass filter which transmits most visible light, but strongly absorbs near-infrared light, unlike other types of glass which are nearly as transparent in the near-infrared as they are in the visible.

Heat treating The process of heating material (generally a metal) to change its characteristics. For example, the surfaces of certain steels may be heated by a laser beam to alter the crystalline structure to a harder form that would protect the surface from wear, but which might be too brittle for use for the whole component.

Heterojunction A junction between semiconductors that differ in doping levels and also in their atomic compositions. For example, a junction between layers of GaAs and GaAlAs. A double-heterojunction laser contains two such junctions; a single-heterojunction laser contains only one.

HF-DF laser A laser in which the active medium is hydrogen fluoride or its isotopic variant, deuterium fluoride. The two types operate very similarly, except the 2.5 to 3 micrometer (μm) output of the HF laser is strongly absorbed by the atmosphere, while the 3.5- to 4-μm output of the DF laser is transmitted well. Functionally, the two gases can usually be used in the same laser device. The excited molecules which emit light are produced by a reaction of hydrogen (or deuterium) and fluorine.

HgCdTe Mercury-cadmium telluride, an infrared detector material. By changing the relative concentrations of mercury and cadmium, and by adjusting operating temperature, this detector can be made sensitive to various infrared wavelengths.

Holographic camera An integrated system for recording holograms.

Holographic diffraction gratings Diffraction gratings in which the pattern of light-diffracting lines was recorded holographically rather than mechanically ruled into the surface. Typically this term is applied to originals, rather than to gratings which were replicated from originals.

Holographic optical elements (HOES) Holograms which have been made to diffract light in the same pattern as other optical components. For example, it is possible to produce (usually by computer synthesis techniques) a hologram which mimics the function of a lens. In some applications, such holographic optical elements are less costly or easier to use than conventional optics.

Holographic sandbox An inexpensive way to minimize vibration in a holographic setup is to mount the components in sand, which tends to isolate them from vibrations. The performance cannot approach that of a high-quality optical table, but neither does the price.

Homojunction A junction between semiconductors that differ in doping levels but not in atomic composition. For example, a junction between *n*-type and *p*-type GaAs.

Honeycomb A mounting surface in which top and bottom surfaces are separated by a latticework structure rather than a solid block of material. In cross section such structures look like a honeycomb, hence the name. A honeycomb structure can offer the rigidity and strength of a solid table, without the massive weight.

Illuminance The amount of visible light incident per unit area.

Image-converter camera A camera which converts images from one wavelength region to another, typically from the infrared to the visible.

Image digitizer A device which measures light intensity at each point in an image and generates a corresponding digital signal which indicates that intensity. It converts an analog image to a digital signal.

Image dissector An early type of video camera tube in which a photoemissive surface produces an electron image, which is focused in the plane of an aperture and scanned by the aperture to record the image. It is still used in certain imaging applications.

Image intensifier A viewing system which functions as a light amplifier, taking a faint image and amplifying it so that it can be viewed more readily. One common application is in viewing faint astronomical images.

Infrared viewer An imaging system which detects infrared radiation and produces a visible-light image of the infrared scene. Typically infrared viewers are handheld devices operating at near-infrared wavelengths, not throughout the entire infrared.

Injection laser See Diode laser.

Integrating sphere A sphere used in optical measurements which generally has two openings, one for light to enter, another for the light to leave. The openings are arranged so the entering light is reflected within the sphere, effectively integrating it over the output aperture to provide uniform illumination, as required for some types of measurements.

Intensity The amount of light incident per unit area. For human viewing of visible light, the usual term is illuminance; for electromagnetic radiation in general, the term is usually radiant flux.

Interference filter An optical filter which selectively transmits certain wavelengths of light because of interference effects which take place in dielectric coatings applied to the surface of the material. Multilayer interference coatings may include metallic layers.

Interferometer A device which divides a single beam of light into two (or sometimes more) components, then recombines them to produce interference. In general, the distances light travels along the different arms will differ; the difference is proportional to the wavelength of the light times the number of interference fringes. Typically interferometers are used for precise measurements of distance.

Intrinsic (germanium or silicon) detectors Detectors in which the semiconductor material is purified and has natural response of germanium or silicon to incident radiation.

Inversion See Population inversion.

Iodine laser A laser in which excited iodine atoms produce laser light at 1.3 micrometers (μm). Early versions derived their energy from flashlamps, which photodissociated molecules containing iodine to produce the excited atoms. More recently, the "chemical oxygen-iodine laser" has been developed, in which a chemical reaction produces excited oxygen atoms, which transfer their energy to iodine atoms—emitting light at 1.3 μm.

Ion laser A laser in which the active medium is an ionized gas, typically one of the rare gases argon or krypton, or a mixture of the two emitting on the wavelengths of both elements. Xenon ion lasers are also available but are much less common. The helium-cadmium laser also can be called an ion laser because the cadmium is ionized, but laser specialists rarely do so.

IR Infrared.

IRED An infrared-emitting diode, or infrared LED. Generally the term LED is used even for devices which emit invisible infrared radiation.

IRTRAN One of a family of synthetic infrared-transmitting materials developed by the Eastman Kodak Co.

Jack A positioning device which raises or lowers an optical component.

Jacket A plastic layer applied over the coating of an optical fiber, or sometimes over the bare fiber. Jackets may be used for color-coding in optical cables, to make handling easier, or for protection of the fiber against mechanical stresses and strains.

Kerr cell A device in which the Kerr effect is used to modulate light passing through a material. The modulation depends on rotation of beam polarization caused by the application of an electric field to the material; the degree of rotation determines how much of the beam can pass through a polarizing filter.

Kerr effect A birefringence that is induced in a material not normally birefringent, by the application of an electric field. The degree of birefringence—that is, the difference in refractive indexes for light of orthogonal polarizations—is proportional to the square of the electric field. The Kerr effect thus causes a polarization rotation which can be combined with polarizers to modulate light.

Ladar An acronym for "laser detection and ranging". (See Laser radar.)

Laser An acronym for "light amplification by the stimulated emission of radiation," applied to a wide range of devices which produce light by that principle. Compared to other light sources, laser light covers a narrow range of wavelengths, tends to be coherent, and is emitted in a directional beam.

Laser diode array A device in which the output of several diode lasers is brought together in one beam. The lasers may be integrated on the same substrate, or discrete devices coupled optically and electronically. In some laboratory devices, individual diode lasers in an array can emit light independently of other lasers in the array.

Laser glass An optical glass doped with a small concentration of a laser material, typically neodymium. When the impurity atoms are excited by light, they can be stimulated to emit laser light. Glass is easier to grow than YAG, especially in large chunks, but its thermal properties are not as good as the synthetic crystal. (See YAG.)

Laser gyroscope A device in which changes in the wavelength of light going in different directions around a ring is used to measure rotation speed. Laser gyroscopes are used in Boeing 757 and 767 aircraft and in some military systems to measure changes in bearing.

Laser head A laser tube packaged in a housing ready for use, generally requiring an external source of the voltage needed to drive the laser.

Laser line filter A filter which transmits light in a narrow range of wavelengths centered on the wavelength of a laser. Light at other wavelengths is reflected. Such filters are used to remove light from nonlaser sources, which could interfere with operation of a laser system.

Laser profilometer A device for measuring the cross-sectional profile of a laser beam—i.e., the intensity pattern of the beam.

Laser radar A laser equivalent of radar, in which the transit time of a laser pulse is measured to determine the distance to a remote object. There are similar systems in which the wavelengths of the returned light are also observed and used to detect air pollutants or other materials, but these often go under the name lidar. Laser radar is sometimes known as ladar, for "laser detection and ranging."

Laser simulator A light source which simulates the output of a laser. In practice, the light source is a 1.06-micrometer LED which simulates the output of a neodymium laser at much lower power levels.

Laser tube A tube of glass (or other material) filled with laser gas and often containing integral optics. Requires external packaging and a voltage supply.

LED A light-emitting diode. A semiconductor device in which light is produced when current carriers combine at a *pn* junction. The emission is spontaneous and there are no feedback mirrors, unlike diode lasers. Output is lower in power than from diode lasers, reflecting the use of lower operating currents. Generally LEDs are less expensive than diode lasers and can operate at shorter wavelengths without the rapid degradation which occurs with visible-wavelength laser diodes.

Lidar (light detection and ranging) Lidar is analogous to radar. Like radar, lidar systems can be used for distance measurement by measuring the round-trip time of a light pulse. Many lidar systems also observe the wavelengths of light returned, and use the spectroscopic information collected to detect air pollutants or other materials. The term is often interchangeable with laser radar or ladar.

Light-emitting diode A semiconductor diode which emits visible or infrared light. Light from an LED is incoherent spontaneous emission, as distinct from the coherent stimulated emission produced by diode lasers and other types of lasers.

Linear polarization Light in which the electric-field vector points in only a single direction is considered linearly polarized. There are two possible linear polarizations, each orthogonal (perpendicular) to the other.

Linear position-sensing detector An optical detector which can measure the position of a light spot along its length.

Liquid-crystal light valve A device used in optical processing to convert an image in incoherent light into an image in coherent light. For example, an image of a gear obtained in ordinary light can be transformed into coherent light for optical processing. The liquid-crystal light valve allows this to be done continually.

Liquid laser In broad terms, a laser in which the active medium is a liquid. In practice this term is largely obsolete, and when used, generally refers to tunable dye lasers, although nontunable liquid lasers have been demonstrated in the laboratory.

Longitudinal mode Laser oscillation mode defined by wavelength and distance between resonator mirrors, occurring when twice the cavity length equals an integral number of wavelengths. A single longitudinal mode includes only a narrow range of wavelengths. A laser can operate on several longitudinal modes at slightly different wavelengths (within its gain curve), yet still emit a single transverse mode.

Lumen The unit for measuring the flux or power of light visible to the human eye; the photometric equivalent of the watt, which is a radiometric unit.

Luminous An adjective applied to measurement terms to indicate that the quantity being measured is light visible to the human eye. For example, luminous energy (measured in lumen seconds) corresponds to joules in radiometric measurements. There is no luminous energy present if the radiation is not visible to the human eye.

Manual positioners Slides, translation stages, and other types of positioning devices in which the turning of a knob or other adjustment by the human hand directly moves the component being shifted.

Maser An acronym for "microwave amplification by the stimulated emission of radiation," the microwave equivalent and predecessor of the laser. It produces coherent microwaves by the stimulated emission of microwave radiation.

Meniscus lens A lens with one concave surface and one convex surface.

Metallic coating A thin layer of metal applied to an optical surface to enhance reflectivity.

Microdensitometer A device for measuring the density of photographic films or plates on a microscopic scale; the small-scale version of a densitometer.

Micromanipulator A positioning device for making small adjustments to the position of an optical component or other device.

Mode A stable condition of oscillation in a laser. A laser can operate in one mode (single mode) or in many modes (multimode). Lasers have both transverse and longitudinal modes.

Modelocker A device used to lock together the various oscillating modes in a laser cavity. This produces a series of ultrashort laser pulses regularly spaced at roughly the transit time of the laser cavity. It can be visualized as a process producing a large clump of photons which bounces back and forth within the laser cavity, producing very short but intense and regularly spaced pulses.

Modulation transfer function (MTF) A measurement of image quality, used in evaluation of lenses and optical systems.

Monochromatic Of a single wavelength or frequency. In reality, light cannot be purely monochromatic and actually extends over a range of wavelengths. The breadth of the range determines *how* monochromatic the light is.

Monochromator An optical device which uses a prism or diffraction grating to spread out the spectrum, then passes a narrow portion of that spectrum through a slit. Thus it produces monochromatic light from a non-monochromatic source.

MOPA Acronym for "master oscillator power amplifier," a configuration in which energy stored in a laser medium is released when a short, low-power pulse from a master oscillator is passed through it. Typically used to extract high energies in short pulses, often (but not always) from solid-state lasers.

Mounting platform A base upon which optical components, mounts, or other equipment can be placed. A complex subassembly might be bolted together on a mounting platform, then picked up and moved as an entity when needed.

MTF Modulation transfer function.

Multichannel spectral analyzer A measurement system which sorts signals into a number of different channels, then counts and analyzes the signals channel by channel.

Multilayer coatings Optical coatings in which several layers of different thicknesses of different materials are applied to an optical surface. Interference effects on light passing through the layers influence reflection and transmission differently at different angles of incidence and different wavelengths. Also known as interference coatings and dielectric coatings, although in some cases some of the layers are metallic rather than true (i.e., nonconductive) dielectrics.

Multimode fiber An optical fiber capable of carrying more than one mode of light in its core. Although strictly speaking this could be two or more modes, in practice it generally means many.

***N* (usually lower case)** Symbol for index of refraction.

NA Numerical aperture.

Negative lens A lens which causes light rays initially parallel to its optical axis to spread out or diverge. Its focal length is a negative number.

NEP Noise-equivalent power.

Neutral density filter A filter which has uniform transmission throughout the part of the spectrum where it is used. For example, a neutral-density visible filter would transmit the same fraction of red and blue light.

Noise-equivalent power The amount of optical power which must be incident on a detector to produce an electrical signal equal to the level of noise inherent in the detector. This is a measure of the sensitivity of a detector. Abbreviated NEP.

Nonlinear effects Optical interactions which are proportional to the square or higher powers of electromagnetic field intensities, not simply to the first power of intensity. Passive optical transmission of light, such as by a lens, is linear transmission. Nonlinear effects generate harmonics of optical frequencies, and sum and difference frequencies when two light waves are mixed. Nonlinear materials are materials in which nonlinear effects are likely to occur at moderate power levels.

Nonsilica glasses A glass in which the primary constituent is a material other than silica (silicon dioxide). The term is sometimes applied to mean nonoxide glasses, those which do not contain oxygen compounds. In fiber optics, some of these materials are used for fibers transmitting mid-infrared wavelengths.

Numerical aperture The sine of the half-angle over which an optical fiber or optical system can accept light rays, multiplied by the index of refraction of the medium containing the rays (which is 1 for air). Abbreviated NA.

Off-axis mirrors Mirrors in which the center of the mirror does not correspond to the axis of the optical figure of the mirror.

Optical attenuation meter A device which measures the loss or attenuation of an optical fiber, fiber-optic cable, or fiber-optic system. Measurements generally are made in decibels.

Optical comparator An inspection instrument which produces a composite showing a reference contour and the image of an actual contour for comparison. Typically this is a projection device.

Optical density A measurement of transmission equal to the base-10 logarithm of the reciprocal of transmittance (the fraction of the incident light transmitted). An object with optical density of zero is transparent; an optical density of one corresponds to 10 percent transmission.

Optical flat An optical component with a surface that is precisely flat, used in measurements comparing flatness of other surfaces.

Optical glass A high-purity, high-quality glass made especially for use in optical components by melting special types of sand. There are several varieties of optical glass, with properties that differ somewhat.

Optical paints Special paints which have well-known optical properties. Typical examples would have broad, uniform reflectance for white paint, or broadband uniform absorption for black paints.

Optical time domain reflectometer A device which sends a very short pulse of light down a fiber-optic communication system and measures the time history of the pulse reflection. The reflection indicates fiber dispersion and discontinuities in the fiber path, such as breaks and connectors. The time it takes for the light pulse to travel to and from the discontinuity indicates how far it is from the test set.

Oscillator (laser) A laser cavity equipped with mirrors that allows laser light to *oscillate* within it, stimulating further emission from the active medium. Such a device is capable of producing a laser beam without requiring an external laser operating at the same wavelength. (Note that the term can be used for devices which are pumped by light at a *different* wavelength from a separate laser—that is, for laser-pumped lasers.)

Parametric oscillator A nonlinear device which, when pumped by light from a laser, can generate tunable output. The beam produced by the parametric oscillator relies on oscillation within the nonlinear material.

PCS Plastic-clad silica, a type of optical fiber.

Peak power The maximum power delivered instantaneously during a pulse of laser light. Typically this is much greater than the power averaged over many laser pulse cycles.

Pellicle An extremely thin, tough membrane which is stretched over a frame. Because of its thinness, it transmits some light and reflects other light, and hence can serve as a beamsplitter. Its thinness avoids the problem of ghost reflections sometimes produced by other types of beamsplitters.

Penta prism A five-sided prism which deviates a beam of light by 90° without inverting or reversing the image. All light rays entering such a prism are turned by the same 90° angle regardless of where they enter the prism or at what angle they enter.

PEV Pyroelectric vidicon, an imaging tube with a pyroelectric target.

Phase matching Alignment of a nonlinear crystal with respect to the incident laser beam in the proper way to generate a harmonic of the laser frequency in the material.

Photoacoustic cell A gas cell which is designed so pulses of incident light will change the gas pressure rapidly, creating acoustic waves. This generally requires a gas which absorbs the light energy, and is typically used in conjunction with photoacoustic spectroscopy.

Photoconductor A type of detector which changes its resistivity when illuminated by light; the changes in resistance can be measured to determine the amount of incident light.

Photodarlington A detector in which a phototransistor is fabricated on the same chip with a second transistor which amplifies the signal from the phototransistor. The circuit formed is a "Darlington" circuit, thus the name. This is a simple and inexpensive type of detector with limited performance.

Photodiode A diode which detects light. Vacuum photodiodes are tubes in which detection relies on the photoelectric effect producing free electrons which are collected by a positively charged electrode. Silicon photodiodes are solid-state devices in which illumination of the semiconductor junction releases current-carrying electrons and holes.

Photometer A device which measures the intensity of visible light as it is perceived by the human eye (in photometric units). Often this term is loosely and incorrectly applied to other types of light-measurement instruments which measure optical flux in other units, and whose response is not limited to the range and nonlinear sensitivity of the human eye. (See Radiometer, for contrast.)

Photomultiplier A type of electron tube in which photons incident on a photocathode produce electrons by photoemission. These electrons are then amplified (or multiplied) by passing them through an electron multiplier, which increases their numbers. Electrons passing through the multiplier are accelerated by high voltages and hit metal screens, from which they free more electrons.

Photon counting A measurement technique used for measuring low levels of radiation, in which each photon (at least ideally) generates a signal, which can be "counted" in some way.

Photon drag detector A room-temperature infrared detector in which infrared photons passing through a doped germanium crystal cause a potential drop across the material, which can be used to determine light intensity.

Photosensitive paper A paper with a coating which responds to light, typically by darkening or changing color.

Phototransistor A transistor in which one of the two junctions is illuminated by light, which releases current carriers. The transistor treats this current as an input, which it amplifies, making it a simple detector-amplifier, albeit one of limited performance.

Piezoelectric effect The generation of an electric potential when pressure is applied to certain materials or, conversely, a change in shape when a voltage is applied to such materials. The changes are small, but piezoelectric devices can be used to precisely control small motions of optical components.

Pinhole aperture A small-diameter hole in a sheet of opaque material. Such apertures usually transmit the central part of a laser beam and produce an output beam with an abrupt edge, in contrast to the gradual falloff in intensity found with unmodified laser beams.

***pin* photodiode** A semiconductor diode light detector in which a region of intrinsic silicon separates the *p*- and *n*-type materials. It offers particularly fast response and is often used in fiber-optic systems.

Plane polarized light Light in which the oscillation of the electric field is only in a single plane, equivalent to linear polarization. There are two possible plane polarizations, each one perpendicular (orthogonal) to the other, but any individual light wave can be plane polarized in any direction.

Plasma tube A laser tube—that is, a tube of glass (or other material) filled with laser gas that may contain integral optics as well. The plasma tube requires external packaging and a voltage supply before it can be used as a laser.

Plastic-clad silica A step-index optical fiber in which a silica core is covered by a transparent plastic cladding of lower refractive index than the core. The plastic cladding is usually a soft material, although hard-clad versions have recently been introduced.

Pockels cell A device in which the Pockels effect is used to modulate light passing through the material. The modulation relies on rotation of beam polarization caused by the application of an electric field to a crystal; the beam then has to pass through a polarizer, which transmits a fraction of the light dependent on its polarization.

Pockels effect An increase in the birefringence of materials which normally are birefringent, caused by the application of an electric field. The change in birefringence is linearly proportional to the applied electric field and causes a rotation of polarization which can be combined with polarizers for use in modulators and other active optical devices.

Polarimeter A device for measuring the degree and type of polarization of a light source.

Polarization analyzer A polarizing component which can be rotated about its axis, either to control the amount of incident polarized light which is transmitted, or to determine the plane of polarization of incident light.

Polarization compensator A device which can be adjusted to produce different amounts of polarization retardation—i.e., relative phase shifts between light or orthogonal linear polarizations. The commonest type is the Soleil-Babinet compensator. The term "compensator" comes from its use in measuring the polarization of light.

Polarization-maintaining fiber A single-mode optical fiber which maintains the polarization of the light which entered it. Normal single-mode fibers, and all other types, allow polarization to be scrambled in light transmitted through them.

Polarization retarder An optical device which causes light of one linear polarization passing through it to be retarded relative to light of the orthogonal linear polarization. The result is a relative phase shift or retardation.

Polarizer A filter which transmits light of only a single polarization.

Polarizing coatings Coatings which influence the polarization of light passing through them, typically by blocking or reflecting light of one polarization and passing light that is orthogonally polarized.

Population inversion An energy distribution in which more atoms or molecules are in an upper energy state than in a lower one. If a transition is possible between the two energy levels, a population inversion is said to exist—meaning that the population is the inverse of the situation at thermodynamic equilibrium where more atoms or molecules are in the lower energy level than in the upper one.

Positive lens A lens which focuses rays initially parallel to its optical axis down to a point, causing them to converge. Its focal length is a positive number.

Power/energy meter An instrument which measures the amount of optical power (typically in watts) or energy (joules). It can operate in the visible, infrared, or ultraviolet region, and detect pulsed or continuous beams. This general term can encompass most types of light-measurement instruments regardless of their calibration.

Preform A cylinder of glass which is made to have a refractive index profile that would be desired for an optical fiber. The cylinder is then heated and drawn out to produce a fiber.

Pulse-forming network Electrical circuitry used to generate high-voltage pulses of particular shapes, and to modify the shapes of pulses generated by other sources.

Pumping, nuclear The use of nuclear reactions to produce the excitation needed for laser action. Nuclear pumping has been demonstrated in the laboratory, but remains far from practical.

Pumping, optical The use of light to raise atoms or molecules to higher energy levels. In a laser the goal is to produce a population inversion which can generate laser emission.

Pyroelectric detectors Detectors of visible, infrared, and ultraviolet radiation which rely on the absorption of radiation by pyroelectric materials. Heating of such materials by the absorbed radiation produces electric charges on opposite sides of the crystal, which can be measured to determine changes in the amount of radiation incident on the detector. Such detectors are sensitive to a broad range of wavelengths and operate at room temperature, but have a relatively inefficient response and require beam chopping to detect a CW signal.

Quadrant detectors Detectors which are divided up into four angularly symmetric sectors or quadrants. The amounts of radiation incident on each quadrant can be compared to one another for applications such as making sure that a beam is centered on the detector.

Quantum efficiency The fraction of incident photons which elicit a response from a detector or other electro-optic equipment. The term is sometimes used with lasers that are optically pumped by other lasers, or with frequency-conversion accessories.

Quantum noise Noise due to the discrete nature of light—i.e., its quantization into photons.

Quarter-wave plate A polarization retarder which causes light of one linear polarization to be retarded by a quarter wavelength (90°) relative to the orthogonal polarization.

Quartz A natural transparent form of silica, which may be marketed in its natural crystalline state, or crushed and remelted to form fused quartz.

Quaternary semiconductor A semiconductor compound made up of four elements, such as InGaAsP. In quaternary semiconductor lasers, two of the elements normally come from one group of the periodic table and two from another group. An example is $In_{1-x}Ga_xAs_{1-y}P_y$, where x and y are numbers less than one.

***Q* switch** An optical device which changes the Q (quality factor) of a laser cavity, typically raising it from a value below laser threshold to one well above threshold. This technique produces a short, intense pulse, known as a "*Q*-switched" pulse. Q switches can be based on acousto-optic or electro-optic devices, rotating mirrors (mechanical Q switching), frustrated internal reflection, or saturation of absorption in a dye.

Radiant fluxmeter A device which measures the amount of radiant flux (electromagnetic radiation) emitted or absorbed. The units typically would be watts per unit area.

Radiometer An instrument which measures optical power in radiometric units (watts). A radiometer measures electromagnetic power linearly over its entire spectral range, while the response of a photometer is calibrated to reflect the response of the human eye over the visible spectrum only. Thus a radiometer will respond to ultraviolet or infrared radiation as well as visible light, while a photometer will not (or should not).

Radius In common optical use, the radius of a lens or mirror generally refers to the radius of curvature of its surface. The physical size of the optical element generally is measured as diameter.

Rails, optical Long, linear rods or rails which are attached to an optical bench. Optical mounts can be affixed to the rail, creating an optical bench mounted on an optical table.

Raman shifter A device which alters the wavelength of light by inducing Raman shifts in the light passing through it. Raman shifts are changes in photon energy which occur when molecules scatter photons, caused by the transfer of vibrational energy to or from the molecule.

Raman spectroscopy The study of the Raman shift in light, caused by molecules absorbing light, then re-emitting the light after changing their vibrational quantum state. The result is a shift in wavelength, which is a characteristic of the molecule involved.

Ramp generator An electrical power supply which generates a voltage that increases at a constant rate. A plot of voltage vs. time shows a ramplike waveform; thus the name. The ramp typically can be repeated if desired.

Range finder A laser system which measures the distance to a target by timing how long it takes a laser pulse to travel to the target and back. The term is generally used with military equipment, although there are analogous civilian systems.

Ranging The measurement of distance, by timing how long it takes a laser pulse to make a round trip from the laser to a distant object. Military systems for finding the range to a distant target are known as range finders. While time-of-flight ranging is the most frequently encountered technique, some quasi-continuous-wave, phase-sensitive techniques have been demonstrated.

Rare gas halide A molecule containing one atom of a rare gas such as xenon, and one atom of a halogen such as chlorine or fluorine. These molecules are called excimers, which can exist only in their excited state, breaking up once they release their excitation energy. The term excimer covers a broader category than rare gas halide, but in describing lasers the two terms are sometimes used interchangeably.

Ratiometer An instrument which measures the ratio of two signals.

Real image An image produced by a lens or optical system which has a real physical existence and can be projected onto a screen as well as seen by the eye.

Receiver A device which detects an optical signal, converts it into an electronic form, then processes it further so it can be used by electronic equipment. From the standpoint of components, it can be viewed as a combination of detector and signal-processing electronics.

Reflectance The fraction of incident light which is reflected by a surface.

Refractometer An instrument which measures the refractive index of a substance.

Repetition rate The rate at which pulses are repeated. For lasers, typically measured as pulses per second or hertz. Sometimes lasers are said to be "rep rate" or "repetitively pulsed" when they produce pulses regularly at fixed intervals, rather than producing single pulses.

Replica (or replicated) gratings Diffraction gratings which were produced from originals by mechanically transferring the pattern of lines ruled into the original to a plastic material. The plastic material replicates the diffraction grating, hence the name. Their prime attraction is low cost and reasonable performance (if carefully produced).

Replicated optics Optical components formed by transferring a master pattern to a roughly machined substrate, using an epoxy layer to form a final optical surface. The epoxy layer is then coated with a reflective layer to form the final component. The process allows mass production of complex optical surfaces much less expensively than conventional polishing techniques. In certain cases, it is possible to use transparent replicating materials to form transparent replicated optics.

Resistor trimming The use of a pulsed laser to remove material from a resistor on a hybrid circuit, thus changing the value of the resistance. This technique is used to adjust the value of the resistance to a precise value not produced by the thin-film deposition process.

Resonant scanner A mechanical beam scanner which vibrates back and forth on a mechanical resonance.

Resonator Generally a pair of mirrors located at either end of a laser medium, which cause light to bounce back and forth between them while passing through the laser medium. One of the mirrors lets some of the light leak through or around it, to produce a laser beam. Sometimes three or more mirrors define a ring-shaped resonator.

Resonator, stable A laser resonator in which a ray of light bouncing back and forth between the mirrors will eventually move toward the laser axis and stay there. This is the commonest type for commercial lasers, particularly low-power types, but is not always the most efficient way to extract energy from a laser medium.

Resonator, unstable A type of laser resonator in which a ray of light bouncing back and forth between the mirrors will continually move away from the laser axis. Commercial versions produce beams with a doughnut-shaped distribution of energy across them and can extract energy from certain types of lasers more efficiently than standard stable resonators.

Retardation plate See Polarization retarder.

Reticle A glass window on which is etched or printed a pattern, typically for use in measurement or alignment. The simplest type of reticle is the crosshairs of an alignment telescope.

Retroreflector An optical device which reflects an incident beam of light back in precisely the direction from which it came. The corner-cube prism is the most familiar example, although three mirrors can be joined together to form a hollow retroreflector (or corner cube) as well.

Right-angle prism A prism in which two of the faces meet at right angles. These are often used to perform the function of a mirror in reflecting light at a 90° angle.

Roof prism See Amici prism.

Rotational transition A change in the rotational state of a molecule. Rotational transitions involve less energy than either electronic or vibrational transitions, and typically correspond to wavelengths in the far-infrared, longer than about 20 micrometers.

Ruled diffraction gratings Diffraction gratings in which the light-scattering lines are mechanically ruled by an etching machine. These gratings are normally replicated and the replicas are sold for most applications; original ruled gratings are very expensive.

Scribing The use of a laser to drill a row of holes across a brittle wafer of ceramic or semiconductor material. When the wafer is bent, the material will break along the row of holes. In essence, it is the same as perforating paper so the paper will tear along a certain line.

Sealed laser A gas laser in which the laser gas is sealed in a tube. Gas may flow through the tube or remain stationary within it, but there is no requirement for a continuous supply of new gas. However, in some models periodic replacement and/or replenishment of the gas may be required.

Selfoc lenses Segments of optical fibers specially designed to function as lenses.

Semiconductor laser A laser in which stimulated emission is produced at a *pn* junction in a semiconducting material. Sometimes called a diode or injection laser. Laser action has been demonstrated from junctionless semiconductors in the laboratory, but such devices have not proved commercially viable. LEDs are similar in structure to diode lasers, but lack feedback-producing cavity mirrors and are generally operated at lower drive currents.

Semiconductors, II–VI Semiconductors composed of elements from group II and group VI of the periodic table—a term sometimes extended to cover elements with valences of 2 and 6. Typical II–VI compounds are cadmium telluride and cadmium selenide.

Semiconductors, III–V Semiconductors composed of atoms from groups III and V of the periodic table, such as gallium (group III) and arsenic (group V), which form gallium arsenide.

Signal-to-noise ratio The ratio of the power of the signal to that of background noise, usually measured in decibels. This is a common measure of the quality of analog electronics or transmission systems. Abbreviated SNR.

Single-mode fiber An optical fiber which can guide light in only a single mode.

Solar blind (detector) A detector which contains filters to block sunlight, so the detector becomes essentially "blind" to the sun. In almost all cases, this involves blocking wavelengths longer than approximately 300 nanometers. For photomultipliers, the filters may be the materials of the tube itself.

Soleil-Babinet compensator A pair of plates which can be continuously adjusted to retard polarization of a beam of light by any amount within the range of the device. Essentially an adjustable polarization retarder.

Spark gap A pair of electrodes separated by an air gap, to which a high voltage can be applied. Normally they do not conduct electricity, but they become conductive if a spark appears between them. They may be used for high-voltage switches; that is, to become conductive when the voltage exceeds a certain level (by air breakdown), although some trigger variability is likely. Breakdown can also be triggered by sending a laser pulse through the air between the electrodes.

Spatial coherence The coherence of light over an area of the wavefront of a beam—for example, where the beam hits a surface.

Spatial filter An aperture in an opaque material which blocks part of the incident light and lets the rest through, depending on position of the light in the beam. Typically spatial filters are used to remove stray light on the fringes of the beam, or to cut off low-intensity parts of the beam to produce a uniform beam with sharp cutoff.

Spatial light modulator A light modulator in which the degree of modulation of light passing through it depends on the position of the light on the modulator. In its simplest form, it might block part of the incident light on its top half, and transmit all incident light on its lower half. Ideally spatial light modulators should have high enough resolution to turn diffuse white light into an image, but that has yet to be achieved practically.

Speckle camera An instrument designed to observe speckle patterns produced by coherent light, and obtain measurements from the speckle patterns.

Spectral bandwidth The range of wavelengths emitted by a laser.

Spectrometer A spectroscope (an instrument which spreads out light into a spectrum) which includes an angular scale for measurement of the angular deviation and wavelengths of the components of the spectrum.

Spectroradiometer An instrument which measures the amount of power in various portions of the spectrum; that is, power as a function of wavelength. A spectroradiometer could be used to plot a blackbody curve, for example, while a radiometer would produce only a single value for total power from the blackbody.

Spectroreflectometer A device which measures the reflectance of a surface as a function of wavelength.

Spectroscope A device which spreads out the spectrum for analysis. Conceptually the simplest type would use a prism or diffraction grating to spread out the spectrum on a piece of paper or ground glass.

Spectrum analyzer A device which breaks a signal up into its various spectral components, which can then be analyzed. A prism or a diffraction grating can serve as an optical spectrum analyzer, although instruments sold as spectrum analyzers are more complex.

Spherical Optical elements are considered spherical if their surfaces are portions of spheres. Typically the term is applied to lenses and mirrors.

Spherical aberration An effect which causes light rays passing through the edge of a lens to come to a focus at a different distance from the lens than those which pass through its center.

Spontaneous emission The emission of a photon by an atom or molecule when it spontaneously (i.e., without outside intervention) drops from a high energy level to a lower one.

Step-index fiber An optical fiber in which there is a discontinuous (step-function) change in refractive index at the boundary between fiber core and cladding.

Steradian The solid angle subtended at the center of a sphere by an area on the surface equal to the square of the radius of the sphere. This is a standard unit of solid angular measurement and is widely used in optics.

Stimulated emission The process that makes laser light. It occurs when an atom or molecule in an excited state is stimulated to emit a photon and drop to a lower energy level by another photon of just the right energy for the transition. Typically the photon doing the stimulating has come from an atom or molecule that has just made the same transition.

Streak camera A device used for measuring ultrafast events. It gets its name because it sweeps a linear image across a light-sensitive surface, creating a record of the intensity of light as a function of time.

Sum frequency generation The addition of the operating frequencies of two lasers to produce a third, higher frequency. From the standpoint of wavelength, the process produces a shorter wavelength than that of either of the laser wavelengths which were added together.

Superluminescent diode A compromise between a diode laser and an LED, which is operated at the high drive currents characteristic of diode lasers, but lacks the cavity-mirror feedback mechanisms that produce stimulated emission. It is used when high-power output is desired, but coherent emission is not wanted.

Target designation The use of a coded series of pulses from a laser to "mark" a target for a smart bomb or missile. A soldier equipped with a designator aims the laser at the target. Light reflected from the target is detected by a homing device in a smart bomb, which looks for the particular code of pulses used by that designator.

TEA (type of laser) TEA is an acronym for "transversely excited, atmospheric pressure." It commonly refers to carbon dioxide lasers in which electrical excitation pulses are applied transverse to the beam direction, and in which the gas pressure is somewhere in the range of one atmosphere, rather than the much lower pressures in other CO_2 lasers. The TEA design can produce short, powerful pulses, but is not adaptable to continuous output because of the high gas pressure.

Temporal coherence The coherence of light over time. Light is temporally coherent when the phase change during an interval *t* remains constant regardless of when the interval is measured.

Ternary semiconductor A semiconductor compound made up of three elements, with the sum of the concentration of two elements equalling the concentration of the third, such as $Ga_{1-x}Al_xAs$.

Thermopile An infrared detector in which rods of antimony and bismuth are connected alternately in series. Light-induced heating of the junctions produces a thermoelectric current, which can be used to measure the amount of incident radiation.

Threshold A minimum level of excitation energy needed to produce laser action. In a typical laser diode, for example, tens or hundreds of milliamperes of drive current must be flowing through the semiconductor junction before laser action starts.

Thryatron An electronic switching tube filled with gas. Application of a voltage to the control grid turns on the current. Normally thryatrons are used for high-voltage switches.

Transition The shift of an atom or molecule from one energy level to another, involving the gain or loss of energy.

Transverse mode Oscillation mode of a laser defined by size and shape of the resonator mirrors, visible as distribution of energy across the beam. The lowest-order transverse mode is TEM_{00}, but a TEM_{00} beam can contain several longitudinal modes at different wavelengths.

Tunable laser A laser which can have its output wavelength changed by some tuning process. The best-known tunable laser is the dye laser, but many other types can be tuned over a limited range.

Tuning In the laser world, tuning generally refers to adjusting the wavelength of a laser.

UV Ultraviolet (wavelengths less than 400 nanometers).

Vacuum monochromator A monochromator in which the beam path is through a vacuum. Such monochromators are usually used to isolate lines in the vacuum ultraviolet spectrum.

Vacuum photodiode A vacuum tube in which light incident on a photoemissive surface (the cathode) frees electrons, which are collected by the positively biased anode. Most photocathode materials are sensitive to visible light. Photodiodes do not include the electron multipliers used in photomultipliers.

Vacuum ultraviolet Ultraviolet wavelengths which are not transmitted by the atmosphere because of strong absorption. There is no firm standard definition, but usually this is taken as wavelengths shorter than about 200 nanometers, extending to the borders of the x-ray region.

Variable-iris aperture The type of adjustable iris found on cameras, which can be mechanically adjusted to change its diameter.

Velocimetry The measurement of velocity. Laser velocimetry relies on detection and measurement of the Doppler shift of the wavelength of light caused by motion of the object being studied. See Doppler shift.

Vibration isolation equipment Systems which minimize the transfer of vibrations from the floor and surrounding environment to the surface of an optical table or other equipment mounted upon them.

Vibrational transition A change in the vibrational state of the atoms within a molecule, resulting in a transition between energy levels. Generally vibrational transitions involve less energy than electronic transitions, but more energy than rotational transitions. In practice, vibrational and rotational transitions usually occur simultaneously as vibrational-rotational transitions.

Vibronic transition A simultaneous change in both vibrational and electronic energy state of a molecule, with the amount of energy involved similar to that for electronic transitions.

Vidicon A video camera imaging tube.

Virtual image An image formed by an optical system which can be seen by the human eye, but which does not have a physical existence and cannot be projected onto a screen.

VUV Vacuum ultraviolet.

Wavefront A surface normal to a bundle of light rays as they travel from a source, passing through the parts of the rays that are in phase with each other. For a diverging bundle of rays, the wavefront is a portion of a sphere.

Waveguide A device which guides light along its length; e.g., an optical fiber.

Waveguide carbon dioxide laser A carbon dioxide laser in which the laser gas is excited within a narrow bore that functions as a waveguide for the infrared radiation produced. Generally the gas flows rapidly through the tube. Using this approach, compact CO_2 lasers can produce several watts of output.

Wavenumber A measurement of frequency in inverse centimeters, commonly used in spectroscopy. Wavenumber equals the wavelength in centimeters divided into one; that is, the inverse of the wavelength in centimeters.

Waveplate A polarization retarder.

Wedge prism A prism with a wedgelike cross section which is used to bend a beam of incident light through a small, known angle. The bending is caused by refraction and is dependent on wavelength. Two wedge prisms can be combined to steer an incident laser beam to any direction within a narrow cone centered on the beam axis.

White light A mixture of colors of visible light that appears white to the eye. In theory, a mixture of three colors is sufficient to produce white light.

X-ray laser A laser which emits x rays. In practice, the definition is fuzzier than it sounds because the borderline between x-rays and the extreme ultraviolet is a hazy one. Some researchers tend to push the x-ray definition to longer wavelengths so they can claim they are working on an x-ray laser; others try to call the same wavelengths extreme ultraviolet, either because they are conservative or because they want to avoid security restrictions on x-ray laser research.

X-ray preionizer An x-ray source which "preionizes" a laser gas, preparing it for the excitation pulse which will stimulate laser emission. Preionization has been shown to increase the amount of energy which can be extracted from a laser medium.

YAG Yttrium aluminum garnet, a crystalline host used in solid-state neodymium lasers. Its excellent thermal and optical properties make it the commonest host for neodymium, but crystal growth remains something of a black art.

YALO Yttrium aluminate, a crystalline host for solid-state lasers, which may be doped with neodymium or other elements.

YLF Yttrium lithium fluoride, a crystalline host used for solid-state lasers. For a variety of reasons, it has never attained the popularity of YAG. May be doped with neodymium or other elements.

APPENDIX

B

types of lasers

Types of Commercial Lasers, Organized by Wavelength

Wavelength, μm	Type	Chapter	Output type and power
0.152	Molecular fluorine ("excimer")	13	Pulsed, average power to a few watts
0.192	ArF excimer	13	Pulsed, average power to several watts
0.222	KrCl excimer	13	Weaker than ArF, pulsed
0.248	KrF excimer	13	Pulsed, average power to tens of watts
0.266	Quadrupled Nd	20	Pulsed, under 1 W average power
0.308	XeCl excimer	13	Pulsed, to tens of W
0.325	He–Cd	9	CW, a few milliwatts
0.337	Nitrogen	14	Pulsed, under 1 W average power
0.347	Doubled ruby	21	Pulsed, under 1 W average power
0.35	Argon or Kr ion	8	CW, to 2.5 W
0.351	XeF excimer	13	Pulsed, tens of watts average power
0.355	Tripled Nd	20	Pulsed, to a few watts
0.3–1.0	Pulsed dye	17	Pulsed, to tens of watts average (flashlamp pumped)
0.4–0.9	CW dye	17	CW, to a few watts
0.442	He–Cd	9	CW, tens of milliwatts
0.45–0.52	Argon ion	8	CW, milliwatts to tens of watts
0.48–0.54	Xenon ion	16	Pulsed, low average power

Types of Commercial Lasers, Organized by Wavelength *(Continued)*

Wavelength, μm	Type	Chapter	Output type and power
0.51	Copper vapor	12	Pulsed at several kilohertz, average power tens of watts
0.532	Doubled Nd	20	Pulsed or CW, to several watts
0.543	He–Ne	7	CW, under 1 mW
0.578	Copper vapor	12	Pulsed at several kilohertz, average power to tens of watts
0.628	Gold vapor	12	Pulsed at several kilohertz, average power to 10 W
0.6328	He–Ne	7	CW, to about 50 mW
0.647	Krypton ion	8	CW, to several watts
0.694	Ruby	21	Pulsed, to a few watts
0.7–0.8	Alexandrite	22	Pulsed, to several watts average power (CW in lab)
0.75–0.9	GaAlAs diode	18	CW or pulsed, under 1 W
0.85	Erbium	24	Pulsed, under 1 W
1.06	Nd–YAG and glass	20	CW or pulsed, to hundreds of watts
1.15	He–Ne	7	CW, milliwatts
1.1–1.6	InGaAsP diode	18	CW or pulsed, milliwatts
1.3	Iodine	16	Pulsed
1.32	Nd–YAG	20	CW or pulsed, to a few watts
1.4–1.6	Color center	23	CW, 100 mW
1.523	He–Ne	7	CW, milliwatts
1.54	Er-glass	24	Pulsed
1.73	Erbium-crystalline	24	Pulsed
2–4	Xe–He	16	CW, milliwatts
2.06	Holmium crystalline	24	Pulsed
2.3–3.3	Color center	23	CW, milliwatts
2.6–3.0	HF chemical	11	CW or pulsed, to hundreds of watts
2.7–30	Lead salt diode	19	CW, milliwatt range
3.39	He–Ne	7	CW, milliwatts
3.6–4.0	DF chemical	11	CW or pulsed, to hundreds of watts
5–6	Carbon monoxide	16	CW, to tens of watts
9–11	CO_2	10	CW or pulsed, to tens of kilowatts
10–11	N_2O	16	CW, tens of watts
40–1000	Far-infrared gas	15	CW, generally under 1 W, (also pulsed or chopped)

CW = continuous wave.

Index

ABOUT THE AUTHOR

Jeff Hecht, an electronics engineer, is cofounder of and contributing editor to *Lasers & Applications* magazine and is former managing editor of *Laser Focus* magazine. He has written articles for *New Scientist*, *High Technology*, *Computers & Electronics*, *Omni*, *Electronic Times*, and other magazines. Mr. Hecht is now a full-time writer and consultant on lasers and their applications. He obtained his B.S. from the California Institute of Technology in 1969.